STREAM FISH COMMUNITY DYNAMICS

STREAM FISH
COMMUNITY DYNAMICS

A Critical Synthesis

WILLIAM J. MATTHEWS
EDIE MARSH-MATTHEWS

JOHNS HOPKINS UNIVERSITY PRESS

BALTIMORE

Johns Hopkins University Press
2715 North Charles Street
Baltimore, Maryland 21218-4363
www.press.jhu.edu

Library of Congress Cataloging-in-Publication Data

Names: Matthews, William J. (William John), 1946–, author | Marsh-Matthews, Edie, author.
Title: Stream fish community dynamics : a critical synthesis / William J. Matthews, Edie
    Marsh-Matthews.
Description: Baltimore : Johns Hopkins University Press, 2017. | Includes bibliographical
    references and index.
Identifiers: LCCN 2016026377| ISBN 9781421422022 (hardcover : alk. paper) | ISBN 9781421422039
    (electronic) | ISBN 1421422026 (hardcover : alk. paper) | ISBN 1421422034 (electronic)
Subjects: LCSH: Freshwater fishes—Ecology—United States. | Stream ecology—United States. | Fish
    communities—United States.
Classification: LCC QL627 .M38 2017 | DDC 597.176—dc23
    LC record available at https://lccn.loc.gov/2016026377

A catalog record for this book is available from the British Library.

*Special discounts are available for bulk purchases of this book. For more information,
please contact Special Sales at 410-516-6936 or specialsales@press.jhu.edu.*

Johns Hopkins University Press uses environmentally friendly book materials, including
recycled text paper that is composed of at least 30 percent post-consumer waste,
whenever possible.

All color photographs are by W. J. Matthews unless otherwise indicated.

CONTENTS

PREFACE

This is a book about the spatial and temporal dynamics of warm-water stream fish communities, based largely on our individual and combined research since the 1970s. Since 1994 we have shared our research on stream fish communities, using broad geographic sampling, continued long-term sampling in numerous systems, and experiments to help us understand mechanisms underlying the phenomena we have measured in the field. We use a substantial number of our own large data sets throughout the book to summarize community structure and dynamics, in what is, we hope, a "critical synthesis." These data sets range from spatially very broad, such as our collecting fish at more than 80 stream sites throughout the midwestern United States in one month of one year, to several that span four to five decades of sampling at fixed sites within watersheds. With publication of this book, we make all of these data sets available publicly and invite all who might be interested to use them in further analyses. Data are available from the Dryad Digital Repository: http://dx.doi.org/10.5061/dryad .2435k. We ask only that we be acknowledged as the source of any data that are used.

The book relies heavily on new analyses of fish communities that we have surveyed at multiple sites upstream to downstream at least four times on 9 streams (which we refer to as "global" communities) and analyses of the "local" communities at 31 individual sites that we also have sampled four or more times. The experiments we address in the book range in dimensions from "tabletop" tests (in units as small as 1 L boiling flasks), to large outdoor stream mesocosms that mimic riffles and pools in small streams, to enclosures in actual streams.

The goals of the book are to combine field observations and related experiments to provide a synthesis of "the big picture" that our own work reveals, all in the context of historical questions or previous research on stream fishes by ourselves or many other workers who have made stream fish ecology the energetic and vibrant science that it is today. In each chapter we briefly review the key concepts and previous publications that have set the tone for research in stream fish ecology in particular or for ecology in general, and, where appropriate, we test conventional wisdom or theoretical concepts with our empirical data sets.

The data sets and experiments were obtained across more than 40 years, for many different reasons and across many different individual projects. No single hypothesis drove the collection of these data, although we have always had an underlying question about why stream fish are where they are, and how they do what they do there. Looking back across these individual projects, we now combine them into a single large synthesis, or "workshop," in which we use these data sets and experiments as tools to test ideas across systems to look for pervasive factors in the dynamics of stream fish communities.

We address topics including comparative structure of these fish communities; taxonomic and emergent properties of fish communities and variability; traits of individual species such as body form, environmental tolerances, or resource user patterns; species interactions, including competition, predation, and facilitation; effects of extreme disturbance events; temporal and spatial dynamics of fish communities with emphasis on both short-term and long-term change; effects of fish species or communities in ecosystems; and an overview with thoughts about future projects. Zoogeography of North American stream fishes as related to our study region was covered in detail in Matthews's (1998) *Patterns in Freshwater Fish Ecology* (Chapman and Hall, New York) and in a large chapter, "Evolution and Ecology of North American Freshwater Fish Assemblages," by Ross and Matthews (2014) in *Freshwater Fishes of North America*, volume 1 (Johns Hopkins University Press, Baltimore), so it is not covered in any detail in the present book. We also do not address other important topics for stream fishes, such as internal physiology, genetics, phylogenetic analyses, or modeling, as these areas of research are outside our own, and we wish to keep the topics in the book within our areas of expertise or experiences. The book is not an attempt to update previous comprehensive works on stream fishes, such as Matthews's 1998 book or Ross's (2013) *Ecology of North American Freshwater Fishes* (University of California Press, Berkeley). Accordingly, readers should refer to the books above for thorough literature surveys. Here, we focus on our own work, using selected examples of papers by other authors to introduce or to summarize particular topics.

We do not use this book to make statements about our personal views on scientific nomenclature for any of the fishes we have worked with. We recognize that, during the writing of this book or while it was in press, nomenclatural changes continued to be forthcoming in peer-reviewed literature, but for standardization of names we followed *Common and Scientific Names of Fishes from the United States, Canada, and Mexico*, 7th edition, 2013 (American Fisheries Society) for all taxonomy, rather than to attempt to incorporate every change that could apply as new papers were published.

The specific questions that we address with our data are diverse, and many have been the subject of ecological studies for a long time. The fundamental questions about factors that influence distribution and abundance patterns of species, which date to the founding of the discipline, remain as important as ever given that the effects of global climate change, human perturbations, and invasive species, among other issues, on fish community structure are just beginning to be understood. We believe that the strength of this book is that we can ask many of those questions or test hypotheses using data sets for samples that we have personally collected, across many systems, giving us the insight that comes from knowing the sampling and the environments of the systems in intimate detail.

The book is organized into 10 chapters. Chapters 1 and 2 describe our methods and the study systems in detail. In chapter 3 we characterize the stream fish communities with respect to emergent properties, and then we use those properties in comparisons of community dynamics and the factors that affect dynamics. Chapters 4 and 5 explore mechanisms related to species traits and interspecific interactions, respectively, that are important for understanding community structure and dynamics. In chapters 6, 7, and 8, we address different aspects of community change over time. In chapter 6, we examine effects of extreme events (e.g., floods and droughts), and in chapter 7, we take a more general approach and relate variation in community structure over time to the model of loose equilibrium. In chapter 8, we add a spatial component to our study of community change and explore spatiotemporal community dynamics. In chapter 9, we examine effects of fishes on ecosystem properties and relate change in community structure to potential effects in ecosystems. Chapter 10 summarizes our conclusions and thoughts about future studies.

Colleagues who kindly reviewed chapters or parts of chapters, and greatly helped us to make improvements, include Katie Bertrand, Mark Eberle, Tony Echelle, David Edds, Aaron Geheber, Melissa Gibbs, Keith Gido, Chad Hargrave, Andrew Marsh, Mary Power, Steve Ross, Jake Schaefer, Art Stewart, Larry Weider, and Jeff Wesner.

We especially thank the late Robert S. (Bob) Matthews, who helped WJM locate field sites and took part in all collecting on Piney Creek from 1972 to 1981. Without Bob Matthews, the Piney Creek project could not have been started, and there might have been no long-term studies (or this book). We also offer special thanks to Bob Cashner and Fran Gelwick, who were instrumental in long-term field studies on Brier Creek and in experimental work; to Henry Robison for sharing data and for many years of collegial support; and to Betty Crump, Keith Gido, Mary Power, Steve Ross, Jake Schaefer, and Art Stewart for their key roles in collaboration on major projects.

We thank many others for their help in the field or laboratory, for sharing data, or with analyses, on various projects that form the basis for this book, including Ginny and Reid Adams, who now collaborate with us on Piney Creek, Ken Asbury, Judith Barkstedt, Hank Bart, Robert Bastarache, Jeff Bek, Barry Bolton, Melody Brooks, Rich Broughton, Carol Brown, Linda Byrd, Irene Camargo, Curtis Campbell, Sara Cartwright, Alan Clingenpeel, Ken Collins, Danny Crump, James Cureton, Cari Deen, Mike Douglas, Alice Echelle, Tony Echelle, David Edds, Mike Eggleton, Jack Feminella, Daniel Fenner, Courtney Franssen, Nate Franssen, Bryan Frenette, Aaron Geheber, David Gillette, Steve Golliday, Joe Gryzbowski, Chad Hargrave, George Harp, Bret Harvey, Kim Hauger, Tom Heger, Loren Hill, Kiki Hiott, Jan Hoover, Dan Hough, Dennis House, Clark Hubbs, Donald A. Jackson, Pete Johnson, Bob Kinneburgh, Jim Lauchman, Ricky Lehrter, Mike Lodes, Frank McCormick, Roland McDaniel, David McNeely, Amber Mackowitz, Bill Magdych, Jody Maness, Andrew Marsh, Rebecca Marsh, Amy Matthews, Bob Matthews, Scott Matthews, Woody Matthews, Mike Meador, Maria Miller, Vernon Miner, Ray Moody, Bob Nairn, Susan Orosz, Zolton Orosz, Larry Page, Mitzi Pardew, Randy Parham, David Partridge, Brian Pounds, Kerri Pratt, Rhea Putnam, Mike Putz, Mark Pyron, John Robinson, Gary

Schnell, Mark Schorr, Mike Schwemm, Bill Shelton, Bill Shepard, Nick Shepherd, C. Lavett Smith, Dave Soballe, Rich Standage, Bruce Stewart, Jeff Stewart, Marsha Stock, Jim Stout, John Styron, Eric Surat, Katy Sutherland, Barbara Taylor, Chris Taylor, Jacob Thompson, Michi Tobler, Jona Tucker, Brent Tweedy, Caryn Vaughn, Bruce Wagner, Peter Wainwright, Mel Warren Jr., Jeff Wesner, Matt Winston, Zach Zbinden, and Earl Zimmerman. Coral McAllister and Andy Vaughn did the final figures, and the cover photograph was graciously provided by Garold Sneegas. Special thanks go to Richard Page, Donna Cobb, Malon Ward, Wendal Porter, and George Martin for many years of logistical support at the University of Oklahoma Biological Station or on the University of Oklahoma main campus. Our experimental streams would not have worked without their efforts.

We are grateful to all the families who have allowed us access to field sites on private property, including (for Brier Creek) Ricky Coleman, Jim Martin, the Parrish family, Jim Williams, and John Woody, and (for Piney Creek) the Byler family, the Calvin, Marie, and Arthur Jones families, the Killian family, the Kever family, Mike and Sue Richardson, Mickey and Tamara Bevill, the Marge Cooper family, Morris and Mavis Jones, the Gameliel and Melissa Jones family, Frederick and Sharon Wright, Albert Smith, and Dr. Adam Gray. For providing a wealth of information on Piney Creek and floods in the watershed, we particularly thank Adam Gray, Arthur Jones, and the late Calvin Jones. We also thank Camp Egan, Tahlequah, Oklahoma, for support and access to Baron Fork.

We thank the University of Oklahoma Biological Station and Department of Biology at the University of Oklahoma for logistical support and the Sam Noble Oklahoma Museum of Natural History for archiving specimens. For collecting permits, we thank the Arkansas Game and Fish Commission and the Oklahoma Department of Wildlife Conservation for many years of continuous approval, and the fish divisions of other states throughout the Midwest and in Virginia. All collecting (since IACUCs were established) has been by approval of the IACUC of the University of Oklahoma. Over the years our work has been funded at various times by the National Science Foundation; US Environmental Protection Agency; US Department of Agriculture Forest Service; US Fish and Wildlife Service; Oklahoma Department of Wildlife Conservation; Oklahoma Department of Environmental Quality; US Department of Defense; North Texas Municipal Water District; the Warren County R-III School District of Warrenton, Missouri; and a Raney Fund Award to WJM from the American Society of Ichthyologists and Herpetologists.

Finally, we wish to acknowledge our executive editor at Johns Hopkins University Press, Vince J. Burke, for his excellent advice and assistance in all phases of this book. We also thank Meagan Szekely and Debby Bors at Johns Hopkins University Press for helping us through the process of book submission and production and Ashleigh McKown and Deborah Tourtlotte for skilled copyediting and indexing, respectively.

STREAM FISH COMMUNITY DYNAMICS

CHAPTER 1

Studying Stream Fish Communities

Our Approach to a Critical Synthesis

The goal of this book, as indicated by the title, is a critical synthesis of stream fish community structure and dynamics. This synthesis arises from our own field and experimental research across five decades, to test what we consider to be important questions or concepts in ecology writ large. We also place our work in the context of key papers and historical studies by others working in stream fish ecology. We have attempted to draw together much of the information we have gained across decades in numerous field and experimental studies to see where there are patterns in common or broad explanations and (by way of being "critical") to identify issues where we have not delved as deeply as we might have, and so point to potentially useful future studies that the next generation of graduate students might consider. Specifically, in this book we have used our own long- and short-term data sets and spatially broad data (described in chap. 2) to look for general patterns across different kinds of streams and fishes, to confront theory with real-world empirical data, or to test conventional wisdom ("everybody knows that . . .") about many aspects of stream fish ecology. In this effort we have used a combination of our published and unpublished field data, most gathered with the assistance of students or colleagues. We have complemented our analyses of field data with experimental assessments of mechanisms or processes. And we have freely drawn on our approximately 40 years each of observing fish in the field to bring our subjective impressions into any discussion. Instead of attempting in this initial chapter to list each theory, idea, or concept tested in the rest of the book, we address these in each chapter, beginning with simple but important concepts about how to characterize a fish community, culminating in thoughts in later chapters about long-term changes in fish communities and the roles of fish in ecosystems. But first, let's consider what a stream fish community is.

Assemblages and Communities

The nature of animal communities has long been debated (Shelford 1931; Allee et al. 1949; McIntosh 1995), and even the definition of a "community" is quite variable. Mittlebach (2012) began his textbook with the definition of a community as "a group of species that occur together in space and time," following Begon et al. (2006) and numerous previous authors. In his textbook, Morin (2011) recounted numerous definitions of community that have been published in books or in the primary literature for the past century, many of which emphasize not only co-occurrence but also interaction of community members. In studies of stream fishes, we (Ross et al. 1985; Matthews et al. 1988; Matthews 1998; Matthews and Marsh-Matthews 2006a,b) and numerous others (e.g., Smith and Powell 1971; Grossman et al. 1982; Schlosser 1982; Moyle and Vondracek 1985; Meffe and Sheldon 1990; Fausch and Bramblett 1991; Winemiller 1991; Davey and Kelly 2007; Hoeinghaus et al. 2007) have variously used "assemblage" or "community" to describe the fishes found together at a particular place and time, as reviewed by Winemiller (2010). Editors of two influential books in stream fish ecology (Matthews and Heins 1987; Gido and Jackson 2010) used the term "community," whereas Matthews (1998) used "assemblage" for fish in one stream reach at one time and "fauna" for the fish of entire watersheds.

In this book, we use the term "community" in a context that suggests that there is predictability about those species found together, that interspecific interactions (McIntosh 1995; Winemiller 2010) occur among community members, and that those interactions (along with abiotic factors and traits of individual species) act to shape community structure. To understand community structure and dynamics, it is necessary to repeatedly sample the assemblage of species present at the same time and place and use those samples to reveal patterns of representation of different community members and potential interactions among them. In addition, large-scale experiments in the field or in artificial stream mesocosms can dissect the myriad complex interactions responsible for community dynamics.

We cannot overemphasize the importance of long-term data for understanding community dynamics. Simple models of community dynamics and their underlying mechanisms date to the origin of the discipline itself (e.g., Lotka-Volterra models developed in the 1920s), and theoretical modeling in the 1970s (e.g., papers in May 1976) yielded predictions relative to more complex communities, but it is only in recent decades (e.g., papers in Cody and Smallwood 1996) that long-term data from real communities (with which to test these predictions with empirical data or to refine the models) have become available. Particularly important are long-term studies at the scale of decades (Kiernan and Moyle 2012; Matthews et al. 2013), because longer spans of time may incorporate rare "pulse" (extreme floods; Matthews 1986) or persistent "press" disturbances (e.g., long-term drought; Lake 2011) that may not be detected over shorter periods. And Bêche and Resh (2007) underscored conclusions from shorter studies that may well be reversed after accumulation of longer data series.

Here we use our own data sets with at least four samples across time from thirty-one creek or river systems to address major questions about stream fish ecology. Two are very-long-term data sets (Piney Creek, Arkansas, and Brier Creek, Oklahoma), with a dozen or more surveys over a span of five decades. Others span months or years, not decades, but provide a wide array of study systems in which to evaluate shorter-term dynamics of fish communities. Finally, we have two spatially broad data sets, with one encompassing fish assemblages throughout the southern Great Plains from Nebraska to south Texas, and the other including samples at 143 sites across the Red River basin in Oklahoma.

Sampling Fish

With apologies to Theodosius Dobzhansky, nothing in (fish) biology makes sense except in the light of adequate sampling. Samples used in this book were made by at least one of the authors (with a few exceptions, as noted for two of the study systems in which colleagues made one or two surveys without us). In this section we describe the sampling methods that produced our databases that are the basis of much of this book, going into more details than are found in the relatively terse "methods" sections of our published papers. We address advantages or challenges of using seining as a primary tool for assessing stream fish communities, and consider effectiveness and amount of effort in seining various kinds of habitats as well as the kinds of species for which seining is most effective or for which it presents challenges. We address repeatability of seining samples and seining crew size, and we consider situations in which other methods may be equally or more efficient. We also describe our typical sampling at a site as to length of time seining, number of seine hauls, the length of reach we sample, and the actual portion of a stream reach through which a seine passes.

Most of our sampling of fish communities on which this book is based has been in daylight hours in wadeable streams not more than about 1.5 m deep or in the shallows of deeper rivers with small-meshed seines (fig. 1.1). We normally use a seine that is 4.6 m long and 1.2 m deep, with 4.8 mm bar measure mesh (usually of the Ace or Delta fabric as commonly marketed by net companies). Probably 90% of our samples were collected with nets of the dimensions and mesh size above. In riffle-pool-structured streams we often used a second, shorter net (1.8 or 2.4 m long, but with the same mesh size) for kick-seining in riffles. Occasionally, we have used a mesh size of 3.2 mm when we wished to capture juveniles or young-of-year of small species, and at some of our widest sites we have supplemented collections by using a bag seine 7.6 m long, with 6.4 mm mesh, in addition to the usual 4.6 m seines described above.

The numbers of persons involved in the collections have varied greatly. Many of the collections have been made by just the two authors (see plate 31), for example, our samples throughout the southern Great Plains in June 1995, or at many other sites, or by WJM and a single helper in the earlier years. For instance, the earliest Piney Creek collections, in July and December 1972, were made with only Bob Matthews and WJM seining and measuring water quality (fig. 1.2).

FIG. 1.1 William Matthews and Nathan Franssen examine fishes collected by seining in Buncombe Creek, Oklahoma.

FIG. 1.2 The late Robert S. (Bob) Matthews making water-quality measurements at site P-2 in Piney Creek in 1973.

Often three to four individuals were involved in seining when we worked with graduate students or colleagues on various projects. Many of our samples from 1981 to the present—including some in Brier Creek, the Kiamichi River, and southern Oklahoma—were made by our ichthyology classes or by two classes supervised by Bob

FIG. 1.3 Robert C. (Bob) Cashner, who supervised the University of Oklahoma Biological Station classes that sampled all the Kiamichi River sites in 1986 and 1987 and the Brier Creek sites in 1995.

Cashner (fig. 1.3), at the University of Oklahoma Biological Station (UOBS) or on the main campus in Norman.

Regardless of the number of persons involved, the goal has always been to thoroughly sample for fish in all kinds of microhabitats within a stream reach, attempting to detect as many species as possible and to adequately represent their relative abundances. We make the assumption that sampling was thorough, and comparisons we have made for Brier Creek surveys (Matthews et al. 2013), for which we have the greatest variation in numbers of persons (from two to three professionals to entire classes), indicated no substantial effect of numbers of persons in the field with regard to effectiveness of fish collection. Nine of sixteen surveys in Brier Creek were made by professionals with one or two field assistants, and seven were made by classes. There was no difference in numbers of species taken in surveys by professionals and classes (mean species per survey was 20.4 for professionals and 20.6 for classes). A few experienced professionals are thus likely to make equally as good or even better samples than an entire class of inexperienced students (although, with supervision by us and our graduate students, the classes seem to do a good job of sampling the fish community).

Our seining techniques (including factors that influence collecting efficiency) are described in Matthews (1986), Matthews et al. (1988), and Matthews et al.'s (2013) supplementary online materials. A typical "whole-community" sample takes from 45 to 90 minutes or more. In Piney Creek, Arkansas, for example, surveys in recent decades have averaged (based on our field notes) from 66 to 86 minutes of sampling at each of a dozen sites, from small headwaters about 5 m wide to large, 30 m–wide lower mainstem sites.

We typically work in a downstream direction, taking a substantial number of sweeps with the fully unrolled 4.6 m seine in pools (fig. 1.4, *top*) and either rolling up that seine or using a shorter one to take kicksets in riffles or rapids. Our field notes show that we typically make about 25 to 40 or more individual seine hauls (including both sweeps and kicksets) for collections at a site (see plate 17). In open areas where sweeps with the seine are unobstructed, we may make fewer but longer sweeps but usually cover not more than about 10 m in a haul, before closing to the bank to trap fish. Longer hauls actually give fish a better chance of escaping capture, so we emphasize a large number of short hauls, each within one kind of microhabitat (open water, edges of weed beds, around brush, and so on). It is particularly important to sample shallows in pools as well as deeper areas, because juveniles or small adults of many species will occur more in the shallow water at pool edges (Matthews et al. 1994). When there are obstructions such as wood debris, large snags, dead trees, root wads, or undercut banks, we surround these structures with the seine and then vigorously kick into the structure to frighten fish into the net. In streams with substantial structure we take many more but shorter seine hauls, making judgments haul by haul as to the most likely way to capture fish efficiently in all of the different microhabitats.

In riffles or rapids we use the shorter seine held stationary by one to two persons, with the lead line held firmly on the substrate. Helpers then vigorously kick the gravel or rubble substrate upstream of the seine to dislodge benthic fishes (fig. 1.4, *bottom*). In a riffle or rapids we often make 5 to 10 kicksets, depending on the size of the area or the variety of substrates and current speeds. In a riffle it is important to sample at the edges as well as midriffle and to try to sample in all substrates from small gravels (for juvenile darters or sculpins) to cobbles or slabs of shale, where larger darters, madtom catfish, or adult sculpins are more often taken. In swift or deeper riffles or rapids we sometimes also run downstream with the seine unrolled, trying to exceed the current speed, to capture minnows or other water-column fishes that may be in the riffle.

A sample at a site in riffle-pool-structured streams often included a stream reach of 200 to 300 m or more, with usually three to four or more individual pools sepa-

FIG. 1.4 Methods of fish collecting used by the authors. *Top*, standard technique using a 4.6 m seine. *Bottom*, kickset collecting with short seine.

rated by riffles or rapids. In "channel"-type streams, with more uniform widths and flows and no rapids or riffles, we ideally sample a reach of 200 m or more. We advocate sampling more than 100 m of stream where possible. In 1988, WJM and an ichthyology class compared fish collections in adjacent measured (and block-netted) 100 m segments of Brier Creek at five sites. The spatially adjacent segments had percentage similarity in composition (based on relative abundance) of only 60% to 70%, and some species were detected only in one segment or the other (Matthews 1990). A different and more complete view of the local assemblage emerged when all 200 m of stream were combined per site, relative to the structure that would have been shown by only one or the other of the 100 m segments.

In some simple habitats, like shallow, sandy streams in the Great Plains, samples of 100 m or less may provide an adequate sample of the community—that is, when repeated seine hauls continue to produce "more of the same species"—and no new species or potential microhabitats seem likely. In June 1995, sampling small creeks to wide rivers in Kansas, Nebraska, and Missouri, our field notes indicate that our sampling included a range of stream reaches from 30 to 200 m, with a mean reach of 96.8 m sampled at 31 sites. Of those, several of the shortest reaches plus one "backwater only," where the Platte River was in snowmelt flood, were omitted from subsequent analyses (Marsh-Matthews and Matthews 2000), as we judged them to be inadequate to represent the structure of the fish community.

One also must make value judgments about the adequacy of a sample, post hoc, and in some cases accept that, in spite of best efforts, some sites are not sampled sufficiently to allow characterization of the fish assemblage. Collecting across Kansas and Nebraska in 1995, we devised, a bit tongue in cheek, a quality of data quotient (QDQ), scoring each site after completing the sampling, on a scale from 1 (for horrible conditions for seining, full of snags, many hauls aborted) to 10 (for the "perfect" place to take a sample, with no obstructions, gradually sloping stream bottoms of firm sand or gravel, and gentle currents). For 16 sites we recorded a QDQ ranging from a 1 to two 9s, with a mean of 5.15. The sites with the extremely low QDQs, corresponding to virtually impossible collecting conditions, were omitted from future analyses. While decisions to include or omit a site from analyses are subjective, we suggest that judgments based on experience are probably a fair way to assess whether a sample is adequate. We note, however, that some samples of relatively short reaches, or where there is limited water or few fish, may be completely valid, especially if part of a longer-term study at fixed sites in headwaters where it can be important to document paucity of fish or habitats during conditions such as a drought.

Another consideration is whether to remove and preserve fishes. For whole-community surveys, it is essential to take the full collection of fishes (albeit possibly measuring and releasing large individuals of easily recognized species such as gamefish or gars and duly recording in field notes), especially for smaller-bodied species or juveniles of any species for which identification can be difficult in the field. With careful examination of whole-community samples in the laboratory, we almost always find one or two individuals of species that we failed to see in the field, and these are likely to be locally rare species that were overlooked in making the field samples. So, most of our collections have been taken as whole collections in the field and sorted later in the laboratory. We make exceptions to this practice when we are making repeated, frequent sampling at fixed sites, for example, monthly samples in small streams. There, once vouchers have been collected and we are familiar with all of the species likely present, we use waterproof field sheets to count and release fish, retaining for the laboratory any individuals of doubtful identity or newly detected species. Thus, while we advocate full sampling for community studies, we balance that with the need to take care to not bias a fish community by our own removal of fish in small streams over short periods, such as sampling month to month.

In relatively large streams, with pools 10 to 20 m wide and complex habitats, removal of a sample of fishes has little impact on the local assemblage, and what we take is a subsample and is in no way intended to remove all of the fish in a reach. Evidence for this practice comes from times when we sampled the same reach of streams over short periods with removal. In July 1995, we sampled the most downstream long-term site in Piney Creek, Arkansas (see chap. 2), using our usual techniques. Two days later, we resampled the site with two other colleagues and found approximately the same species, in similar proportions to those detected previously. At a midreach site (BR-5) in Brier Creek, Oklahoma, we obtained two removal samples a week apart. The two collections produced totals of 1559 and 1235 individual fish (for more detail, see the paragraph on repeatability, below). In other words, in spite of having thoroughly sampled both of these sites in a first collection, we found extremely similar fishes present from days to a week later in the second sample. Finally, in repeated sampling over longer periods, we see no reduction in the numbers of fish taken. In Blaylock Creek of the Little Missouri River headwaters in southwest Arkansas, for example, WJM made seining collections on seven dates from November 1989 to August 1992. Numbers of fish taken on those dates were 643, 454 (in high water with difficult sampling), 989, 645, 704, 677, and 679. Thus there was no evidence for reduction in numbers of fish present over that period despite removal of individuals from the community.

Repeatability of sampling is an important issue. As indicated above, the goal of community sampling is to take all species present at each site in proportion to their relative abundance as a representative subsample. The ability of our sampling to meet this criterion is reflected in two examples. In 1981, WJM directed class collections at six sites in Brier Creek in June and again in July. In June we collected 1167 individuals, and in July we collected 1025 individuals at the same sites, although by July the headwater sites had partially dried, so less water was available. Faunal resemblance (Ross et al. 1985) between the two surveys was very high (Morisita's index = 0.93). On 19 June 2001, WJM and two assistants sampled a 300 m reach at BR-5 for 90 min (field number WJM 2972). On 26 June 2001, WJM and EMM with one assistant sampled the same reach with similar effort (WJM 2978), but without having data from WJM 2972, as that sample had not yet been counted; that is, the second collection was made "naively," without detailed information from the first. The two collections produced similar numbers of fish, as noted above. And we took 15 species in the first sample and 14 species in the second, with a percentage similarity based on abundance of 95.9% between the two collections. These comparisons suggest that our seining techniques provide a repeatable sampling of the fish community at individual sites or throughout a watershed.

Detectability of different fish species, or of different-sized individuals within a species, is also important. Some species, such as water-column minnows or topminnows, are easy to capture in pools with a seine. Benthic fishes in riffles are also relatively easy to capture in kicksets, if substrate is disturbed deeply enough to dislodge madtom catfishes, which (during the daytime) burrow more deeply into the substrate than darters and sculpins. There are caveats, however. After kicking riffles in the Roanoke River drainage for two years, I had a relatively small sample of the Roanoke Logperch (*Percina rex*). Then Bob Jenkins, of Roanoke College, told me that in spite of being darters, Roanoke Logperch usually swim above the substrate of riffles, in the water column, and are most reliably captured by seining rapidly downstream through riffles. Some other difficult fish to detect by seining in typical plains streams are bullhead catfishes (*Ameiurus* spp.), because they often (as observed while snorkeling in Brier Creek) occupy holes in stream banks or hide under larger stones, where a seine can easily pass over them. Larger individuals of large-bodied species such as black basses (*Micropterus* spp.) or suckers (Catostomidae), with greater ability to avoid a seine, also are probably underrepresented. We do capture adults of these species, but we more reliably collect their young-of-year or juveniles, sometimes in large numbers. But seining is probably not a reliable way to measure total biomass of large-bodied species such as game fish. Electrofishing (often used by management agencies) probably gives a more reliable estimate of biomass for large species. For simple detection (presence) of most species, however, we suspect that thorough seining by experienced operators is as likely to give a representative assessment of their presence and relative abundance as any other technique, short of a total kill of a stream reach by rotenone (which takes a huge effort by a large field crew and is now used only rarely in stream fish assessments).

Another issue related to detectability is simply that numerous species can be present but truly rare in a watershed. Within a reach we sample only perhaps 40% of the surface area, and even in well-surveyed systems like Piney and Brier Creeks, our sampling reaches represent only a tiny fraction of the total length of all the streams in the watershed. Thus, in any survey, most of a watershed is not sampled, and especially for relatively rare, mobile species, their detection or lack thereof in any particular sample may be due to the chance of their moving in or out of the space we actually sample on a given day. For example, in Piney Creek we occasionally capture Ozark Shiner (*Notropis ozarcanus*). In many surveys we take none, but in one recent survey we took 24 individuals, the most we had ever found at one time. We doubt that the species is actually disappearing and reappearing in the Piney Creek watershed and instead at-

FIG. 1.5 William Matthews counting fish by snorkeling in a pool of Brier Creek, Oklahoma.

tribute its occasional appearance in our samples merely to its overall rarity and the chance of finding it or not at a particular place and time.

Finally, how does detection of the composition of a community by seining compare to results we would get from snorkeling or backpack electrofishing? Snorkeling is a good way to detect and count fish in clear streams (Matthews et al. 1994; Matthews and Marsh-Matthews 2006a), but it is only useful in waters sufficiently clear that "bank-to-bank" visibility is possible in small pools (fig. 1.5).

Across most systems in our region of the United States, water clarity will not allow snorkeling as a general survey tool. Backpack electrofishing is possible in essentially all of the streams we have included in this book, but it requires a team of at least several persons, with operators well trained to recover small fish as well as the more obvious large ones. In October 1992, a team composed of Betty Cochran (US Department of Agriculture Forest Service), Fran Gelwick, and WJM (unpub. data) compared seining, snorkeling, and backpack electroshocking at seven sites with clear water on Blaylock Creek in Ouachita National Forest, Arkansas. Some of the sites were easy to seine, consisting of open pools and relatively unobstructed gravel riffles, and others were challenging to seine, with upthrust shale rock, large cobbles, or other obstructions. Under those conditions, snorkeling provided the highest total counts of fish. In comparing results for seining and electrofishing for the more common species, seining produced more individuals than electrofishing for three minnow or topwater species (Bigeye Shiner, *Notropis boops*; Redfin Shiner, *Lythrurus umbratilis*; and Brook

Silversides, *Labidesthes sicculus*), electrofishing produced more individuals for High-land Stoneroller (*Campostoma spadiceum*), which tend to hide in substrates when pursued by seine, and also for Creek Chub (*Semotilus atromaculatus*), Northern Hog Sucker (*Hypentelium nigricans*), Yellow Bullhead (*Ameiurus natalis*), Green Sunfish (*Lepomis cyanellus*), Longear Sunfish (*Lepomis megalotis*), and Smallmouth Bass (*Micropterus dolomieu*). For four other common species, including Striped Shiner (*Luxilus chryso-cephalus*), Northern Studfish (*Fundulus catenatus*), Greenside Darter (*Etheostoma blen-nioides*), and Orangebelly Darter (*Etheostoma radiosum*), captures were approximately equal between the two techniques. In general, the three species for which seining produced more individuals were small-bodied pool dwellers, whereas those for which electrofishing produced more individuals included the larger-bodied centrarchids, a sucker, and two relatively large-bodied minnows.

Overall, both seining and electrofishing were effective in this comparison in one stream but gave somewhat different results in detection or relative abundance of species, as would be expected. The most important point, especially for long-term surveys of fixed sites or for spatially broad surveys across a wide geographic area, is that different techniques might give different pictures of the fish communities, but it is important to use the same technique throughout. Thus, having started our careers with small-meshed seines as the primary tool for sampling fish communities, we have continued this approach across five decades, and for all of the reasons provided above, we make the assumption that seining has given a reasonably comparable view, with "adequate sampling," of these fish communities across time and space.

Experimental Approaches

The field studies on which we have drawn for most of this book (see chap. 2 for details on sites and systems) were often complemented by field or laboratory experiments to test mechanisms related to our observations in natural habitats. These experiments, conducted over many years and for different projects, ranged in size and complexity from experimental habitats as small as a simple 1 L flask to compare critical thermal maxima (CTMs) of different species, to in-stream pens or divisions of whole pools in natural streams, to large, modular outdoor experimental stream systems that we designed and built (Matthews et al. 2006). The history of use or development of many of our experimental approaches includes chambers developed by WJM to determine selectivity of different fish species in gradients of temperature, oxygen, pH, or other water-quality factors (Matthews and Hill 1979; Matthews 1987); dividing whole Brier Creek pools with construction plastic to determine effects of algae-grazing Central Stonerollers (*Campostoma anomalum*) on stream ecosystems (Power et al. 1985); construction of arrowhead-shaped open or closed pens in Baron Fork of the Illinois River to additionally determine effects of fish on ecosystem processes (Gelwick et al. 1997); and use of troughs or small circulating experimental mesocosms in a greenhouse to test effects of the stonerollers under controlled conditions with more replication (Partridge 1991; Vaughn et al. 1993).

FIG. 1.6 Experimental mesocosms used in numerous studies of stream fish communities. These are located at the Aquatic Research Facility at the University of Oklahoma in Norman.

Finally, after having seen many of our laboriously constructed in-stream pens washed away by floods before or during experiments, we decided to construct a large array of experimental riffles and pools, initially at the University of Oklahoma Biological Station, then a second set at UOBS, and, more recently, a third set on the main university campus in Norman. The system, described in detail in Matthews et al. (2006), consists of interspersed pool and riffle units, with circular pools approximately 1.8 m in diameter and 0.5 m deep, and riffles 1.8 m long, with depths and current speeds controlled by various combinations of pumps that recirculate water from a footbox to a headbox (fig. 1.6). The design allows isolation of experimental units as small as a single riffle and pool, connection of two or three pools and riffles as an experimental unit, or even using rows of six to twelve pools and intervening riffles as a larger stream unit.

Former graduate students, colleagues, and students of our students have now replicated the system at other facilities from South Dakota to central Texas to southern Mississippi, providing great opportunity for future coordinated experiments across latitudes. By the time of Matthews et al. (2006), at least 39 species had been used in experiments in these systems, 20 of which successfully spawned in the experimental streams, and comparisons had been made between physical conditions in the natural Kings Creek and a set of these streams built by Keith Gido at the Konza Prairie Reserve of Kansas State University. By 2006, at least 10 graduate theses had been based all or in part on these stream systems across numerous universities. We continue to use our artificial streams at the University of Oklahoma to test factors such as the effects of piscivores on colonization success of minnows (Marsh-Matthews et al. 2011, 2013).

But setting up biologically meaningful experiments with applicability to the real systems (Benton et al. 2007) can be tricky. For example, Skelly and Kiesecker (2001) showed that across a large number of experiments on larval anurans, the choice of size or kind of experimental arena biased the likelihood of finding that competition was important. Experiments can be simple to design sitting at a desk indoors. They are much more difficult and labor intensive to conduct successfully either in the field or in mesocosms. We have had good success in general using separated sections of pools, large pens in streams, or fencing to isolate pools to carry out various ecological experiments (e.g., Power et al. 1985; Gelwick and Matthews 1992; Gelwick et al., 1997), but at least half of the instream pens that were built laboriously at various times and in different streams by WJM in collaboration with Mary Power, Marsha Stock, Fran Gelwick, or Caryn Vaughn were washed away by floods before experiments were started or completed! Field manipulations are golden when they work and can provide strong inference about many questions in stream fish ecology (e.g., Gelwick and Matthews 1992), but be prepared for failures unless you get lucky. That is a major reason the modular experimental streams of Matthews et al. (2006) were designed, in an effort to incorporate some degree of reality in fully replicated stream units that did not wash away. The trade-off between the greater reality of conditions in the field and the safer bet on controlled conditions in laboratory or mesocosm experiments is a choice each experimenter has to make based on the question, the potential to actually build and maintain pens or other structures throughout an entire experiment, and the degree of reality that is demanded to answer the question. And setting up any controlled arena and testing responses of fish makes the assumption that what we see in experimental units can translate to the field (Fausch 1988).

Finally, any experimenter using artificial units must be vigilant that nothing is wrong with any of the species in the experiment. Based on our field observations and experiments on the effects of stonerollers on algae in streams (Matthews et al. 1987; Gelwick and Matthews 1992), for example, WJM wanted to compare the effects of stonerollers to those of another algivore, Southern Redbelly Dace (*Chrosomus erythrogaster*). So, we put a lot of effort into obtaining enough dace from the field for an experimental trial and set them up in greenhouse stream units to compare the effects of the two species over several weeks. But the dace did not tolerate the experimental conditions well and began to look sickly and behave erratically. We hoped they would get well and the experiment could proceed, but they did not, and we had to scrap the entire effort.

It should be obvious that one does not base an experiment on sick fish, but more subtle divergence from normal behavior could be hard to detect. In our experiments (e.g., Matthews and Marsh-Matthews 2006b; Marsh-Matthews and Matthews 2010) we have usually worked with fish species with which we have much experience from field observation (Matthews et al. 1994). Familiarity with the typical behaviors of one's fish can help any experimenter to design good experiments that have biological reality and helps in detecting any troubles with behavior of the fishes. If they don't look right, they probably are not, and you have to give serious consideration to abandoning the experiment.

Summary

Throughout this book we use the term "community" to describe species that predictably occur together in stream reaches or across whole watersheds and that have the potential to interact. Our definition includes a temporal component, based on repeated sampling across time. In subsequent chapters we will examine communities at different spatial scales. The communities described in this book were (with one exception) sampled by seining with small-meshed seines, targeting all possible microhabitats and kinds of fish in a reach in proportion to their abundance. In chapter 2 we describe a large number of local sites or whole streams where we have used these techniques to collect fish during the last 40 years. One of the strengths of the book is that we have sampled the fishes in all of these systems in the same way, so we can make broad comparisons between streams or syntheses across widely different fish communities. We have also noted our approach to experimental investigations of mechanics and to performing trials to help us understand, through controlled conditions, what may be underlying our findings about communities derived from our fieldwork.

References

Allee, W. C., O. Park, A. A. Emerson, T. Park, and K. P. Schmidt. 1949. Principles of animal ecology. W. B. Saunders, Philadelphia, PA.

Bêche, L. A., and V. H. Resh. 2007. Short-term climatic trends affect the temporal variability of macroinvertebrates in California "Mediterranean" streams. Freshwater Biology 52: 2317–2339.

Begon, M., C. R. Townsend, and J. L. Harper. 2006. Ecology: from individuals to ecosystems. Blackwell, Oxford, UK.

Benton, T. G., M. Solan, J. M. J. Travis, and S. M. Salt. 2007. Microcosm experiments can inform global ecological problems. Trends in Ecology and Evolution 22:518–521.

Cody, M. L., and J. A. Smallwood, eds. 1996. Long-term studies of vertebrate communities. Academic Press, San Diego, CA.

Davey, A. J. H., and D. J. Kelly. 2007. Fish community responses to drying disturbances in an intermittent stream: a landscape perspective. Freshwater Biology 52:1719–1733.

Fausch, K. D. 1988. Tests of competition between native and introduced salmonids in streams: what have we learned? Canadian Journal of Fisheries and Aquatic Science 45:2238–2246.

Fausch, K. D., and R. G. Bramblett. 1991. Disturbance and fish communities in intermittent tributaries of a western Great Plains river. Copeia 1991:659–674.

Gelwick, F. P., and W. J. Matthews. 1992. Effects of an algivorous minnow on temperate stream ecosystem properties. Ecology 73:1630–1645.

Gelwick, F. P., and W. J. Matthews. 1997. Effects of algivorous minnows (Campostoma) on spatial and temporal heterogeneity of stream periphyton. Oecologia 112:386–392.

Gelwick, F. P., M. S. Stock, and W. J. Matthews. 1997. Effects of fish, water depth, and predation risk on patch dynamics in a north-temperate river ecosystem. Oikos 80: 382–398.

Gido, K. B., and D. A. Jackson, eds. 2010. Community ecology of stream fishes: concepts, approaches, techniques. American Fisheries Society Symposium 73. American Fisheries Society, Bethesda, MD.

Grossman, G. D., P. B. Moyle, and J. O. Whitaker Jr. 1982. Stochasticity in structural and functional characteristics of an Indiana stream fish assemblage: a test of community theory. American Naturalist 120:423–454.

Hoeinghaus, D. J., K. O. Winemiller, and J. S. Birnbaum. 2007. Local and regional determinants of stream fish assemblage structure: inferences based on taxonomic vs. functional groups. Journal of Biogeography 34:324–338.

Kiernan, J. D., and P. B. Moyle. 2012. Flows, droughts, and aliens: factors affecting the fish assemblage in a Sierra Nevada, California, stream. Ecological Applications 22:1146–1161.

Lake, P. S. 2011. Drought and aquatic ecosystems: effects and responses. Wiley-Blackwell, Oxford, UK.

Marsh-Matthews, E., and W. J. Matthews. 2000. Geographic, terrestrial and aquatic factors: which most influence the structure of stream fish assemblages in the midwestern United States? Ecology of Freshwater Fish 9:9–21.

Marsh-Matthews, E., and W. J. Matthews. 2010. Proximate and residual effects of exposure to simulated drought on prairie stream fishes. Pages 461–486 in K. B. Gido and D. A. Jackson, eds. Community ecology of stream fishes: concepts, approaches, and techniques. American Fisheries Society Symposium 73. American Fisheries Society, Bethesda, MD.

Marsh-Matthews, E., W. J. Matthews, and N. R. Franssen. 2011. Can a highly invasive species re-invade its native community? The paradox of the Red Shiner. Biological Invasions 13: 2911–2924.

Marsh-Matthews, E., J. Thompson, W. J. Matthews, A. Geheber, N. R. Franssen, and J. Barkstedt. 2013. Differential survival of two minnow species under experimental sunfish predation: implications for re-invasion of a species into its native range. Freshwater Biology 58:1745–1754.

Matthews, W. J. 1986. Fish faunal structure in an Ozark stream: stability, persistence, and a catastrophic flood. Copeia 1986:388–397.

Matthews, W. J. 1987. Physicochemical tolerance and selectivity of stream fishes as related to their geographic ranges and local distributions. Pages 111–120 in W. J. Matthews and D. C. Heins, eds. Community and evolutionary ecology of North American stream fishes. University of Oklahoma Press, Norman, OK.

Matthews, W. J. 1990. Fish community structure and stability in warmwater midwestern streams. US Fish and Wildlife Service Biological Report 90:16–17.

Matthews, W. J. 1998. Patterns in freshwater fish ecology. Chapman and Hall, New York, NY.

Matthews, W. J., and D. C. Heins, eds. 1987. Community and evolutionary ecology of North American stream fishes. University of Oklahoma Press, Norman, OK.

Matthews, W. J., and L. G. Hill. 1979. Influence of physico-chemical factors on habitat selection by Red Shiners, *Notropis lutrensis* (Pisces: Cyprinidae). Copeia 1979:70–81.

Matthews, W. J., and E. Marsh-Matthews. 2006a. Persistence of fish species associations in pools of a small stream of the southern Great Plains. Copeia 2006:696–710.

Matthews, W. J., and E. Marsh-Matthews. 2006b. Temporal changes in replicated experimental stream fish assemblages: predictable or not? Freshwater Biology 51:1605–1622.

Matthews, W. J., R. C. Cashner, and F. P. Gelwick. 1988. Stability and persistence of fish faunas and assemblages in three midwestern streams. Copeia 1988:945–955.

Matthews, W. J., K. B. Gido, G. P. Garrett, F. P. Gelwick, J. G. Stewart, and J. Schaefer. 2006. Modular experimental riffle-pool stream system. Transactions of the American Fisheries Society 135:1559–1566.

Matthews, W. J., B. C. Harvey, and M. E. Power. 1994. Spatial and temporal patterns in the fish assemblages of individual pools in a midwestern stream (U.S.A.). Environmental Biology of Fishes 39:381–397.

Matthews, W. J., E. Marsh-Matthews, R. C. Cashner, and F. Gelwick. 2013. Disturbance and trajectory of change in a stream fish community over four decades. Oecologia 173:955–969.

Matthews, W. J., A. J. Stewart, and M. E. Power. 1987. Grazing fishes as components of North American stream ecosystems: effects of *Campostoma anomalum*. Pages 128–135 in W. J. Matthews and D. C. Heins, eds. Community and evolutionary ecology of North American stream fishes. University of Oklahoma Press, Norman, OK.

May, R. M., ed. 1976. Theoretical ecology: principles and applications. W. B. Saunders, Philadelphia, PA.

McIntosh, R. P. 1995. H. A. Gleason's "individualistic concept" and theory of animal communities: a continuing controversy. Biological Reviews 70:317–357.

Meffe, G. K., and A. L. Sheldon. 1990. Post-defaunation recovery of fish assemblages in southeastern blackwater streams. Ecology 71:657–667.

Mittlebach, G. G. 2012. Community ecology. Sinauer Associates, Sunderland, MA.

Morin, P. J. 2011. Community ecology, 2nd ed. Wiley-Blackwell, Oxford, UK.

Moyle, P. B., and B. Vondracek. 1985. Persistence and structure of the fish assemblage in a small California stream. Ecology 66:1–13.

Partridge, W. D. 1991. Effects of *Campostoma anomalum* on the export of various size fractions of particulate organic matter in streams. MS thesis. University of Oklahoma, Norman, OK.

Power, M. E., W. J. Matthews, and A. J. Stewart. 1985. Grazing minnows, piscivorous bass and stream algae: dynamics of a strong interaction. Ecology 66:1448–1456.

Ross, S. T., W. J. Matthews, and A. A. Echelle. 1985. Persistence of stream fish assemblages: effects of environmental change. American Naturalist 126:24–40.

Schlosser, I. J. 1982. Fish community structure and function along two habitat gradients in a headwater stream. Ecological Monographs 52:395–414.

Shelford, V. E. 1931. Some concepts of bioecology. Ecology 12:455–467.

Skelly, D. K., and J. M. Kiesecker. 2001. Venue and outcome in ecological experiments: manipulations of larval anurans. Oikos 94:198–208.

Smith, C. L., and C. R. Powell. 1971. The summer fish communities of Brier Creek, Marshall County, Oklahoma. American Museum Novitates 2458:1–30.

Vaughn, C. C., F. P. Gelwick, and W. J. Matthews. 1993. Effects of algivorous minnows on production of grazing stream invertebrates. Oikos 66:119–128.

Winemiller, K. O. 1991. Ecomorphological diversification in lowland freshwater fish assemblages from five biotic regions. Ecological Monographs 61:343–365.

Winemiller, K. O. 2010. Preface: stream fish communities from patch dynamics to intercontinental convergences. Pages 23–28 in K. B. Gido and D. A. Jackson, eds. Community ecology of stream fishes: concepts, approaches, and techniques. American Fisheries Society Symposium 73. American Fisheries Society, Bethesda, MD.

CHAPTER 2

The Stream Fish Community Study Systems

Descriptions of the Study Systems

The databases used in this book (table 2.1) range temporally, from long term in ecological time (across decades) to short term (months to a few years), and spatially, with many sites throughout the southern Great Plains or in the Red River basin in Oklahoma (fig. 2.1). In this chapter we describe these databases, the stream systems, and the projects that generated the data. There are 11 named databases in table 2.1, including 9 streams with multiple sample sites and 22 streams with a single sample site, all sampled by seining at least four times. One other long-term data set is from snorkel counts of fish in 14 consecutive pools in Brier Creek, spanning 30 years from 1982 to 2012, which we treat separately from the data sets from seine sampling. We also have two spatially broad databases. One is for streams sampled once each in 1978 and again in 1995 across a 1500 km north–south gradient in the southern Great Plains from Nebraska and Iowa to south Texas. The other is a data set for sites sampled once each, across many years, at 143 localities throughout the Red River basin in Oklahoma. All of the samples described in this chapter were made by one or both of the authors, with a few exceptions in which reliable colleagues or students took some of the samples for a data set (noted in those specific cases) using the same techniques that we did.

Descriptions of the temporal data sets are organized generally from long- to short-term studies, followed by the two spatial data sets. Information based on these primary databases is supplemented as appropriate by various other field collections we have made, which are described in the applicable chapters. But here we describe the primary databases used in the book—where, when, how, or why the collections were made—and we also provide detailed descriptions of the systems.

Table 2.1. Global and local streams included in this book

Database	Stream	County	Date Range	Number of Surveys	Number of Sites	Map Site
Piney Creek, Arkansas	Piney Creek	Izard	1972–2012	12	12	1
Brier Creek, Oklahoma	Brier Creek Seine Collections	Marshall	1976–2012	16	6	2
	Brier Creek Snorkel Surveys		1982–2012	22	14	
Kiamichi River, Oklahoma	Kiamichi River	Pushmataha and LeFlore	1981–2014	5	6	3
Roanoke River, Virginia	Roanoke River	Roanoke and Montgomery	1978–1979	6	10	
Ouachita National Forest streams, Arkansas	Bread Creek	Saline	1989–1991	4	6	4
	South Fork Alum Creek	Saline	1989–1991	4	6	5
	Crooked Creek	Montgomery	1989–1991	4	6	6
	Blaylock Creek	Montgomery and Polk	1989–1991	4	6	7
Grand River tributaries, northeast Oklahoma	Little Elm Creek	Ottawa	2005–2007	13	1	8
	Garrett Creek	Ottawa	2004–2007	15	1	9
	Cow Creek	Ottawa	2004–2014	27	1	10
	Coal Creek	Ottawa	2004–2014	22	1	11
North Canadian River tributaries, central Oklahoma	Mustang Creek	Canadian	2010–2011	4	1	12
	Crutcho Creek	Oklahoma	2010–2011	4	13	13
	Choctaw Creek	Oklahoma	2010–2011	4	1	14
	Gar Creek	Seminole	2010–2011	4	1	15
Canadian River	Canadian River	McClain and Cleveland	1976–1977	4	1	16

(continued)

Table 2.1. Global and local streams included in this book (*continued*)

Database	Stream	County	Date Range	Number of Surveys	Number of Sites	Map Site
Other Oklahoma streams	Ballard Creek	Adair	1981–2010	5	1	17
	Illinois River, Lake Frances	Adair	1981–2010	5	1	18
	Tyner Creek headwater	Adair	1998–2010	5	1	19
	Baron Fork, Illinois River	Cherokee	1998–2004	4	1	20
	Salt Fork Red River	Greer	2000–2013	5	1	21
	Blue River	Johnston	1992–2013	5	1	22
	Morris Creek	LeFlore	1998–2006	5	1	23
	Hickory Creek	Love	1981–2009	5	1	24
	Hauani Creek	Marshall	1982–2008	4	1	25
	Little Glasses Creek	Marshall	2005–2013	10	1	26
	Borrow ditch, US 70	McCurtain	1982–2006	7	1	27
	Glover River	McCurtain	1976–2000	4	1	28
	Lukfata Creek, Borrow pit	McCurtain	1976–2004	5	1	29
	Chigley Sandy Creek	Murray	1993–2008	4	1	30
Interior Plains, central United States	sites in 13 drainages from Iowa to Texas		1978 and 1995	2	65	
Red River and Arkansas River basins in Oklahoma and Arkansas			1976–2014		143	

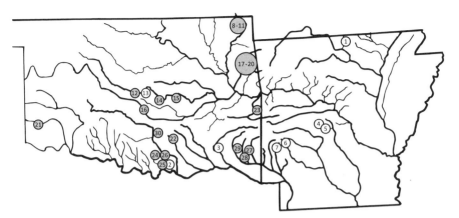

FIG. 2.1 Map of 30 of the 31 study systems used in the book. The Roanoke River is in Virginia and not on this map. Numbers correspond to map sites in table 2.1. Study systems indicated with open circles are those with multiple collecting sites within a stream (global), and those indicated by shaded circles represent single site systems (local).

Piney Creek, Arkansas

Our longest collections began in 1972, when WJM started an MS thesis at Arkansas State University under the direction of George Harp (Matthews 1973; Matthews and Harp 1974). Graduate students with George, or his colleague Ken Beadles, surveyed fish in the late 1960s and 1970s in whole watersheds or regions of Arkansas that had never before been sampled in detail. These surveys produced information on fish species presence or abundances in at least eight large creeks to small rivers in the Ozark uplands or nearby parts of north Arkansas and south Missouri (Matthews 1982). For his MS thesis, WJM sampled Piney Creek, Izard County, Arkansas (site 1 in fig. 2.1), at 15 fixed sites across the watershed in August 1972, December 1973, and April 1973 (Matthews 1973). He subsequently sampled 12 of those sites from August 1982 to August 1983 (a period including a "flood of the century"; Matthews 1986). WJM and EMM made seasonal samples at the 12 sites from December 1994 to July 1995 and again in summer 2006 and 2012 (with Ginny Adams and Reid Adams in 2012), which bracketed a second flood of the century in spring 2008 (Funkhouser and Eng 2009; Matthews et al. 2014). The Piney Creek data set used in most analyses in this book includes 11 complete surveys of all 12 long-term sites (fig. 2.2). Collections made by WJM in August 1972 were omitted from most analyses because in this first survey, riffle habitats were not adequately sampled (as WJM had not yet learned that kicksets are more effective in riffles or rapids than pulling seine hauls). Supplementary collections at five of the sites were made to provide more detail on variation among summers, or to track changes after the great flood in spring 2008, with G. and R. Adams (Matthews et al. 2014). Matthews and Marsh-Matthews (2016) provide

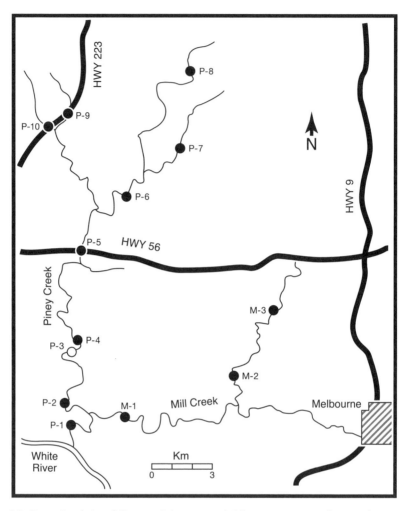

FIG. 2.2 Piney Creek, Izard County, Arkansas, with 12 permanent sample sites shown by numbered solid dots.

an overview of all Piney Creek surveys since inception of the project, with assessment of long-term community trajectories in a loose equilibrium.

The Piney Creek watershed, described in Matthews and Harp (1974), Matthews et al. (1978), Ross et al. (1985), Matthews (1986), Matthews et al. (1988), Matthews et al. (2014), and Matthews and Marsh-Matthews (2016), is dendritic, with a mainstem approximately 47 km long, draining 450 km², and joining the White River east of Calico Rock, Arkansas. From our uppermost mainstem headwater site (P-8) to the lowermost mainstem site (P-1), gradient averages 2.29 m/km. The watershed is in the rural southern Ozark Mountains, with rugged hill slopes and tall bluffs of limestone and dolomite lower in the watershed (see plate 1), grading to more gently sloping landscapes in the headwaters. There are no incorporated cities, no heavy manufacturing, and no large areas of row crops. Grazing is largely limited to cattle, with no large commercial livestock operations. Downstream, much of the creek is bordered by dense forests with limited acreage devoted to pasture. In headwaters the terrain is less steep and more land is in pasture, but at all sites there is substantial riparian forest of oak-hickory or pine. All our study sites are on private land, and we owe a great debt to the families (some now in the third generation) who have allowed us access to their land.

Ross et al. (1985) showed that Piney Creek is benign on the harsh–benign gradient of Peckarsky (1983). Many small springs or spring seeps provide perennial flow in the lower mainstem. Headwaters also flow continuously in all but extreme drought conditions. Even during the "exceptional" drought of 2012 (Matthews et al. 2014), flow was interrupted only at two headwater sites. Our measurements of physicochemical conditions in the 1970s and in 2012 (during one of the worst drought periods on record for north Arkansas) never showed water-quality conditions potentially stressful to warm-water fishes (Ross et al. 1985; Matthews et al. 2014). Overall, Piney Creek is a clear, warm-water, upland stream with excellent water quality and high diversity of habitat for native fishes, with conditions essentially unchanged since the first surveys in the 1970s.

Brier Creek, Oklahoma

Brier Creek, Marshall County, south Oklahoma (site 2 in fig. 2.1), has a long history of fish community studies. The first comprehensive ichthyological surveys were by C. Lavett Smith and C. R. Powell, who seined weekly at seven sites in summer 1969 (Smith and Powell 1971). In summer 1976, WJM was in a class taught by Tony Echelle at the University of Oklahoma Biological Station (UOBS) that sampled fish at six of the Smith and Powell sites (fig. 2.3). Starting in 1981, WJM made collections at those same sites with classes he taught at the UOBS, with colleagues including Bob Cashner and Fran Gelwick, and all with EMM after 1996. Early collections of fish at five of the six sites on Brier Creek (excluding one extreme headwater site) were described in detail in Ross et al. (1985) and Matthews et al. (1988), and Matthews et al. (2013) summarized collections across all six sites. The collections and collectors from 1981 to 2008, and their collecting efforts, are described in detail in Matthews et al. (2013) or in online supplementary material. Here we expand that data set to include our survey in 2012.

In addition to long-term seining collections, WJM made 22 snorkeling surveys in 14 pools of a 1 km reach of Brier Creek from 1982 to 2012 (Power and Matthews 1983; Matthews et al. 1994; Matthews and Marsh-Matthews 2006). The reach surveyed by snorkeling was just upstream of seining site BR-5 (fig. 2.3), with pools about 15 to 90 m long, separated by gravel or cobble riffles, and pool substrates varying from sand and gravel to cobble or bedrock, typically with clear water and good visibility.

Brier Creek has also been the site of general ecological studies on trophic cascades, predator–prey relationships, long-term change in minnow species; water–land

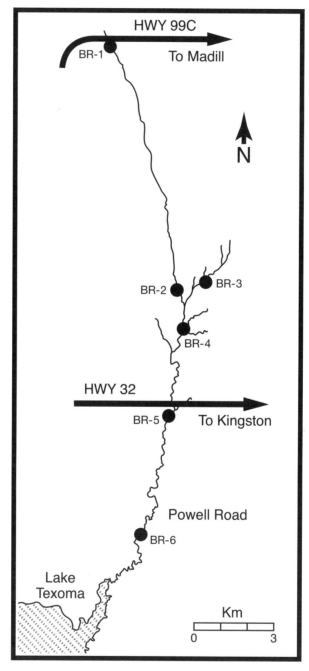

FIG. 2.3 Brier Creek, Marshall County, Oklahoma, with permanent sample sites shown by numbered solid dots.

interactions, or effects of floods on algae (Power and Matthews 1983, Power et al. 1985, Harvey 1987, 1991; Gelwick and Matthews 1992; Matthews and Marsh-Matthews 2006, 2007; Marsh-Matthews and Matthews 2010; Marsh-Matthews et al. 2011; Wesner 2011, 2013; Gillette 2012).

Brier Creek flows into the Red River arm of Lake Texoma, an impoundment of the Red and Washita rivers on the Oklahoma-Texas border, completed in 1946. The watershed is rural, with no incorporated communities, and land use is dominated by pasture or forest (Matthews et al. 2013). The creek is approximately 22 km long, draining 59.6 km^2, with only 1 small tributary. From the narrow pasture channel in the extreme headwaters (BR-1) to our lowermost site (BR-6) (see plate 2), which is 3.7 km upstream from Brier Creek's confluence with Lake Texoma at normal lake level, the mainstem is 18 km long with an average gradient of 3.9 m/km. The creek has mostly riffle-pool structure with bedrock-gravel substrate, although erosional deposits of sand or silt cover the substrate near the confluence with Lake Texoma (Matthews and Marsh-Matthews 2007). Flow is primarily from runoff, but spring seeps above our site BR-5 maintain flow in the lower reaches of the creek most of the year. Brier Creek is reduced to isolated pools in dry years, however, and in extreme drought, long reaches of the creek dry completely (Matthews 1987; Marsh-Matthews and Matthews 2010). Brier Creek also has had numerous extreme, erosional flood events during the span of our studies, with stage rises of 4.5 m or more, and with the worst flood of the century in October 1981 due to 66 cm of rain in 3 days (Matthews et al. 2013). Brier Creek is environmentally harsh (Ross et al. 1985), particularly in the headwaters, where temperatures can fluctuate as much as 10°C in a day, and dissolved oxygen in isolated pools is often below 1 ppm in the summer (Matthews 1987).

Kiamichi River, Oklahoma

The Kiamichi River (site 3 in fig. 2.1) occupies a long, narrow valley, bordered on the north and south by steep slopes of the Ridge and Valley Belt of the Ouachita Mountain Province (Johnson 2006). It arises as small runoff channels that converge near the Arkansas–Oklahoma border, then flows southwest through parts of four Oklahoma counties (LeFlore, Latimer, Pushmataha, and Choctaw) to its confluence with the Red River, south of Fort Towson, Oklahoma. Fishes were sampled comprehensively throughout the drainage by Jimmie Pigg, as summarized in Pigg and Hill (1974). Echelle and Schnell (1976) used multivariate factor analysis to define fish groups within the drainage. Matthews et al. (1988) assessed temporal change in local Kiamichi fish assemblages relative to those in Piney Creek and Brier Creek, based on sampling in all three systems from the 1970s to mid-1980s.

Four surveys at six mainstem sites (fig. 2.4) on the Kiamichi River were by classes taught by WJM or Bob Cashner at the UOBS in the summers of 1981, 1985, 1986, and 1987, with WJM participating in the first three trips. These six sites were again surveyed by WJM and graduate students in summer 2014. The lower mainstem, downstream of all our sample sites, is impounded by Hugo Lake, and the largest tributary

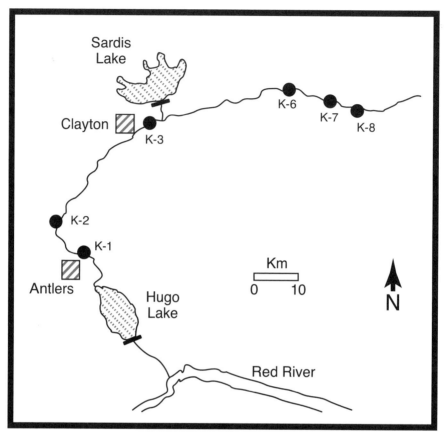

FIG. 2.4 Kiamichi River, southeast Oklahoma, with permanent sample sites shown by numbered solid dots.

to the upper river, Jackfork Creek, was impounded by Sardis Lake in 1983. The impoundment has the potential to affect fish at our three lowermost collecting sites, and by 1992, Pyron et al. (1998) showed distinct changes in fish in the Kiamichi River near the confluence of Jackfork Creek. But in our sampling in 1985–1987 and in 2014 we detected no substantive changes in fish at these mainstem sites relative to the first survey in 1981.

The Kiamichi River watershed drains 4719 km^2, with a total mainstem length of 278 km. Above our most downstream sampling site, at Antlers (K-1), the drainage area includes 2924 km^2, and the distance from our uppermost sampling site at Big Cedar (K-8) to the Antlers site is 164 km. The Kiamichi River is a trellised drainage, reaching only fifth order in spite of its length because there are few large tributaries to the mainstem. Between our upper and lower study sites the river mainstem has an average gradient of only 0.728 m/km. The Kiamichi River flows over mostly sandstones and shales, largely lacking calcareous strata. There are no large springs, and the river is fed mostly by runoff from the steep surrounding slopes (Oklahoma Water Resources Board 2000), which is substantial in wet years. But the entire drainage is vulnerable to drying in drought years, with serious consequences to aquatic biota (Allen et al. 2013; Vaughn et al. 2015).

The headwaters are clear with prominent riffle-pool structure, flowing over stony cobble or boulder substrate or gravel bars with dense stands of water willow (*Justicia americana*) (see plate 3). Farther downstream the river tends toward longer, wide pools interspersed with riffles, and the lowest collecting site (Antlers, K-1) is almost entirely one wide channel, typically rather turbid. The Kiamichi River watershed is mostly rural, with the largest community in our study area, Clayton, having a population of only 1012 in the 2010 census. Land cover on the surrounding slopes is almost entirely forest, with extensive pasturing for livestock in the valley. There is no heavy manufacturing, and there are no known major point sources of pollution.

Matthews et al. (1988) considered the Kiamichi River to be environmentally benign on a harsh-to-benign continuum. But in recent years a combination of drought and water retention in Sardis Lake (which controls 30% of the flow to the river) has made the lower river more vulnerable to dewatering of long reaches (Allen et al. 2013; Vaughn et al. 2015). Our sampling period from 1981 to 1987 included no flood or drought events that were unusual or extreme for the Kiamichi River. Between July 1981 and August 1987 there were numerous substantial stage rises, mostly in winter or spring, recorded at the US Geological Survey (USGS) gage at Clayton, Oklahoma, with 13 events exceeding 10,000 cfs up to 22,000 cfs. But data for gages on the Kiamichi River from the 1920s to the present indicate that such stage rises, especially in the winter, are common (C. Vaughn, pers. comm.) Also, in summer 1982 and 1983 there were extended periods of low flow, with 16 consecutive days of no flow at Clayton, but those low-flow events were not as extreme as some in more recent years (C. Vaughn, pers. comm.). So, the fish surveys from 1981 to 1987 spanned a period with some notable stage rises and substantial nonflowing days at the Clayton USGS gage, but these events were not unusual for the Kiamichi River, and certainly not as extreme for the system as were the two great floods in Piney Creek in 1982 and 2008, the flood in Brier Creek in October 1981, or extreme summer droughts in 1980 (Ross et al. 1985) or in recent years in Brier Creek (Marsh-Matthews and Matthews 2010; Matthews et al. 2013). Our collections in 2014 were made at a time with substantial flow.

Roanoke River, Virginia

The South Fork of the Roanoke River in Roanoke and Montgomery Counties, Virginia, arises on the slopes of the Blue Ridge Mountains at the junction of Goose and Bottom Creeks, and is joined by the North Fork of the Roanoke River at Lafayette, Virginia, forming the Roanoke River proper. In 1977, WJM joined the faculty of the Biology

Department at Roanoke College, Salem, Virginia, and began sampling fish in the upper Roanoke River drainage (e.g., Matthews and Styron 1981). He then selected nine sites on the South Fork of the Roanoke River or its tributaries and two sites on the Roanoke River below the junction of the North Fork and South Fork for monthly sampling (fig. 2.5) from May to November 1978, plus additional sampling in cold weather in March 1979. Roanoke College students Jeff Bek and Eric Surat carried out much of the field sampling in summer months, with WJM setting up the project and taking part in sampling in spring, fall, and winter. This work was stimulated by research on comparative use of habitat and foods by minnows (Surat et al. 1982) and darters (Matthews et al. 1982), but whole-community collections were taken at all times, separated by pool, riffle, or run. These samples also provided specimens for the ecomorphology study by Douglas and Matthews (1992).

Above our lowermost site the Roanoke River has a drainage area of about 754 km^2. The mainstem of the Roanoke River and South Fork upstream of our lowermost site is 27 km long from our lowest site (R-1, at Dixie Caverns; see plate 4) to the uppermost site on Bottom Creek. After originating on slopes of the Blue Ridge Mountains, the system flows through the Ridge and Valley Physiographic Province, with an overall gradient of 3.58 m/km between the most upstream and downstream sites. The drainage is dendritic, as the South Fork originates from the junction of two large creeks and the mainstem is joined farther downstream by substantial tributaries like Elliot Creek and many smaller branches or "runs." Tributaries range from second to fourth order, and the mainstem is fifth order at our fixed sites (Surat et al. 1982). Individual sites varied from 1 m wide in a small tributary to 25–30 m at 4 sites on the mainstem. Tributaries had locally higher gradients (approximately 12–22 m/km) than the mainstem, where local gradients ranged from about 2 to 9 m/km (Surat et al. 1982). Tributaries had distinct riffle-pool variation with bottoms ranging from sand to rubble or bedrock, whereas the mainstem locations had large riffles or rapids, deep pools, and long reaches of relatively uniform channel at some sites (Surat et al. 1982). The upper parts of the watershed were heavily forested, with cleared pastures near the river mainstem.

The spring-fed South Fork had clear water and perennial flow at even the smallest sites. At the time of our studies the South Fork Roanoke River drainage was mostly rural, with no large cities or heavy manufacturing. There were 3 small communities in the drainage (Lafayette, Elliston, and Shawsville) with a combined population less than 2000. The rest of the drainage was characterized by small farms or steep forested mountain slopes. At the time of our studies no serious sources of pollution were known (Surat et al. 1982), and stream channels were not modified in any significant manner. In March 1978 the Roanoke River had a major flood (Matthews 1998, 334), and in April 1978, just before we began our monthly samples, the Roanoke River had the largest flood in the previous six years (Matthews 1998, 334–335). During both floods we documented behaviors of fish relative to refuge use and their subsequent return to normal habitats postflood (Matthews 1998). From May to autumn 1978 (the

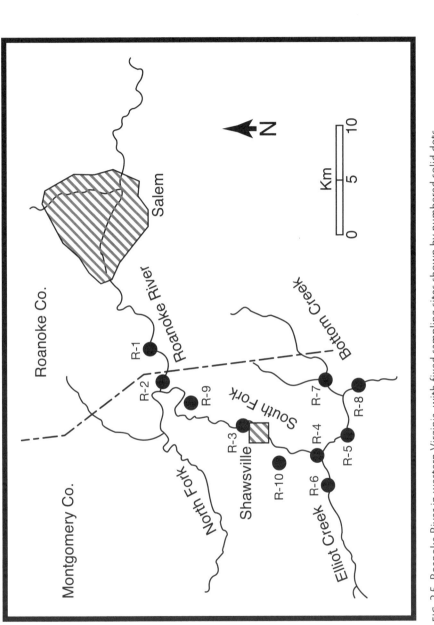

FIG. 2.5 Roanoke River in western Virginia, with fixed sampling sites shown by numbered solid dots.

period of our monthly samples) there were no more major floods or severe drought, but there was 1 large stage rise (to 5200 cfs) in February 1979, between our autumn 1978 and winter 1979 samples.

Ouachita National Forest Streams, Southwest Arkansas

The US Department of Agriculture Forest Hydrology Laboratory, Oxford, Mississippi, contracted with WJM to study effects of timber harvest by USDA Forest Service protocols in the upper parts of two drainages in the Ouachita National Forest of southwest Arkansas. Four creeks were sampled: two in the Saline River headwaters in Saline County, north of Hot Springs, Arkansas, and two in the headwaters of the Little Missouri River in Montgomery and Polk Counties, northwest of Langley, Arkansas (sites 4–7 in fig. 2.1). One creek in each drainage had a recent history of timber harvest and the other had had no timber harvest in recent decades. In the Saline headwaters, Bread Creek (site 4 in fig. 2.1) was the "harvested" creek and the South Fork of Alum Creek (site 5 in fig. 2.1) had not been harvested for about 40 years. In the headwaters of the Little Missouri River, Crooked Creek (site 6 in fig. 2.1; see plate 5) had some recent timber harvest, and Blaylock Creek (site 7 in fig. 2.1) was nonharvested. Six fixed sites were selected in each creek, spaced from small headwaters with small pools and riffles, to downstream reaches with large or long pools and swift riffles or rapids.

All 24 sites were sampled by WJM and at least 2 assistants in October–November 1989, May 1990, October 1990, and May 1991, with approximately 1–1.5 hours of sampling at each site. Seining was under good conditions, although some headwater sites presented challenges, such as coarse substrates or upthrust rock, to sampling. During dry weather in autumn 1989 and 1990, some headwater pools in Bread and Alum Creeks were isolated, and few fish were found. In October 1990 and May 1991, no fish were found in the most upstream site on Alum Creek, even though some water remained in pools. But flow was always substantial in the lower parts of each creek with fast riffles and large pools.

A preliminary analysis using a five-way analysis of variance (ANOVA), with harvest as one of the main effects and a multivariate analysis, showed no major effects of timber harvest (Matthews 1993), so the data across all four surveys of the 24 sites are used in this book to assess variation in local and whole-watershed assemblages across space or time. Subsequent funding from the USDA Forest Service allowed examination of 1589 stomachs of 21 fish species (Matthews et al. 2004), which provided basic ecological information for these species and estimates of redundancy in ecosystem roles among taxonomically similar species.

Grand River Tributaries, Northeast Oklahoma

Collections were made in Ottawa County, northeastern Oklahoma, in a study comparing fishes in parts of the Tar Creek watershed and nearby streams (Franssen et al. 2006). Tar Creek is affected by outflow of heavy-metal-contaminated water extruding under artesian pressure from huge underground zinc and lead mines that were abandoned in the early 1970s (Nairn et al. 2014). Collections for the Tar Creek project included numerous sites within the heavily contaminated areas, and four fixed-reference sites in similarly sized streams outside the contaminated area (sites 8–11 in fig. 2.1). The unaffected reference sites that are used in this book included one site each on Little Elm Creek (see plate 7), a direct eastern tributary of Grand River; Garrett Creek, an eastern tributary of Tar Creek; and Cow and Coal Creeks (see plate 8), which are direct western tributaries to Grand River. Although not affected by mining, these creeks are relatively harsh, as they dry substantially at times, markedly reducing available habitat for fish.

Crutcho Creek and Other North Canadian River Tributaries, Central Oklahoma

The North Canadian River arises in New Mexico, flows generally eastward through the Oklahoma panhandle (where it is known as the Beaver River and is often dry), and then turns southeastward though western Oklahoma. In central Oklahoma it resumes an eastward course and joins the Canadian River proper in eastern Oklahoma to form Lake Eufaula, from which the Canadian River then flows east and north into the Arkansas River near the town of Webbers Falls (Matthews et al. 2005). As it flows through central Oklahoma, a number of small tributaries join the North Canadian River from the south and west in Canadian, Oklahoma, Pottawatomie and Seminole Counties. Data on fish communities of five of these tributary creeks were compiled during a project conducted from 2009 to 2011 to compare the fishes of the Crutcho Creek system to those of nearby streams.

The Crutcho Creek watershed in Oklahoma County (site 13 in fig. 2.1) includes two named streams, Crutcho Creek and Soldier Creek, both of which originate as runoff on Tinker Air Force Base in Midwest City, Oklahoma (fig. 2.6; see plate 6). In 2009, Raymond Moody, natural resource biologist for Tinker Air Force Base, contracted with us to conduct a comprehensive survey of fishes of the Crutcho Creek system and to compare the fish community with those of reference streams. Matthews and Gelwick (1990) surveyed the Crutcho Creek system outside the Air Force base. As reference sites we chose similar-sized creeks that were also tributaries of the North Canadian River, and established collection sites on each creek. Two of these streams, Gar Creek (Seminole County, site 15 in fig. 2.1) and Choctaw Creek (Oklahoma County, site 14 in fig. 2.1) were located east of Crutcho Creek, and one other, Mustang Creek (Canadian County, site 12 in fig. 2.1) was located to the west. These streams were generally similar in size and structure. Lengths of the creek mainstems ranged from 9.8 to 19.9 km; elevational gradients along the mainstems ranged from 1.5 to 3.9 m/km. Substrate was primarily mud with incised mud banks (often stabilized with rip-rap or debris). Streams varied with respect to the surrounding landscape: Crutcho Creek, Mustang Creek, and Choctaw Creek flow through urban or suburban areas, while Snake/Gar Creek flows through a more rural area.

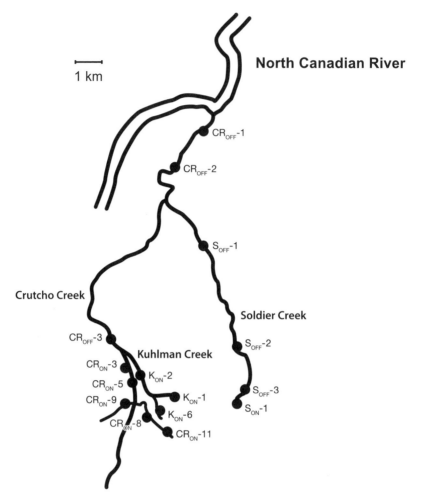

FIG. 2.6 Crutcho Creek sampling sites, Oklahoma County, Oklahoma.

South Canadian River, Central Oklahoma

The Canadian River (also known as the South Canadian River) arises in the southern Rocky Mountains in Colorado and New Mexico and flows generally eastward to join the Arkansas River in eastern Oklahoma. In central Oklahoma, the South Canadian is a sand-bed river with highly unstable substrates, a highly variable flow regime, and frequent changes in location of the flowing channel within the wide riverbed. The mainstem is considered among the harshest of riverine environments (Matthews and Zimmerman 1990; Matthews et al. 2005), and Hefley (1937) considered the South Canadian River bed and adjacent terrestrial habitat as "one of the harshest on Earth." For dissertation research, WJM sampled fish seasonally from May 1976 to January 1977, measured microhabitat features, and estimated mesohabitat type at a large number of individual small seine hauls (Matthews 1977; Matthews and Hill 1979a,b, 1980). The samples all were within one 600 m reach of the river west of Norman and east of Newcastle, Oklahoma (site 16 in fig. 2.1; see plate 12), and the associated mouth of Pond Creek, with additional samples in upper Pond Creek, with 18 to 25 randomized points in the river proper or creek mouth, depending on seasonal water levels, with each point sampled 4 times (once in the morning and once in the afternoon of 2 different days) in May 1976, August 1976, October 1976, and December 1976–January 1977. At each point, WJM made one seine haul 5 m long with a short seine, so that approximately 10 m^2 was included in each seine haul. For each seine haul, WJM recorded type of structure, substrate, current, depth, dissolved oxygen, conductivity, turbidity, water temperature, pH, and shade, and counted and released all fish in each seine haul. Mesohabitat, following designations of Polivka (1999), was also assigned to each seine haul from original field notes. Discharge (at Bridgeport, Oklahoma, the nearest consistent USGS gage) on dates sampled ranged from 7 to 252 cfs, with very low water levels in the river mainstem in August and October 1976.

Other Oklahoma Streams Sampled Four or More Times

In 13 other Oklahoma streams, from the Ozarks in the east to the southwestern corner of the state, we have at least four, and up to ten, samples in one site each, over time. These streams range from small tributaries like upper Tyner Creek in Adair County (site 19 in fig. 2.1; see plate 10), to relatively large but seinable rivers like the Illinois River just west of the Oklahoma–Arkansas state line (site 18 in fig. 2.1; see plate 9), Blue River in southcentral Oklahoma (site 22 in fig. 2.1; see plate 13), and the Salt Fort of Red River near Mangum, Oklahoma (site 21 in fig. 2.1; see plate 11). In addition to normally free-flowing streams, we included two backwater overflow sites, Borrow Pits along US Highway 70 north of Idabel, Oklahoma (site 27 in fig. 2.1), and Lukfata Slough (site 29 in fig. 2.1), both created by borrow pits dug to provide earth for highway construction, that flood frequently, creating swampy habitat with heavy growth of aquatic vegetation. Many of these sites have been sampled by classes under our direction, using similar collecting techniques across years. Although collected by students, class samples were sorted in the laboratory and identifications confirmed by us. Exceptions include Little Glasses Creek, Marshall County, Oklahoma (site 26 in fig. 2.1), where the earlier collections were by classes but we did the more recent collections, counting and releasing fish, and some streams in south Oklahoma where we personally made the collections as part of our studies of Red Shiner (*Cyprinella lutrensis*) declines.

Streams of the Southern Great Plains, Midwestern USA

In June 1978, WJM made more than 100 collections in streams throughout the southern Great Plains, including the Osage Plains and Central Lowlands (Brown and Matthews 1995) from Iowa and Nebraska to south Texas, to examine geographic variation in morphology and nuptial color of male Red Shiners (Matthews 1987a, 1995). At most sites, he also sampled the entire fish community and recorded habitat conditions (Matthews 1985). Seventeen years later, in June 1995, we together revisited 82 of those sites to resample the fish community and record environmental data. The streams sampled in these collections ranged in size from very small pasture streams to large, shallow rivers, and varied in features like width, depth, flow, substrate, riparian cover, and in-stream structure.

Two databases used in this book were compiled from these collections: 1 with both fish and habitat data from 61 sites that had comparable collections in 1978 and 1995 and another with fish and habitat data from 65 sites collected in 1995. For the database with paired sites, we included only collections made when conditions at the site were similar in the two years and when collecting effort was comparable. We omitted several sites where high water in 1995 prevented a comprehensive sample of all microhabitats that had been sampled in 1978. These data are used to compare fish communities at these paired sites in 1978 and 1995, and to ask to what extent different habitat features were correlated with changes.

For the 1995 database, we included only sites at which the entire habitat was thoroughly sampled (omitting sites with areas too deep or too swift to seine effectively or with too many obstructions; see discussion of QDQ in chap. 1). These data were used to examine factors that influence structure of stream fish communities (Marsh-Matthews and Matthews 2000a) and are used in this book to address additional questions related to geographic variation in community structure. Data from 50 sites collected in 1995 at which Red Shiners were present were also used to examine the relationship between relative abundance of Red Shiner and the structure of the residual fish community (Marsh-Matthews and Matthews 2000b).

Red River Basin

This data set consists of 143 different sites in streams of the Red River basin, from extreme east to west Oklahoma, crossing many gradients of rainfall, elevation, and natural land cover. All collections were made by seining using our typical methods, by one or both of the authors, by our classes, or with our professional collaborators from 1976 to 2014. A total of 111 species has been detected across these samples. Samples were made in essentially every kind of habitat available within the upper Red River basin, from uplands of the Ouachita Mountains to the flat prairies to the west, from small headwaters barely a meter wide to the Red River proper, and from clearwater upland springs and creeks to low-gradient, muddy streams or sand-bottomed forks of the Red River in southwest Oklahoma.

Spatial Scales of Stream Fish Communities

The spatial scale at which a community is defined depends on several factors, including: (1) the scale at which it was sampled, (2) proximity of sample sites within the stream (which relates to probability or strength of interactions among fishes found at different sites), and (3) the scope of the community function or property of interest (such as the impacts of migratory or invading species). Nine of our study systems allowed us to examine community structure and dynamics at both global and local spatial scales: Piney Creek (Matthews and Marsh-Matthews 2016), with 12 sites spanning the watershed; Crutcho Creek with 13 sites, and the Roanoke River with 11 tributary and mainstem sites; Brier Creek in southern Oklahoma, Kiamichi River in east Oklahoma, and 4 creeks in western Arkansas in the Ouachita National Forest, each with 6 sites from headwater to lower mainstem. For these systems it is possible to pool data from all collection sites to characterize these communities at the global scale of the whole watershed (Matthews et al. 2013, Matthews and Marsh-Matthews 2016).

For these systems, consideration of a whole-creek global community is reasonable because these systems are small enough that fishes can move readily throughout. For example, Matthews (1987) recorded recolonization of headwaters of Brier Creek by numerous species within one month following rewetting after drought. Ross et al. (1985) also noted downstream movement of some species in Brier Creek during a severe drought year, and we have often found shoals of minnows to move from pool to pool within 1 km of Brier Creek in our snorkel surveys (Matthews et al. 1994; Matthews, unpub. data). On the other hand, longitudinal variation in local community structure (i.e., at a given site) is common and generally persistent in all the streams we have studied (Matthews and Marsh-Matthews 2016 and other references). Regardless of the factors responsible for this longitudinal variation (Matthews 1998), some of which we address in later chapters, these streams obviously contain headwater species like Southern Redbelly Dace (*Chrosomus erythrogaster*) that occur only in small upstream tributaries, and that for some species pairs, such as two sculpin and two darter species in Piney Creek, differences between upstream and downstream species can be real. Thus communities sampled at the local scale will likely differ from those at the global scale.

In one example, we compared the global and local communities of our nine multisite study systems with respect to number of species detected per sample. For each of the study systems, we chose one site at random from among all sampling sites to represent the local community for that system. For every global-local pair, the number of species detected was lower for the local community (fig. 2.7).

Because the scale of community analysis has such a profound impact on community properties, hereafter we analyze our nine global communities separately from communities sampled at the local scale. There were 22 streams sampled at only the local scale (1 site per stream). To these we added the randomly selected local community from each of the 9 multisite systems, for a total of 31 local communities. In the chapters that follow it will be made clear whether analyses focus on the 9 global communities pooled within multisite streams, the 31 local communities, or a comparison of the global with the local.

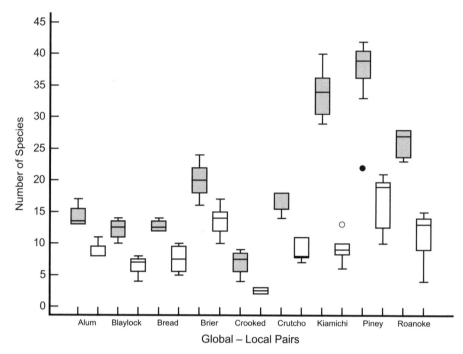

FIG. 2.7 Comparison of species captured per collection at the global (gray) versus local (open) scale for each of the nine study systems with multiple collection sites. In every case, the number of species captured at the global scale was greater than that at the local scale.

Summary

Over the course of more than 40 years, we have made collections of stream fish communities throughout the central United States (plus Virginia), with most collections in Arkansas and Oklahoma. These streams vary from small headwater creeks to large rivers, across many kinds of environments. Over this span of time and space, collections have been made at varying intervals, from months to years to decades. This wide variety of data sets provides a compelling opportunity to explore numerous aspects of stream fish community structure and dynamics at varying spatial scales.

References

Allen, D. C., H. S. Galbraith, C. C. Vaughn, and D. E Spooner. 2013. A tale of two rivers: implications of water management practices for mussel biodiversity outcomes during droughts. Ambio 42:881–891.

Brown, A. V., and W. J. Matthews. 1995. Stream ecosystems of the central United States. Pages 89–116 in C. E. Cushing, K. W. Cummins, and G. W. Minshall, eds. Ecosystems of the world. Vol. 22. River and stream ecosystems. Elsevier, Amsterdam, The Netherlands.

Douglas, M. E., and W. J. Matthews. 1992. Does morphology predict ecology? hypothesis testing within a freshwater stream fish assemblage. Oikos 65:213–224.

Echelle, A. A., and G. D. Schnell. 1976. Factor analysis of species associations among fishes of the Kiamichi River, Oklahoma. Transactions of the American Fisheries Society 105: 17–31.

Franssen, C. M., M. A. Brooks, R. W. Parham, K. G. Sutherland, and W. J. Matthews. 2006. Small-bodied fishes of Tar Creek and other small streams in Ottawa County, Oklahoma. Proceedings of the Oklahoma Academy of Science 86:9–16.

Funkhouser, J. E., and K. Eng. 2009. Floods of selected streams in Arkansas, spring 2008. US Geological Survey Fact Sheet 2008–3103. US Geological Survey, Reston, VA.

Gelwick, F. P., and W. J. Matthews. 1992. Effects of an algivorous minnow on temperate stream ecosystem properties. Ecology 73:1630–1645.

Gillette, D. P. 2012. Effects of variation among riffles on prey use and feeding selectivity of the Orangethroat Darter *Etheostoma spectabile*. American Midland Naturalist 168:184–201.

Harvey, B. C. 1987. Susceptibility of young-of-the-year fishes to downstream displacement by flooding. Transactions of the American Fisheries Society 116:851–855.

Harvey, B. C. 1991. Interactions among stream fishes: predator-induced habitat shifts and larval survival. Oecologia 87:29–36.

Hefley, H. M. 1937. Ecological studies on the Canadian River floodplain in Cleveland County, Oklahoma. Ecological Monographs 7:345–402.

Johnson, K. S. 2006. Geomorphic provinces. Pages 4–5 in C. R. Goins and D. Goble, eds. Historical atlas of Oklahoma, 4th ed. University of Oklahoma Press, Norman, OK.

Marsh-Matthews, E., and W. J. Matthews. 2000a. Geographic, terrestrial and aquatic factors: which most influence the structure of stream fish assemblages in the midwestern United States? Ecology of Freshwater Fish 9:9–21.

Marsh-Matthews, E., and W. J. Matthews. 2000b. Spatial variation in relative abundance of a widespread, numerically dominant fish species and its effect on fish assemblage structure. Oecologia 125:283–292.

Marsh-Matthews, E., and W. J. Matthews. 2010. Proximate and residual effects of exposure to simulated drought on prairie stream fishes. Pages 461–486 in K. B. Gido and D. A. Jackson, eds. Community ecology of stream fishes: concepts, approaches, and techniques. American Fisheries Society Symposium 73. American Fisheries Society, Bethesda, MD.

Marsh-Matthews, E., W. J. Matthews, and N. R. Franssen. 2011. Can a highly invasive species re-invade its native community? The paradox of the Red Shiner. Biological Invasions 13: 2911–2924.

Matthews, W. J. 1973. The fishes of Piney Creek—an Ozark Mountain stream in northcentral Arkansas. MS thesis. Arkansas State University, Jonesboro, AR.

Matthews, W. J. 1977. Influence of physico-chemical factors on habitat selection by Red Shiners, *Notropis lutrensis* (Pisces: Cyprinidae). PhD dissertation. University of Oklahoma, Norman, OK.

Matthews, W. J. 1982. Small fish community structure in Ozark streams: structured assembly patterns or random abundance of species? American Midland Naturalist 107:42–54.

Matthews, W. J. 1985. Distribution of midwestern fishes on multivariate environmental gradients, with emphasis on *Notropis lutrensis*. American Midland Naturalist 113:225–237.

Matthews, W. J. 1986. Fish faunal structure in an Ozark stream: stability, persistence, and a catastrophic flood. Copeia 1986:388–397.

Matthews, W. J. 1987. Physicochemical tolerance and selectivity of stream fishes as related to their geographic ranges and local distributions. Pages 111–120 in W. J. Matthews and D. C. Heins, eds. Community and evolutionary ecology of North American stream fishes. University of Oklahoma Press, Norman, OK.

Matthews, W. J. 1993. Fish community composition and biomass as related to physical stream environments, seasons, and forest management in the Ouachita Mountains of southwestern Arkansas. Final Report to USDA Forest Service. Agreement No. 19-89-043.

Matthews, W. J. 1995. Geographic variation in nuptial colors of Red Shiner (*Cyprinella lutrensis*; Cyprinidae) within the United States. Southwestern Naturalist 40:5–10.

Matthews, W. J. 1998. Patterns in freshwater fish ecology. Chapman and Hall, New York, NY.

Matthews, W. J., and F. P. Gelwick. 1990. Fishes of Crutcho Creek and the North Canadian River in central Oklahoma: effects of urbanization. Southwestern Naturalist 35:403–410.

Matthews, W. J., and G. L. Harp. 1974. Preimpoundment ichthyofaunal survey of the Piney Creek watershed, Izard County, Arkansas. Arkansas Academy of Science Proceedings 28:39–43.

Matthews, W. J., and L. G. Hill. 1979a. Age-specific differences in the distribution of Red Shiners, *Notropis lutrensis*, over physicochemical gradients. American Midland Naturalist 101:366–372.

Matthews, W. J., and L. G. Hill. 1979b. Influence of physico-chemical factors on habitat selection by Red Shiners, *Notropis lutrensis* (Pisces: Cyprinidae). Copeia 1979:70–81.

Matthews, W. J., and L. G. Hill. 1980. Habitat partitioning in the fish community of a southwestern river. Southwestern Naturalist 25:51–66.

Matthews, W. J., and E. Marsh-Matthews. 2006. Persistence of fish species associations in pools of a small stream of the southern Great Plains. Copeia 2006:696–710.

Matthews, W. J., and E. Marsh-Matthews. 2007. Extirpation of Red Shiner in direct tributaries of Lake Texoma (Oklahoma-Texas): a cautionary case history from a fragmented river-reservoir system. Transactions of the American Fisheries Society 136:1041–1062.

Matthews, W. J., and E. Marsh-Matthews. 2016. Dynamics of an upland stream fish community over 40 years: trajectories and support for the loose equilibrium concept. Ecology 97:706–719.

Matthews, W. J., and J. T. Styron Jr. 1981. Comparative tolerance of headwater versus mainstem fishes for abrupt physicochemical change. American Midland Naturalist 105:149–158.

Matthews, W. J., and E. G. Zimmerman. 1990. Potential effects of global warming on native fishes of the southern Great Plains and the Southwest. Fisheries 15:26–32.

Matthews, W. J., J. R. Bek, and E. Surat. 1982. Comparative ecology of the darters *Etheostoma podostemone*, *E. flabellare* and *Percina roanoka* in the upper Roanoke River drainage, Virginia. Copeia 1982:805–814.

Matthews, W. J., R. C. Cashner, and F. P. Gelwick. 1988. Stability and persistence of fish faunas and assemblages in three midwestern streams. Copeia 1988:945–955.

Matthews, W. J., B. C. Harvey, and M. E. Power. 1994. Spatial and temporal patterns in the fish assemblages of individual pools in a midwestern stream (U.S.A.). Environmental Biology of Fishes 39:381–397.

Matthews, W. J., E. Marsh-Matthews, G. L. Adams, and S. R. Adams. 2014. Two catastrophic floods: similarities and differences in effects on an Ozark stream fish community. Copeia 2014:682–693.

Matthews, W. J., E. Marsh-Matthews, R. C. Cashner, and F. Gelwick. 2013. Disturbance and trajectory of change in a stream fish community over four decades. Oecologia 173:955–969.

Matthews, W. J., A. M. Miller-Lemke, M. L. Warren, D. Cobb, J. G. Stewart, B. Crump, and F. P. Gelwick. 2004. Context-specific trophic and functional ecology of fishes of small stream ecosystems in the Ouachita National Forest. Pages 221–230 in J. M. Guildin, ed. Ouachita and Ozark Mountains Symposium: ecosystem management research. General Technical Report SRS-74, US Department of Agriculture, US Forest Service, Southern Research Station, Washington, DC.

Matthews, W. J., W. D. Shepard, and L. G. Hill. 1978. Aspects of the ecology of the Dusky-stripe Shiner, *Notropis pilsbryi* (Cypriniformes: Cyprinidae) in an Ozark stream. American Midland Naturalist 100:247–252.

Matthews, W. J., C. C. Vaughn, K. B. Gido, and E. Marsh-Matthews. 2005. Southern Plains Rivers. Pages 283–325 in A. C. Benke and C. C. Cushing, eds. Rivers of North America. Elsevier Academic Press, Amsterdam, The Netherlands.

Nairn, R. W., J. A. LaBar, K. A. Strevett, W. H. Strosnider, D. Morris, A. E. Garrido, C. A. Neely, and K. Kauk. 2014. Initial evaluation of a large multi-cell passive treatment system for net-alkaline ferruginous lead-zinc mine waters. Pages 635–649 in R. I. Barnhisel, ed. Bridging reclamation, science and the community. American Society of Mining and Reclamation, Champaign, IL.

Oklahoma Water Resources Board. 2000. Kiamichi River Basin water resources development plan. Final Report. Oklahoma Water Resources Board, Oklahoma City, OK.

Peckarsky, B. L. 1983. Biotic interactions or abiotic limitations? a model of lotic community structure. Pages 303–323 in T. D. Fontaine III and S. M. Bartell, eds. Dynamics of lotic ecosystems. Ann Arbor Science, Ann Arbor, MI.

Pigg, J., and L. G. Hill. 1974. Fishes of the Kiamichi River, Oklahoma. Proceedings of the Oklahoma Academy of Science 54:121–130.

Polivka, K. M. 1999. The microhabitat distribution of the Arkansas River Shiner, *Notropis girardi*: a habitat-mosaic approach. Environmental Biology of Fishes 55:265–278.

Power, M. E., and W. J. Matthews. 1983. Algae-grazing minnows (*Campostoma anomalum*), piscivorous bass (*Micropterus* spp.), and the distribution of attached algae in a small prairie-margin stream. Oecologia 60:328–332.

Power, M. E., W. J. Matthews, and A. J. Stewart. 1985. Grazing minnows, piscivorous bass and stream algae: dynamics of a strong interaction. Ecology 66:1448–1456.

Pyron, M., C. C. Vaughn, M. R. Winston, and J. Pigg. 1998. Fish assemblage structure from 20 years of collections in the Kiamichi River, Oklahoma. Southwestern Naturalist 43: 336–343.

Ross, S. T., W. J. Matthews, and A. A. Echelle. 1985. Persistence of stream fish assemblages: effects of environmental change. American Naturalist 126:24–40.

Smith, C. L., and C. R. Powell. 1971. The summer fish communities of Brier Creek, Marshall County, Oklahoma. American Museum Novitates 2458:1–30.

Surat, E. M., W. J. Matthews, and J. R. Bek. 1982. Comparative ecology of *Notropis albeolus*, *N. ardens* and *N. cerasinus* (Cyprinidae) in the upper Roanoke River Drainage, Virginia. American Midland Naturalist 107:13–24.

Vaughn, C. C., C. L. Atkinson, and J. P. Julian. 2015. Drought-induced changes in flow regimes lead to long-term losses in mussel-provided ecosystem services. Ecology and Evolution 5: 1291–1305.

Wesner, J. S. 2011. Shoaling species drive fish assemblage response to sequential large floods in a small midwestern U.S.A. stream. Environmental Biology of Fishes 91:231–242.

Wesner, J. S. 2013. Fish predation alters benthic, but not emerging, insects across whole pools of an intermittent stream. Freshwater Science 32:438–449.

CHAPTER 3

Characterizing the Fish Communities

Community Metrics

The study systems described in chapter 2 represent a broad range of stream morphologies, or physical and climatic characteristics across a wide geographic area. To compare the fish communities among these disparate systems, we need to characterize community structure and dynamics in a way that allows appropriate and meaningful assessments. Here we characterize our study communities with respect to metrics that summarize community composition and emergent community properties. Specifically, we assess family composition, number of species detected, number and importance of core species, evenness, and diversity. Singly or in combination, these metrics provide insight into community structure and dynamics. Such characterization provides a background or baseline against which we can choose appropriate systems to compare and frame testable hypotheses about factors that shape community structure and dynamics. The comparisons and contrasts among these 31 stream fish communities provide a broad overview of the range of variation in warm-water fish communities of stream systems that are typical of much of the central or eastern United States.

Taxonomic Composition

The most basic information about a community is a species list. The species at any given location will reflect biogeographic patterns and evolutionary history, as is well documented in general works like Hocutt and Wiley (1986) and Matthews (1998) or primary papers like Marsh-Matthews and Matthews (2000). Our 31 localities, from Virginia to western Oklahoma, cross numerous biogeographic boundaries, so no species are common to all sites. Taxonomic analyses at the species level would serve to simply confirm the known biogeographic patterns. Taxonomic differences among our communities are actually an asset, however, because any dynamics we detect that apply to differing taxonomic assemblages makes them all the more general. This is in keeping with the thrust of this book, that is, seeking synthesis, general patterns, or mechanisms that are broadly applicable to warm-water stream fish communities.

Although we have different communities from place to place at the species level of taxonomy, the family level of taxonomy can provide useful comparisons of the communities across all these stream systems. For fishes, the taxonomic level of "family" is one of the historically most stable and one of the most potentially important ecologically for any overview or synthesis of communities. Matthews (1998, 43–54) focused on the importance of families in the overall composition of communities, the range of diversity within or among families in a community, and differences in family richness patterns between tropical and temperate streams. Most species in freshwater are more similar morphologically and ecologically within than among families. Most species within a family share a common fundamental shape and positions of mouth, fins, and the like, resulting in many shared ecological traits.

For example, Karr et al. (1986) provided trophic classification of 25 minnow species (Cyprinidae) native to the central United States, of which 17 were insectivores and another 5 were omnivores (but also using insects in their diets). And essentially all darters (Percidae) in Karr et al.'s classification were insectivores. Douglas and Matthews (1992) found for 17 species across five families in the Roanoke River, Virginia, that species within each family clustered separately in a morphological phenogram, and that there was general (but not complete) agreement between morphological and trophic phenograms. In a principal components analysis of stomach contents, Matthews et al. (2004) showed that family differences among 21 species in upland streams of the Ouachita National Forest, Arkansas, were reflected in trophic differences. Darters (Percidae) and madtom catfishes (Ictaluridae) were completely separated from each other and from minnows (Cyprinidae) and sunfish or bass (Centrarchidae), although cyprinids and centrarchids partly overlapped. Gatz (1979a,b) and Douglas and Matthews (1992) also showed the distinctiveness of families in the morphological or ecological structure of fish communities.

Not all species within particular fish families are morphologically or ecologically similar, however. Within the North American Cyprinidae there are large-bodied chubs west of the Rocky Mountains that are very different, ecologically, from their smaller-bodied counterparts to the east. Most of the diverse native cyprinids of eastern North America are water-column insectivores (Karr et al. 1986), but a few genera, such as stonerollers of the genus *Campostoma* and dace of the genus *Chrosomus* are algivorous, and others, like *Hybognathus* and *Pimephales*, are omnivorous, largely feeding on benthic ooze or biofilms. Catfishes in the family Ictaluridae range from large-bodied species such as Flathead Catfish (*Pylodictis olivaris*) and Blue Catfish (*Ictalurus furcatus*) that are mostly piscivorous as adults (Edds et al. 2002) to small-bodied, insectivorous madtom catfishes of the genus *Noturus*.

There was a total of 18 families across our 31 study systems. Among the 9 global communities, the number of families ranged from 4 in Crooked Creek to 12 in the

Kiamichi River. Among the 31 local communities, the number of families ranged from 2 in Crooked Creek to 11 in Morris Creek. Although both the Kiamichi River (global) and Morris Creek (local) are located in the Ouachita Mountain uplift of southeast Oklahoma, other sites outside the uplands also had a large number of families. Communities with ten families each were in five streams, including a swampy, nonflowing backwater overflow along US 70 in southeast Oklahoma, three low-gradient small streams in different parts of Oklahoma (Cow, Hauani, Hickory Creeks), and a complex site on the Illinois River in the Ozark uplands just downstream of a partly broken dam (Lake Frances). Communities with the fewest families (two to four) were in two extremely different systems, including the low-gradient Mustang Creek (with four families in the local community) in central Oklahoma and the high-gradient pool rapids of Crooked Creek in the Ouachita Mountains of west Arkansas (with two families detected at the local scale and four at the global scale). Thus there appears to be no overarching distinction between upland or lowland or small versus large stream systems, with respect to numbers of families that we have found.

Of the 18 families detected across all study systems, Cyprinidae (minnows) was the numerically dominant family in all 9 global communities and in 23 of the 31 local communities. Among the remaining local communities, Poeciliidae (as Western Mosquitofish, *Gambusia affinis*) numerically dominated the community in four small stream sites that had highly variable environments. Sunfish and black basses (Centrarchidae) were the numerical dominants in three other systems, and Esocidae (as Grass Pickerel, *Esox americanus*) was the numerical dominant in one swampy backwater system. At the other extreme, across all of our collections we found three families (Petromyzontidae, Moronidae, and Sciaenidae) represented only once, by one individual each.

The Oak-Hickory Forest Analogy

We borrow from the vegetation literature, using the oak-hickory forest analogy as a basis for naming our community groups. Phytosociologists and foresters have long used either professional judgment and consensus (Eyre 1980) or more formal standards for naming based on highly structured quantitative models (Federal Geographic Data Committee 2008; Jennings et al. 2009) to classify vegetation. Here, our approach for fish community names is similar to that of Eyre's (1980) combined names for forest types, like "Post Oak–Blackjack Oak" or "Yellow Poplar–White Oak–Northern Red Oak" or the familiar "Oak-Hickory Forest" within the Eastern Deciduous Forest (Bailey 1983), based on dominant natural climax tree associations.

Several previous authors have used either family-level names or the names of the most abundant species to characterize fish communities or associations. One of the earliest and still commonly used examples is that of the fish "faunal zones" for European streams by Huet (1959), with trout, grayling, barbel, and bream zones from cold headwaters to warmer, low-gradient downstream reaches. Echelle and Schnell (1976) used a factor analysis to distinguish six groups of associated species, named for the dominant species (e.g., "White Bass group," "Steelcolor Shiner group," etc.). Rahel (1984) referred to three distinct groups in northern bog lakes, including a "centrarchid assemblage," a "cyprinid assemblage," and an "*Umbra-Perca* assemblage." Rahel and Hubert (1991) noted a cold-water "Salmonidae assemblage" that was replaced downstream by a "Cyprinidae-Catostomidae assemblage."

To provide a formal structure for naming our 31 local and 9 global fish communities, we followed the 75% rule. If the individuals in one family comprised 75% or more of all fish detected in the system, we named the community for that taxon, for example, a Cyprinid (Cyp) or a Poeciliid (Poe) community (table 3.1). For all others, we summed the proportion represented by the dominant and successive subdominant families until in combination they reached or exceeded 75% of the total community, resulting in combined names such as Cyprinid-Percid (Cyp-Per), Poeciliid-Centrarchid (Poe-Cen), or Centrarchid-Poeciliid-Elassomatid (Cen-Poe-Elas) communities (table 3.1). Across all nine global communities, four were Cyprinid, two each were Cyprinid-Centrarchid (Cyp-Cen) and Cyprinid-Percid, and one was Cyprinid-Fundulid (Cyp-Fun). Across the 31 local sites (table 3.1), 23 were dominated by cyprinids, with 8 Cyprinid, 8 Cyprinid-Percid, 2 Cyprinid-Poeciliid, 4 Cyprinid-Centrarchid communities, and 1 Cyprinid-Fundulid community. Four communities were poeciliid-dominated, with one Poeciliid, two Poeciliid-Cyprinid, and one Poeciliid-Centrarchid. Three communities were centrarchid-dominated, one each as Centrarchid-Cyprinid-Percid, Centrarchid-Cyprinid-Poeciliid, and Centrarchid-Poeciliid-Elassomatid. One community, at a site in Bread Creek, actually required summing four families to reach the criterion of comprising 75% of the total individuals, resulting in its unwieldy naming as a Percid-Catostomid-Cyprinid-[Centrarchid or Aphredoderid] community.

Naming communities by the 75% rule may be a bit cumbersome in a few cases, with four sites needing three or more family names (table 3.1). But for most of the global and local systems, providing such a family overview is a convenient way to refer to a community in a general sense, and we use these names to compare emergent properties of communities in sections that follow. We (and most ichthyologists) readily relate to differences among communities characterized as, for example, Cyprinid-Percid, Poeciliid-Cyprinid, or Centrarchid-Poeciliid-Elassomatid, with one characterization suggesting an upland stream dominated by minnows and darters, another suggesting a sluggish low-gradient stream with mosquitofish and minnows, and the latter suggesting a low-gradient or vegetated backwater area characterized by sunfish, mosquitofish, and Pygmy Sunfish.

Because many species within a family share characteristics of food or habitat use, we can suggest ecological generalizations based on our named family types. The Cyprinid or Cyprinid-Percid communities generally had relatively low percentages of piscivores like sunfish or black bass. In all of the Cyprinid (>75%) local sites, the proportion of cyprinids ranged from 79% to 95% of all fish captured, whereas at those sites, centrarchids (the family with the most piscivorous species) ranged from <1% to 7%. This finding suggests that at Cyprinid sites, centrarchids were low in relative

Table 3.1. Community characteristics of global and local communities

Stream	Dominant Family	# Fam	S	C	AVG %CORE	CV %CORE	AVG S	CV S	AVG E	CV E	AVG H'	CV H'	O
GLOBAL													
ALUM5	Cyp	9	19	11	97.9	1.4	14.3	13.3	0.577	19.0	1.53	19.4	5
BLAY7	Cyp	6	16	9	98.6	0.8	12.3	13.9	0.754	7.5	1.88	3.8	3
BREAD4	Cyp-Per	9	14	10	88.4	6.4	12.8	7.5	0.780	15.3	1.99	15.8	0
BRIER2	Cyp-Cen	7	30	8	85.9	11.1	20.1	12.0	0.679	11.6	2.00	12.3	4
CROOK6	Cyp-Per	4	9	4	98.8	1.1	7.0	30.8	0.548	0.5	1.02	1.1	1
CRUT13	Cyp-Cen	8	26	12	98.5	2.2	16.8	10.7	0.540	8.5	1.52	11.1	8
KIAM3	Cyp	12	48	19	94.8	2.0	33.8	12.4	0.622	11.1	2.18	10.4	7
PINEY1	Cyp-Fun	9	48	20	95.1	3.6	37.0	15.1	0.719	5.2	2.59	8.8	1
ROAN31	Cyp	6	33	19	99.1	0.5	26.0	9.0	0.707	8.0	2.30	7.1	5
LOCAL													
ALUM5	Cyp	7	15	3	90.0	5.2	8.8	17.0	0.467	25.0	1.01	29.3	6
BALL17	Cyp-Per	9	36	5	49.4	45.9	18.4	33.8	0.619	17.8	1.79	26.9	6
BARON20	Cyp-Per	9	29	9	93.8	3.3	19.0	18.2	0.604	12.8	1.77	13.7	8
BLAY7	Cyp	5	10	4	80.5	20.0	6.5	26.6	0.807	9.9	1.47	10.6	4
BLUE22	Cyp-Per	6	28	10	74.5	14.5	19.8	6.6	0.839	7.7	2.50	9.0	5
BORR27	Cen-Poe-Elas	10	20	1	15.2	104.4	8.6	49.0	0.705	17.0	1.46	37.6	4
BREAD4	Cyp-Per	7	12	2	33.8	49.7	7.5	31.7	0.885	2.5	1.75	17.5	1
BRIER2	Cyp-Cen	5	24	4	74.2	19.7	13.5	13.9	0.634	18.2	1.64	17.8	3
CHIG30	Cyp-Per	7	20	4	70.2	29.5	11.0	29.7	0.614	28.8	1.48	35.6	8
CHOC14	Cyp-Cen	6	17	4	90.5	9.6	8.6	22.7	0.587	24.4	1.23	25.7	6
COAL11	Poe-Cyp	9	23	1	48.5	64.9	7.1	47.1	0.568	48.7	1.13	53.5	7
COW10	Poe-Cyp	10	29	0	54.3	61.7	6.4	87.4	0.728	40.5	1.23	57.4	3
CROOK6	Cyp-Per	2	3	2	98.2	2.3	2.5	23.1	0.854	17.7	0.74	9.5	0

CRUT13	Cyp-Cen	6	14	5	89.9	16.6	9.0	20.8	0.449	44.5	1.00	49.8	2
GAR15	Cyp-Cen	4	12	5	82.4	21.8	8.8	19.6	0.619	21.6	1.31	28.0	1
GARR9	Poe-Cen	6	13	1	63.8	42.9	4.3	47.5	0.595	47.5	0.84	49.1	3
GLOV28	Cyp-Per	8	33	5	62.9	19.3	15.8	29.5	0.634	8.8	1.73	15.6	13
HAUAN25	Cen-Cyp-Poe	10	23	4	39.6	37.4	13.3	16.7	0.615	13.8	1.59	19.7	7
HICK24	Cyp-Poe	10	21	1	11.5	123.5	10.6	36.9	0.587	12.6	1.35	30.4	9
ILL18	Cyp-Poe	10	44	6	45.4	45.1	21.8	26.4	0.657	10.3	2.01	14.8	16
KIAM3	Cyp	8	16	4	89.5	9.7	9.2	27.1	0.613	16.9	1.34	19.0	5
LELM8	Poe	5	12	1	79.7	31.4	3.9	49.3	0.352	94.4	0.53	98.0	3
LGLASS26	Cen-Cyp-Per	8	20	4	55.2	41.5	11.9	22.6	0.783	10.9	1.92	15.4	2
LUK29	Eso-Cen	9	16	2	48.6	61.3	8.2	38.0	0.708	35.0	1.53	46.7	6
MORR23	Cyp-Per	11	40	9	62.4	38.6	22.0	24.3	0.656	14.5	2.01	16.8	17
MUST12	Cyp	4	11	2	56.6	55.5	7.0	36.4	0.635	9.7	1.19	25.6	2
PINEY1	Cyp-Fun	8	31	4	50.6	35.6	16.6	24.9	0.747	10.5	2.08	15.9	4
ROAN31	Cyp	5	22	1	18.3	121.3	11.3	36.6	0.711	17.3	1.64	20.7	5
SALTF21	Cyp	8	20	4	65.4	27.3	11.0	39.6	0.574	32.8	1.25	17.2	7
SCR16	Cyp	7	19	6	99.8	0.2	12.3	32.9	0.363	24.5	0.88	19.4	3
TYNR19	Cyp	6	16	5	86.8	5.9	10.2	23.4	0.664	13.3	1.51	9.8	4

Note: See text for abbreviation definitions.

abundance, so we would predict lower predator pressure than at Centrarchid sites. The three sites dominated by centrarchids (table 3.1) had 44.8% (Little Glasses Creek), 37.4% (Borrow pit sloughs), and 33.2% (Hauani Creek) centrarchids, and Choctaw and Brier Creeks (both Cyprinid-Centrarchid communities) had 44% and 16%, respectively, of centrarchids.

Emergent Properties

Beyond basic taxonomic composition, communities can be characterized by emergent properties such as species richness, diversity, and evenness (table 3.1), all of which can be assessed independently of taxonomy (Magurran 2011). For example, Marsh-Matthews and Matthews (2000) examined factors associated with emergent properties in 65 stream fish communities throughout the southern Great Plains and were able to relate geographic, terrestrial, and instream characteristics to richness and complexity of those fish communities.

Species Richness

The number of species in a community is a key element of many kinds of studies, including those addressing biodiversity (Vaughn 2010), biotic integrity (Karr et al. 1986; Roset et al. 2007), invasibility of systems (Gido et al. 2004; Moyle and Marchetti 2006), or ecosystem function and services (Worm et al. 2006). The actual assessment of species number has been controversial, however. Beginning with Sanders (1968), as refined or corrected by Hurlbert (1971) and Simberloff (1972), many community ecologists have used rarefaction to adjust the number of species detected in a given sample (i.e., species richness) to the abundance of individuals in the smallest of several samples. This approach has most often been used to allow comparison among different study areas or study systems, but it has also been used for temporal comparisons (e.g., Taylor et al. 2008). Rarefaction to adjust the expected number of species to the smallest sample size—with various modifications such as standardization by sampling effort (Gotelli and Colwell 2001), sample "completeness" (Chao and Jost 2012), or other models (Colwell et al. 2012)—has been used extensively for many taxa (e.g., Beketov et al. 2013; Coelho et al. 2014; Garcia et al. 2014), including fish communities (Marsh-Matthews and Matthews 2000; Gillette et al. 2012; Cheek and Taylor 2016; Miyazono et al. 2015).

Despite its widespread use, rarefaction may not be appropriate for all situations. Gotelli and Colwell (2001), discussing the pitfalls of rarefaction, noted that to be valid, "sample sizes must be sufficient, and that the assemblages . . . have been sampled in the same way." Furthermore, rarefaction has the underlying premise that the target organisms are distributed randomly in the environment (Kobayashi 1982, as discussed by Gotelli and Colwell 2001). Gotelli and Colwell (2011) summarized problems that may make rarefaction inappropriate under a variety of conditions because of violation of any of six fundamental assumptions (like equitable sampling, homogeneous distribution of individuals, etc.). Fish communities sampled in finite stream reaches, including those we have sampled over the years, probably violate at least four of the Gotelli and Colwell (2011) assumptions, because of variability in the habitat within the reach, heterogeneous or clumped distribution of individuals within a site, and the open nature of fish distribution in streams.

One of our sites illustrates why rarefaction is probably undesirable for our samples. Little Elm Creek is a small headwater site in northeast Oklahoma. During our 13 surveys we captured a range of fish from an extreme of only 8 to a total of 834 individuals (but with 5 and 6 species, respectively, at those two times). At another time at that site we captured only 13 individuals, but all of one species (Western Mosquitofish; see plate 15). The point is that, regardless of conditions in the reach (so long as water existed even in small pools), we did sample the fish, and the extremely small numbers (at times), compared to hundreds of fish in the same reach at other times, provide a true reflection of the variability in the fish that can be found in a small stream in the midwestern United States as drought waxes and wanes. Now, our sample sizes were sufficient to characterize the community within the fixed reach of Little Elm Creek during drought—but there simply were not many fish present. Did we "sample in the same way," as required by Gotelli and Colwell? Yes, in that we seined as effectively as possible throughout the same reach, wherever there was water. But also no, in that while we might have moved the seine through 100 m or more of this small pasture creek during wet conditions, the total length of reach seined when it was pooled up was probably not one-third as long. Similar examples could be cited from the nearby Cow Creek, where in sampling at extremely low water during heat and drought we found only two individuals of one species, but on other occasions found 150 to 400 individuals, of 9 to 12 species. The simple fact is that if one follows the same reach of small streams in our part of the world, there will be months or years when water volumes are much reduced, with concomitant reductions in abundance (and numbers of species), but the remnant individuals in isolated pools can be very important when rains rewater the streams (Marsh-Matthews and Matthews 2010), and this natural variability is inherent in many of the streams that contain fish in the central United States. We would argue that any stream reach that harbors fish at all, from large river mainstems to the smallest headwaters, represents real habitat for fish and thus is worthy of study.

Relative to the assumption of homogeneous distribution, fish in any given reach in our study systems are distributed neither at random nor homogeneously, and the same is true for streams in general (Kwak and Peterson 2007). And many species, like minnows, are highly clumped (in schools or shoals) within a stream reach. Any field ichthyologist knows that in a sample reach of 100–300 m of most streams a substantial diversity of subhabitats (pools, riffles) or microhabitats (backwaters, undercut banks, rootwads, etc.) exists and that different fish species at different life stages (Schlosser 1991) use those habitats differently. In sampling a typical stream reach, we often make 25 to 40 or more seine hauls or kicksets (see chap. 1) in all observable microhabitats, recognizing that different microhabitats usually have quite different species. To rarify estimated species numbers to the fewest number of fish collected in several to many surveys at a site (often on the worst date of sampling, e.g., during drought) would negate the real-world efforts to find all possible species across all of the different microhabitats available during normal base flow sampling. Lastly, many of the most abundant fish species in our samples (minnows) are strongly schooling, further making their spatial distributions nonhomogeneous, thus requiring extensive, thorough sampling within the study reach to capture and detect as many species as possible.

For our assessments of number of species in our 31 study systems, we opted to use the raw number of species detected in our collections at a given site and time, without adjustment by rarefaction. The primary reason for using raw species detected is that we are most interested in the dynamics of the fish community within a given stream reach (for single sample sites) or within a whole watershed (for global sampled systems) across a wide array of environmental conditions. The environmental conditions in any fixed sample reach may change dramatically from time to time, and it is our specific goal to capture the variation in the fish community as environmental conditions change.

For each of the 9 global and 31 local communities (table 3.1), we determined the total number of species (TOTAL S) detected in all collections combined (global range = 9–48; local range = 3–44), the average number of species per collection (AVG S) (global range = 7–37; local range = 2.5–22), and the coefficient of variation in number of species detected per collection (CV S) (global range = 7.5%–30.8%; local range = 6.6%–87.4%). Total species number and average number of species detected were highly significantly correlated (global $r = 0.986$, $P < 0.001$; local $r = 0.873$, $P < 0.001$) but the relationship between these values varied with scale. At the global scale, a regression of AVG S on TOTAL S yielded a regression line with a slope of 0.712, while at the local scale the analysis produced a line with a slope of 0.483. These slopes were significantly different by analysis of covariance. The overall effect of this difference was that the expected number of species per collection (AVG S) increased at a faster rate, with an increase in TOTAL S at the global scale as compared to the local scale. This means that the average number of species detected per collection can be much lower than the potential total, particularly in systems with the highest total species count, and that the effect of scale of sampling is significant. This finding has important implications because it suggests that although total species detected can help identify biodiversity "hotspots," the actual number of species detected in any collection is a better indicator of community structure at any given time.

Core Species

With temporal samples of a community, it is possible to identify the individual species that occur predictably, that is, those that constitute the "core" of the community. Magurran (2011) defined core species as those that occur regularly in the community. Matthews and Marsh-Matthews (2016) further defined core species as those that occur in every community sample, and this more restrictive definition has been used by other authors to describe community structure (Petanidou et al. 2008; Dolan et al. 2009). Numbers of collections in our communities varied from 4 to 27, and in all but one stream (Cow Creek, with 27 collections) we are able to identify at least one core species (table 3.1).

The number of core species (CORE S) ranged from 4 to 20 in the 9 global communities and from 0 to 10 in the 31 local communities. At both the global and local scales, the number of core species was significantly positively correlated with the total number of species detected in the system (global $r = 0.870$; local $r = 0.474$), which is not surprising given that core species are a part of the total number detected. There was no correlation between the number of core species in the global communities and the number of core species in the randomly chosen local community from the same system, however, suggesting that the detection of core species at the local scale is very different from that at the global scale. At the global scale there are multiple chances (individual site samples) of collecting any particular species in the system, whereas at the local (single reach) scale there is only one chance to detect that species.

In both our global and local study systems, species that contributed to the core included many that were highly abundant, but the core also included some that typically occurred at low abundance. For example, across the seventeen global surveys of all six sites on Brier Creek, 2 of the 8 core species, Bigeye Shiner (*Notropis boops*) and Central Stoneroller (*Campostoma anomalum*), each made up an average of 22% of the individuals taken in a given sample; 2 others, Green Sunfish (*Lepomis cyanellus*) and Orangethroat Darter (*Etheostoma spectabile*), each made up approximately 10%, while the remaining core species averaged from 7.5% (Longear Sunfish, *Lepomis megalotis*) to 5% (Largemouth Bass, *Micropoterus salmoides*) to as low as about 4% (Blackstripe Topminnow, *Fundulus notatus*; Bluegill, *Lepomis macrochirus*). In the local community at Brier Creek site BR-5, 2 of the 4 core species (Bigeye Shiner and Central Stoneroller) each made up an average of about 25% of each collection, while the other 2 species, Orangethroat Darter and Longear Sunfish, made up averages of 18% and 7.5%, respectively. This type of pattern was evident in almost all of the communities examined at both the global and local scales. Of the core species in all nine global communities, the most abundant species had an average relative abundance one to two orders of magnitude greater than that of the least abundant core species. At the local scale, there were several communities with a single core species, but for those with multiple core species the most abundant typically had an average relative abundance one order of magnitude greater than that of the least abundant species. So, even among the core species that were always present, there were substantial differences in their abundance, that is, their numerical contributions to the total community.

Beyond the numbers of core species in each of the communities, we also determined the overall numerical contribution of core species as a group to the overall community (table 3.1). For each collection at a study site, we summed the relative abundances of all core species. We then calculated the average relative abundance of core species (AVG %CORE) across all collections at a site and the coefficient of variation (CV) of the percentage made up by the core (CV %CORE), based on summed relative abundance. The average serves as a measure of the overall contribution of core species to community structure, and the coefficient of variation quantifies the degree to which that contribution varies over time.

One-Time Species

In contrast to core species (and at the other extreme of occurrence), there were species in each community that were detected only once in the entire study period (table 3.1). The number of one-time species ranged from 0 to 8 among the global communities and from 0 to 17 among the local communities. To examine the relative contributions of core versus one-time species to the total species detected, we calculated the difference (core minus one-time) and plotted that against the total species detected (fig. 3.1). For all global communities (fig. 3.1, *top*) the number of core species exceeded the number of one-time species, and the relative number of core species had a positive relationship with the total number of species detected, indicating that core species contributed more to the total increase in species richness than did one-time species. At the local scale (fig. 3.1, *bottom*) the relative contributions of core and one-time species were reversed. The negative slope of this relationship (fig. 3.1, *bottom*) suggests that the increase in total species richness at the local scale is attributable more to increases in one-time species than in core species. This result has important implications for assessments of community richness and for development of conservation plans. Suppose that the total number of species detected is used as a means of identifying diversity "hot spots." The scale at which richness is calculated can be important, because if the scale of assessment is too small, many of the species counted in the total may occur only occasionally and might not benefit from conservation efforts at that site.

Diversity

Another emergent property of communities is species diversity, which can be measured by a number of indices (Maurer and McGill 2011) that incorporate both richness and relative abundance (as evenness) (Magurran 2011). We used the Shannon index, H′, which Maurer and McGill (2011) note is "probably the most commonly used expression of species diversity" and is relatively independent of sample size (McCune and Grace 2002). We used PC-ORD 6.0 to determine H′ for all collections of each community, and calculated the average H′ (AVG H′) and the coefficient of variation in H′(CV H′) for each community (table 3.1). Among our global communities, AVG H′ varied from 1.02 to 2.59, and among the local communities AVG H′ varied from 0.53 to 2.5. The CV H′ ranged from 1.1% to 19.4% (global) and from 9% to 98% (local). At both spatial scales the value of AVG H′ increased with the total number of species detected at a similar rate. A regression of AVG H′ on TOTAL S had a slope of 0.026 for global communities and 0.031 for local communities (which were not significantly different by analysis of covariance). The intercepts of the regression lines did differ

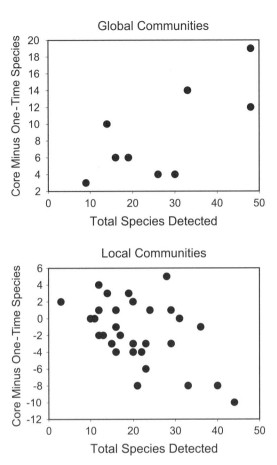

FIG. 3.1 Relative contribution of core versus one-time species to total species detected in global (*top*) and local (*bottom*) communities. The positive slope for the global communities indicates that the increase in total species results from additional core species. The negative slope for the local communities shows that the increase in total species at the local scale depends more on additional one-time species than on additional core species.

significantly, however (global intercept = 1.196; local intercept = 0.793), such that AVG H′ was overall higher in global communities.

Evenness

Evenness is a measure of the relative abundances of species within a sample such that if all species are equally abundant, evenness is maximized. Evenness ranges from approximately 0 to 1 (Maurer and McGill 2011). We used PC-ORD 6.0 to calculate evenness (designated as E in that program) for all samples of all communities. For each community we calculated average E (AVG E) and coefficient of variation in E (CV E) across all samples (table 3.1). Evenness (AVG E) ranged from 0.540 to 0.780 among global communities and from 0.352 to 0.885 among the local communities, but the

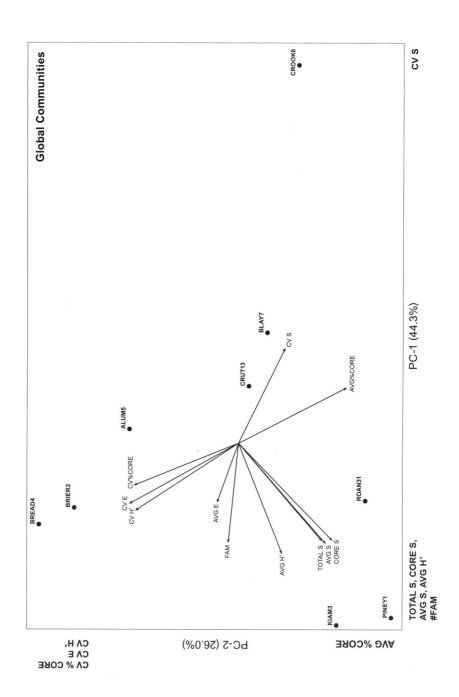

Global Communities

difference in evenness was not significant between the different scales. And observed evenness values did not vary as a function of total species detected at either scale, so the two metrics provide different information on the communities.

Principal Components Analyses of Emergent Community Properties

Eleven of the emergent community characteristics (table 3.1) described above were included in a principal components analysis (PCA) of the communities in character space. Separate PCAs were performed at the global and local scales. The PCAs were done in PC-ORD 6.0 on a correlation matrix of emergent community characteristics, and the number of significant axes was determined by comparing axis eigenvalues to those resulting from a randomization test with 999 runs.

At the global scale the analysis yielded one significant axis that explained 44.3% of the variance and a second axis (marginally significant at $p = 0.073$) that explained an additional 26.0% of the variance (fig. 3.2). Vectors summarizing the correlations of the eleven community characteristics with the two axes are also provided in figure 3.2.

Three community characteristics—total species detected (TOTAL S), average number of species per collection (AVG S), and number of core species (CORE S)—were highly positively intercorrelated, as evidenced by their proximity in multivariate space. Four other characteristics (CV H′, CV E, CV %CORE, and AVG %CORE) were also highly intercorrelated, with the coefficients of variation in diversity (CV H′), evenness (CV E), and percentage core species (%CORE) positively intercorrelated and the average percentage core species (AVG %CORE) negatively correlated with the others. Average evenness (AVG E) and coefficient of variation of species (CV S) detected in a given sample were also negatively correlated with each other.

Among the nine global communities in two-dimensional multivariate space, those with many species were to the left of PC-1, and those for which species numbers were also relatively consistent were also at the lower end of PC-2. Distribution of communities along PC-2 suggests a gradient of community stability in diversity, with evenness and the contribution of core species to community samples with more variable communities at the upper end of PC-2.

FIG. 3.2 (*opposite*) Biplot of PCA of global fish communities in community character space. PC-1 had an eigenvalue of 4.88 and explained 44.3% of the variance; PC-2 had an eigenvalue of 2.86 and explained 26.0% of the variance. Fish communities are identified using a combination of the site name and site number (from fig. 2.1 and table 2.1). In all subsequent plots of this global PCA, the global communities were identified only by site number. Vectors representing each of the 11 community characteristics indicate the direction and strength of correlations of those characteristics with each of the two PC axes. Characteristics with correlations having an absolute value of 0.6 or greater with a given axis are listed adjacent to the axis with negative correlations near the origin and positive correlations at the positive extreme of the axis.

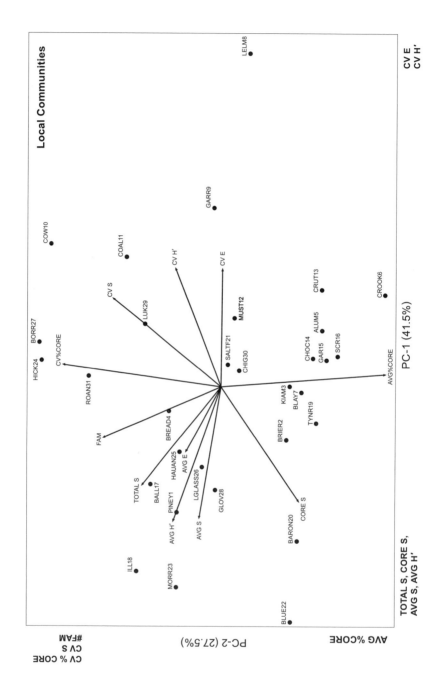

Local Communities

At the local scale the PCA had 3 significant axes that together accounted for 86.2% of the variance (PC-1, 41.5%; PC-2, 27.6%; PC-3, 17.2%). We consider only the multivariate space defined by PC-1 and PC-2 in further analyses. There were fewer strong correlations among the community characteristics that defined the multivariate space for the local communities than for the global communities (fig. 3.3).

In the multivariate space defined by the first two local PC axes, communities with high total species, high average number of species per collection, high diversity, and high evenness were toward the left on PC-1, while those with high variation in those characteristics (CV H, CV E, CV S) were toward the right on that axis. PC-2 was most influenced by characteristics associated with core species, with communities containing high numbers of core species and high percentage of the community attributable to core species near the bottom of that axis and those with high variation in core species near the top. Proximity of communities in multivariate space reflects similarity in characteristics used to define the space. We can therefore examine factors that have the potential to affect similarity among the communities.

Does Geography or Family Composition Predict Emergent Properties?
Geography: USGS Hydrologic Subregions

To examine the possible effect of geography on the location of the communities in emergent property multivariate space, we identified the hydrologic subregion in which each community was located (US Geological Survey Water Boundary Database / four-digit hydrologic unit codes) and separately superimposed convex hulls on biplots of PC-1 versus PC-2 for the global and local scales (fig. 3.4). The nine global communities (fig. 3.4, *top*) spanned six river subregions, with one site in each of five different subregions and four within the Lower Red–Ouachita subregion. The four sites in the same subregion were widely distributed along both axes in multivariate space, indicating that communities within the same subregion showed as much or more variation among sites than did communities from different subregions. So, overall, geography at the scale of subregions was not a major factor in emergent properties for the nine global stream communities.

The 31 local communities comprised 10 subregions, 6 of which had multiple sites where communities were sampled. Convex hulls for five of these subregions occupied

---

FIG. 3.3 (*opposite*) Biplot of PCA of local fish communities in community character space. PC-1 had an eigenvalue of 4.57 and explained 41.5% of the variance, and PC-2 had an eigenvalue of 3.02 and explained 27.5%% of the variance. Fish communities are identified using a combination of the site name and site number (from fig. 2.1 and table 2.1). In all subsequent plots of this local PCA, the local communities were identified only by site number. Vectors representing each of the 11 community characteristics indicate the direction and strength of correlations of those characteristics with each of the two PC axes. Characteristics with correlations having an absolute value of 0.6 or greater with a given axis are listed adjacent to the axis with negative correlations near the origin and positive correlations at the positive extreme of the axis.

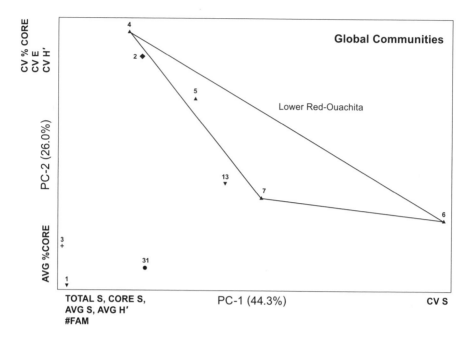

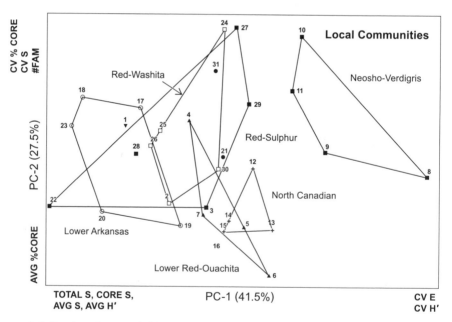

FIG. 3.4 Overlay of convex hulls indicating hydrologic subregions (based on four-digit HUCs) on biplots of PCAs for global (*top*) and local (*bottom*) communities. The global PCA is the same as in figure 3.2, and the local PCA is the same as in figure 3.3.

large areas in the multivariate PCA space (fig. 3.4, *bottom*), indicating that the communities collected within them varied extensively in their emergent community characteristics. Figure 3.4 (*bottom*) also showed that, in numerous cases, local communities from different subregions were close to each other in multivariate space, suggesting overall similarity in emergent community characteristics despite being located in different basins. For example, the closest pairs of local communities in multivariate space included Kiamichi River (Red-Sulphur subregion) and Blaylock Creek (Lower Red–Ouachita subregion); Chigley Sandy Creek (Red-Washita subregion) and Salt Fork of the Red River (Red Headwaters subregion); and South Canadian River (Lower Canadian subregion) and Gar Creek (North Canadian subregion). Similarity among characteristics of communities from different basins was also supported by the extensive overlap of convex hulls for five of the six basins (fig. 3.4, *bottom*) and by the location among the overlapping basins of the four communities that were the only representatives of a given basin: the Roanoke River (Chowan-Roanoke subregion) was within the convex hulls of both the Red-Washita and Red-Sulphur subregions, Piney Creek (Upper White subregion) was located within the convex hull for the Lower Arkansas subregion, the South Canadian River (Lower Canadian subregion) was adjacent to the convex hulls for the North Canadian and Lower Red–Ouachita subregions, and the Salt Fork of the Red River (Red Headwaters subregion) was within the convex hull for the Red-Sulphur subregion in figure 3.4 (*bottom*).

There was one subregion (Neosho-Verdigris) for which the convex hull did not overlap any others (fig. 3.4, *bottom*). All of the communities from this basin were toward the right side of PC-1, indicating that they tended to exhibit higher variation in evenness, diversity, and average number of species per collection than did communities located farther to the left on that axis. All of the sites from which these communities were sampled were similar in being small (maximum widths 4 to 8 m, with one pool 12 m wide; see plate 7) and intermittent, as shown later in the chapter by convex hulls for intermittency. But in spite of this one example of all sites in one subregion diverging from all other sites, the overarching pattern in figure 3.4 for local communities showed no evidence that they consistently differed geographically with respect to emergent properties.

Family Composition

We also examined the distribution of communities in multivariate space with respect to the community type as defined by the dominant families (table 3.1) to determine whether communities with similar family composition displayed similar emergent properties. At the global scale the location of named family types in PCA space, indicated by convex hulls in figure 3.5 (*top*), showed wide variation in community characteristics within a family type and suggested no predictable differences in emergent properties as a function of family types that dominated.

For the 31 local communities, those in which cyprinids were the single most abundant group occupied a large proportion of multivariate space (fig. 3.5, *bottom*), indi-

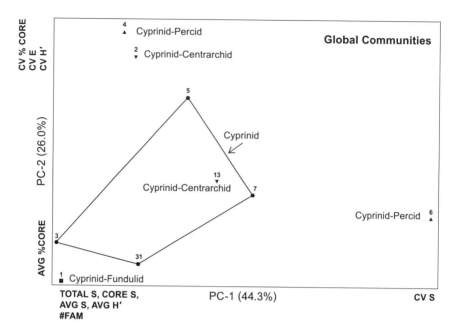

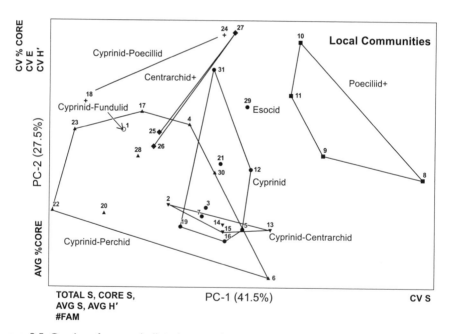

FIG. 3.5 Overlay of convex hulls indicating the community type based on dominant families on biplots of PCAs for global (*top*) and local (*bottom*) communities. The global PCA is the same as in figure 3.2, and the local PCA is the same as figure 3.3.

cating that they exhibited wide variation in emergent community characteristics. Of the more narrowly defined subsets of cyprinid-dominated communities, the Cyprinid-Percid communities exhibited the most variation, followed by the Cyprinid and Cyprinid-Centrarchid types. All of these overlapped in multivariate space, however, indicating shared emergent characteristics among these community types. On the other hand, the poeciliid-dominated communities did not overlap the cyprinid-dominated communities in multivariate space, suggesting that characteristics of those community types differed from the other named types of communities. Although there was substantial variation among poeciliid-dominated communities (as evidenced by the large area occupied by the convex hull for that group), all were displaced to the right of PC-1, indicating that they all showed higher variation in number of species per collection, diversity, and evenness than the cyprinid community types.

Several factors may contribute to the greater variation in these poeciliid-dominated communities. First, the sites where these occur are environmentally variable and the streams are intermittent, both of which can act to limit which species can occur. Western Mosquitofish are highly tolerant of harsh or variable conditions (Pyke 2008) and are good invaders because they can establish populations quickly following recolonization of dry stream reaches. Western Mosquitofish often display boom-and-bust population dynamics (Matthews and Marsh-Matthews 2011), which would contribute to the variation in relative abundance (i.e., evenness).

Do Environmental Factors Predict Emergent Properties?

For the 9 global and 31 local communities, we compared positions on the PCA axes to physical, geographic, or climatic properties of the streams (table 3.2). Properties for comparison included traits that all have potential to relate broadly to the composition of fish communities (Marsh-Matthews and Matthews 2000). Factors we examined included categorical traits (intermittency, location in uplands vs. lowlands, and land use) and continuous traits (width, depth, stream gradient, temperature, and rainfall) for the streams. Categorical environmental variables were determined as follows. Intermittency (following categories of Poff and Ward 1989) was estimated from our field notes or personal knowledge of each site. Global streams or local sites that never were known to cease flowing were scored for intermittency as "none" (0); sites that were nonflowing or dry at least once, but usually were flowing, were scored as "low" (1); and sites that more than once were nonflowing or dry were scored as "high" (2) in intermittency. Local sites and global streams were classified as "upland," "lowland," or "marginal" on the basis of their geographic location within or outside recognized highland areas (Ozark Highland, Ouachita Highland, mountains of western Virginia). Global streams that flowed from uplands to lowlands and local sites at edges of uplands were scored as marginal. Land use for global streams was estimated subjectively on the basis of our knowledge of the entire watershed and checked against Google Earth images. For local sites, land use was determined from Google Earth, classifying by the primary use within a 0.5 km radius around the site. For land use at

Table 3.2. Environmental characteristics of global and local study systems

Stream Abbreviation (Name + Map Site)	Intermittency	Upland vs. Lowland	Land Use (1 km radius)	Max Width (m)	Max Depth (cm)	Elevation (m)	Stream Gradient (m/km)	Annual Mean Air Temperature (°C)	Average Mean High Air Temperature (°C)	Precipitation (cm)
Global										
ALUM5	low	up	forest	20	120	318	13	16.6	23.1	147
BLAY7	none	up	forest	15	125	353	12	15.0	21.3	148
BREAD4	low	up	forest	15	65	299	18	16.6	23.1	147
BRIER2	low	low	mixed	12	150	222	4	17.2	23.9	105
CROOK6	none	up	forest	9	90	371	19	15.0	21.3	148
CRUT13	low	low	urban	15	150	367	2	16.1	22.2	92
KIAM3	low	up	mixed	50	150	160	1	17.2	23.9	128
PINEY1	none	up	mixed	36	150	130	2	13.6	21.0	117
ROAN31	none	up	mixed	50	150	406	3	13.7	19.5	103
Local										
ALUM5	none	up	forest	20	120	318	28	16.6	23.1	147
BALL17	none	up	mixed	15	100	282	2	15.6	21.7	123
BARON20	none	up	mixed	50	100	220	1	15.6	21.7	121
BLAY7	none	up	forest	6	75	309	12	15.0	21.3	148
BLUE22	none	margin	mixed	30	150	291	1	17.2	23.3	106
BORR27	high	low	forest	30	100	104	0	17.2	23.9	132
BREAD4	none	up	forest	15	65	299	14	16.6	23.1	147
BRIER2	low	low	mixed	12	150	222	3	17.2	23.9	105
CHIG30	low	low	pasture	8	30	248	4	17.2	23.3	101
CHOC14	none	low	urban	4	50	337	1	16.1	22.2	92

COAL11	low	margin	pasture	12	50	232	2	15.0	21.1	114
COW10	low	margin	pasture	4	80	237	2	15.0	21.1	114
CROOK6	none	up	forest	4	60	400	23	15.0	21.3	148
CRUT13	none	low	mixed	15	150	347	3	16.1	22.2	92
GAR15	none	low	forest	8	50	269	6	16.7	22.8	104
GARR9	high	margin	mixed	7	100	237	2	15.0	21.1	114
GLOV28	none	up	pasture	40	60	117	1	17.2	23.9	132
HAUAN25	none	low	pasture	19	110	200	2	17.2	23.9	105
HICK24	none	low	mixed	10	50	198	2	17.2	23.9	99
ILL18	none	up	mixed	75	150	272	1	15.6	21.7	123
KIAM3	none	up	forest	12	100	256	6	16.7	23.3	125
LELM8	high	margin	pasture	4	50	247	2	15.0	21.1	114
LGLASS26	high	low	mixed	15	150	195	3	17.2	23.9	105
LUK29	high	low	mixed	30	50	133	0	17.2	23.9	132
MORR23	none	up	pasture	15	150	139	2	16.7	23.3	125
MUST12	low	low	urban	7	150	375	2	15.6	22.2	85
PINEY1	none	up	mixed	10	50	131	3	13.6	21.0	117
ROAN31	none	up	mixed	35	80	359	3	13.7	19.5	103
SALTF21	high	low	pasture	50	50	458	1	16.1	23.9	71
SCR16	low	low	pasture	50	100	342	3	16.1	22.8	95
TYNR19	low	up	mixed	20	50	250	2	15.6	21.7	123

both scales we assigned broad general categories, including pasture, mixed forest and pasture, or urban.

Continuous variables for global streams and local sites were determined as follows. Elevation, latitude, and longitude were taken from Google Earth at the approximate midpoint of the mainstem for each of the 9 global streams and at the 31 individual site locations. Overall stream gradient (m/km) for global streams was between our uppermost and lowermost sample site, as documented in chapter 2. For local sites, stream gradient (m/km) was determined from Google Earth as the difference between the elevation of the streambed 500 m upstream and 500 m downstream of the midpoint of the sampling site. Mean annual temperature, mean daily high temperature, and mean annual precipitation for streams and sites were from the Oklahoma Mesonet or from the National Weather Service for the county or for the nearest reporting weather station. Maximum widths and depths were from our field notes. At sites where any pool was "over the wader tops," we recorded a depth of 150 cm by convention.

Categorical Environmental Variables

For the three categorical variables (intermittency, upland-lowland, and land use) we drew convex hulls around the assigned categories on the PCA biplots of the emergent properties to allow estimation of the importance of those variables in determining positions of communities in the PCAs.

Hydrology (Intermittency)

All global streams were either perennial or occasionally intermittent (low), as none were considered highly intermittent. There was complete separation of perennial and occasionally intermittent global streams on the PCA biplot (fig. 3.6, *top*), suggesting that streams that cease to flow even occasionally can have emergent properties different from those where flow is, as far as we know, uninterrupted. Streams 4, 2, and 5 toward the top of the biplot (South Alum Creek, Brier Creek, and Bread Creek) have all been sampled at times when water was extremely low and all or substantial parts of the streams were nonflowing. These three streams were toward the positive end of PC-2, corresponding to having large coefficients of variation in diversity, evenness, and the percentage of the community made up of core species. The perennial streams all were toward the bottom of the PCA biplot but ranged widely on PC-1, from sites with high total species or consistent diversity and evenness (e.g., stream 1, Piney Creek; stream 31, Roanoke River) to stream 6 (Crooked Creek), which had substantially fewer species and low diversity.

For local communities (fig. 3.6, *bottom*), there was substantial overlap between those with no, low, or high intermittency. But intermittent streams were toward the right side of PC-1 and thus appeared to be more dynamic in emergent properties (CV E, CV H'), whereas perennial streams, toward the left side of PC-1, had more stable community properties (TOTAL S, CORE S, AVG S, and AVG H'). At both the global and local scales, hydrology seemed to matter, and intermittent streams

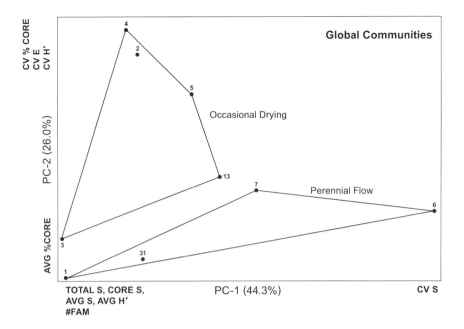

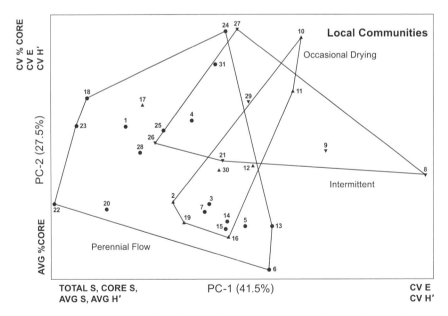

FIG. 3.6 Overlay of convex hulls indicating hydrologic patterns of the collection sites on biplots of PCAs for global (*top*) and local (*bottom*) communities. The global PCA is the same as in figure 3.2, and the local PCA is the same as in figure 3.3.

were suggested to have substantially different emergent community properties overall, compared to perennial streams. These results at both scales are consistent with the findings of Horwitz (1978), Schlosser (1987), and Poff and Ward (1989) that intermittency is one of the most important hydrological variables, with strong potential to affect composition and dynamics of fish communities. Our results are also consistent with older research summarized by Hynes (1972), showing that intermittent streams are more variable in biological properties than perennial streams. In this case, our findings seem to support the conventional wisdom in stream and fish ecology!

Upland versus Lowland Streams

We have often observed that stream fish communities differed greatly between uplands and lowlands, with a preponderance of clear water, stony substrates, riffle-pool morphometry, and stronger flows in the former, contrasted with more turbid water, soft substrates, more channel structure, and lower or inconsistent flows. Our observations are consistent with patterns in distribution of fish or environmental conditions in systems worldwide (e.g., Ibarra and Stewart 1989; Oberdorff et al. 1993; Wolter and Vilcinskas 1997). In many cases the species identity or species distributional limits changes, sometimes sharply even within a drainage, from lowland to upland reaches (Matthews and Robison 1988, 1998; Ibarra and Stewart 1989; Taylor and Lienesch 1996; Matthews 1998, 305). But until now we have not really tested the upland-lowland dichotomy against emergent community properties. Only two global streams (Brier and Crutcho Creeks) were considered to be lowland, as the other seven were all in systems in the uplands. On a biplot of PC-1 versus PC-2 for the global streams (not shown), both Brier and Piney Creeks were well within the convex hull surrounding all of the upland streams, so there was no general differentiation in emergent properties for stream fish communities in the two categories.

For the 31 local communities, figure 3.7 (*top*) showed substantial overlap in the convex hulls surrounding upland, lowland, and marginal sites on the PCA of emergent community properties. There was no simple pattern at either the local or global scale to indicate that position in uplands versus lowlands, alone, had any substantial influence on the emergent properties of warm-water fish communities. Keep in mind that none of our data sets include cold-water systems, in which salmonid communities clearly change to warm-water communities (e.g., Rahel and Hubert 1991).

Land Use

The relationship of land use to stream fish communities is historically a mixed bag. Harding et al. (1998), Meador and Goldstein (2003), and Gido et al. (2010) all found relationships between land use and structure or diversity of local fish communities, but land use in these studies was typically complicated by other natural or anthropogenic factors. Our own assessment for the 65 sites we sampled throughout the central United States in June 1995 showed that near-stream land use loaded strongly only on

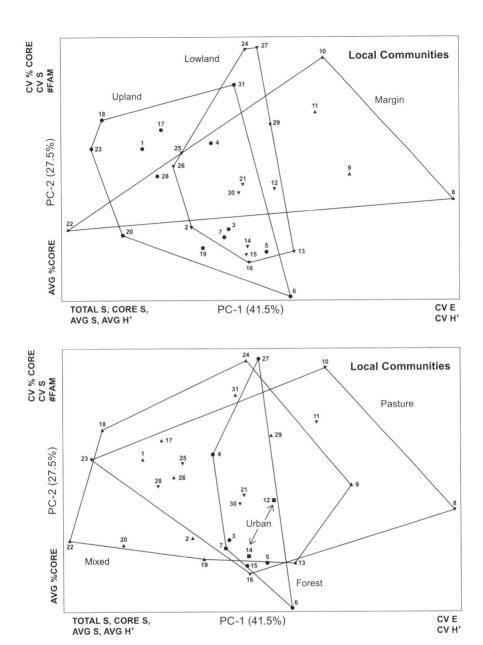

FIG. 3.7 Overlay of convex hulls indicating upland, lowland, and marginal streams (*top*) and land use (*bottom*) of the collection sites on biplots of PCAs for local communities. The local PCA is the same as in figure 3.3.

the third axis of a PCA that included 13 "terrestrial" variables and that this PC-3 axis did not explain significant variation in the local fish assemblages. There seems to be no overarching indication in the literature that "land use X results in fish community change Y," and the large number of case histories of "land use versus fish assemblages" seem to have varying local outcomes.

For our data sets, the convex hulls for land use overlain on the PCA of emergent properties for the global streams (not shown) showed that the four forested watersheds (Alum, Blaylock, Bread, and Crooked Creeks) that were all in the Ouachita National Forest, Arkansas, separated from the other five global stream communities. But these four creeks are known to differ from the others on the PCA for a variety of reasons, so adding land use for the global streams provided little new information. For the 31 local sites, convex hulls overlaid on the PCA of emergent properties (fig. 3.7, *bottom*) also showed strong overlap of forest, pasture, and mixed land-use sites. From a critical perspective, while land use may result in changes in watersheds (Allan 2004) that indirectly affect fish communities, our study systems do not support any strong direct relationship between fish community emergent properties and land use in a basin in general or within a 1 km–diameter area about the local sample sites.

Continuous Environmental Variables

To evaluate the influence of the continuous environmental variables, we first ran product moment correlations between PC-1 and PC-2 for emergent community properties and the abiotic variables to identify any that were significant or nearly so at an uncorrected $p = 0.05$.

Global Communities

To identify environmental variables related to emergent properties of global communities, we used Pearson product moment correlation to examine the relationship of environmental variables to each of the first two emergent community property global PCA axes. We chose those variables with correlations significant at or near $p = 0.05$ for further examination. Although some of these correlations were only marginally significant (likely owing to the low sample size of only nine global communities), we felt that the variables described below were all of potential biological interest.

The first PCA axis (PC-1) for emergent properties of the global communities was related to three environmental variables (maximum stream width, overall stream gradient, and elevation). Stream width was actually more strongly related to the second PCA axis, but because PC-1 accounted for more of the variance in emergent properties than PC-2, we discuss stream width here, as it was related to PC-1.

Interpretation of these three environmental factors relative to PC-1 was based on the associations of individual emergent properties with PC-1. Streams with negative values for PC-1 were more diverse (more total and average number of species and number of core species, more families, and higher values for Shannon diversity). Streams toward the positive end of PC-1 conversely had lower diversity values and were more variable, as shown by the higher coefficient of variation in number of species captured per collection.

PC-1 scores for global communities had at least a marginally significantly positive relationship with both elevation ($R^2 = 0.449$, $p = 0.048$) and stream gradient ($R^2 = 0.405$, $p = 0.066$) and a marginally significantly negative relationship with stream width ($R^2 = 0.394$, $p = 0.070$) (fig. 3.8). Matching these three environmental variables to the emergent properties on PC-1 suggested that global communities in the streams with higher elevation, steeper gradients, and narrower width were more variable in important emergent properties, whereas communities of the streams at lower elevation, with less steep gradients and greater width, tended to have higher numbers of species and families, as well as higher diversity indices, and were less variable overall.

In addition to stream width, two other continuous abiotic variables were associated with global PC-2 with Pearson's $r > 0.60$, including mean annual temperature and mean average high temperature (both highly correlated, so we consider only the mean annual temperature). Emergent properties most associated with PC-2 were the average percentage of the community composed of core species (negative on PC-2) and the CVs for the Shannon diversity, evenness, and proportion of the community consisting of core species (positive on PC-2).

PC-2 scores for global communities were significantly positively related to mean annual temperature ($R^2 = 0.480$, $p = 0.039$) (fig. 3.8). Accordingly, fish communities in streams with higher mean annual air temperatures had higher variability in emergent diversity measures.

The overall impression from comparing emergent properties PC-1 and PC-2 to the suite of categorical and continuous environmental variables was that the global fish communities in smaller (narrower-width) streams at higher elevations with higher stream gradients (e.g., Crooked, Blaylock, and Alum Creeks in the Ouachita National Forest) and in locations with higher environmental temperatures and a tendency for intermittency (Bread and Alum Creeks, and Brier Creek in south Oklahoma) had fewer species and a greater tendency for variation in important emergent properties over time (based on generally larger CVs in numbers of species, diversity, or evenness). These findings are consistent with some established patterns in stream fish ecology, such as Horwitz's (1978) finding that headwaters with more variable flow had lower fish diversity across 15 river basins, or with generalizations by Pires et al. (1999) that temperature, depth, and width were key factors in fish species abundances in intermittent Portuguese streams.

Local Communities

Pearson product moment correlations between emergent property PC-1 or PC-2 and continuous variables for the 31 local sites showed that four correlations were of the most interest, all significant at 0.01 or better without alpha adjustment. For variables thus identified as potentially of interest, we graphed the PCA axis scores as the response variable (y axis) against the independent continuous variable (x axis).

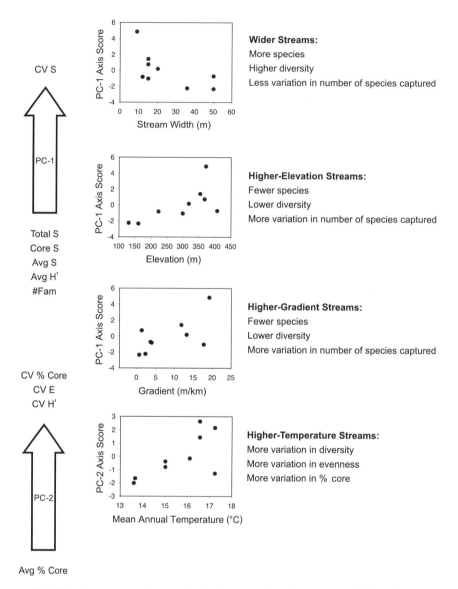

FIG. 3.8 Global community characteristics (as summarized by scores on PC axes) as a function of environmental factors.

On PC-1, negative scores were correlated with the emergent properties of more total species, more core species (i.e., species that were always present), a higher number of average species per survey, and higher Shannon diversity. Conversely, positive PC-1 scores related to greater variability (as indicated by larger CVs in both Shannon diversity and evenness). PC-1 scores for the 31 local communities were significantly negatively related to stream width (fig. 3.9) ($R^2 = 0.198$, $p = 0.012$), indicating that

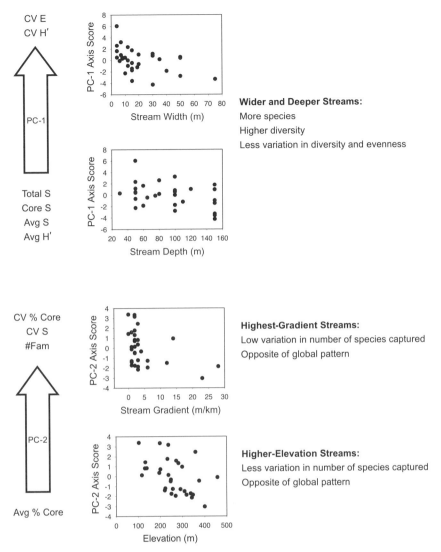

FIG. 3.9 Local community characteristics (as summarized by scores on PC axes) as a function of environmental factors.

wider sites had more overall diversity and narrower sites varied more in emergent properties. There was a similar negative relationship for PC-1 scores and depth across the 31 local sites ($R^2 = 0.156$, $p = 0.028$). Thus deeper sites also had fish communities with higher diversity and more temporal stability, and shallower sites had more variable emergent properties.

On PC-2, communities with positive scores had more families but also higher CVs in number of species and in the contribution of core species to the community, suggesting a lack of stability in emergent properties. Conversely, negative scores on PC-2 were correlated with a higher average percentage of the community made up of core species. Community score on PC-2 was significantly negatively related to both stream gradient ($R^2 = 0.167$, $p = 0.022$) and elevation ($R^2 = 0.215$, $p = 0.009$). Local communities from sites in higher-gradient terrain (often stony bottomed, with riffle-pool structure) and from higher-elevation streams had low variability (CV) in numbers of species and in contribution of core species. These patterns were opposite of those found for the global communities. While we cannot specifically explain any mechanisms causing these differences, they underscore how considering fish communities at different spatial scales can yield important insights.

Based on correlations of the categorical and continuous physical variables with PC-1 from the PCA on emergent properties, wider, deeper, and perennial sites had higher diversity and lower community variability, and smaller or intermittent streams had a tendency toward more community variability. Correlations of physical variables with PC-2 showed a tendency for higher-gradient, upland sites to have more core species and less community variability. Overall, there was substantial correspondence between five physical variables (including intermittency; see above) and the emergent properties of the 31 communities as shown by their positions on axes from the PCA.

The results above all support Schlosser's (1987) conceptual model about factors influencing fish community structure in small streams. For example, Schlosser suggested that in the parts of a stream with larger or deeper pools, local fish communities should be relatively more stable compared to smaller stream reaches with uniform shallows, which were characterized by more temporally variable populations of small fishes. Based on his earlier empirical work in Jordan Creek, Schlosser (1987) showed a smaller coefficient of variation in the downstream, larger parts of the system than in upstream sites. Our findings also support the older but important summary by Sheldon (1968), that deeper stream sites had higher fish diversity and that "in general distribution and diversity of fishes in Owego Creek are controlled by structural features of the habitat." So, our results come as no surprise but help confirm the importance of physical properties to emergent properties of fish communities, across a wide range of stream habitats and environmental settings.

Summary

In this chapter we characterized a wide range of fish communities across many stream types by focusing at the taxonomic level of family (rather than individual species identities) and on emergent properties of the communities. Because emergent properties are independent of species identity, we were then able to use them to compare structure and dynamics across communities that had few or no species in common. Although there were 18 families across our 31 primary study systems, minnows (Cyprinidae) numerically dominated all 9 of the global communities and 23 of 31 local communities. Four other local communities were dominated numerically by Western Mosquitofish (Poeciliidae), three by sunfish and black bass (Centrarchidae), and one by Grass

Pickerel (Esocidae). The family classification can help with predictions about factors like food-web structure or the potential importance of predation (see chap. 5).

Emergent properties—including species richness, number of core species and one-time species, diversity, and evenness, as well as CVs in these traits—were determined for each of the global and local study systems. Individual communities represented a broad range of all of the emergent traits. In PCA biplots based on emergent traits, both global and local communities ranged from having generally higher species richness, diversity, and evenness at one end on the PC axis to having higher CVs in those traits at the other end of the axis.

The positions of global and local communities on PCA biplots based on emergent properties were not specifically related to the hydrologic subregions in which sites were located, except for local communities in the Neosho-Verdigris subregion. Likewise, the identity of the dominant family for the most part did not predict the position of communities in multivariate space based on emergent properties, except for local communities dominated by mosquitofish, which were the same communities as those in the Neosho-Verdigris subregion. These four sites, all in one subregion and all dominated by Western Mosquitofish, were separated from all other local communities toward the end of a PC axis that represented fewer species and greater variability in diversity. In addition, these four sites were characterized by high intermittency (see plate 7). But other sites, dominated by other families, were also intermittent, so all of the available evidence suggests that it is dominance by mosquitofish (see plate 15), with their tolerance for harsh environments and boom-and-bust population dynamics, which separates these four sites from all others. "Family" matters in this case.

Some environmental traits were substantial predictors of emergent properties of the communities. Of three categorical traits (intermittency, location in upland vs. lowland, and land use), only the degree to which streams were intermittent was a good predictor of the position of communities in emergent property PCA space, with those in intermittent habitats shifted toward the ends of PC axes representing higher variation in community characteristics. The lack of an effect of the upland-lowland dichotomy was surprising, given the well-known distinctiveness of these types of communities at the species level. These results suggest that, in spite of the differences in species identities, similar community emergent properties can exist in both upland and lowland streams. Likewise, land use had surprisingly little effect on the emergent properties of global or local communities.

A variety of continuous variables were predictive of community scores on PCA axes. Stream width gave the same answer at both scales, with smaller streams having more variable emergent properties, matching the result of previous studies or the Schlosser (1987) model. But two other continuous variables gave different results at global and local scales. Higher elevation and higher stream gradient suggested more variability for global communities, in contrast to less variability in emergent properties for local communities. At each scale, other, but different, variables explained substantial variation in PCA scores. For global communities, warmer mean annual air temperature corresponded with more variable emergent properties. For local communities, stream depth appeared important, with deeper sites showing a trend toward less variability in emergent properties. Thus our findings for some environmental variables supported conventional wisdom in fish community ecology, but results for other environmental variables were scale dependent, providing different conclusions between local communities and the global communities that were based on the sum of fish across multiple sampling sites in each watershed.

References

Allan, J. D. 2004. Landscapes and riverscapes: the influence of land use on stream ecosystems. Annual Reviews of Ecology, Evolution, and Systematics 35:257–284.

Bailey, R. G. 1983. Delineation of ecosystem regions. Environmental Management 7: 366–373.

Beketov, M. A., B. J. Kefford, R. B. Schafer, and M. Liess. 2013. Pesticides reduce regional biodiversity of stream invertebrates. Proceedings of the National Academy of Sciences USA 110:11,039-11,043.

Chao, A., and L. Jost. 2012. Coverage-based rarefaction and extrapolation: standardizing samples by completeness rather than size. Ecology 93:2533–2547.

Cheek, C. A., and C. M. Taylor. 2016. Salinity and geomorphology drive long-term changes of local and regional fish assemblage attributes in the lower Pecos River, Texas. Ecology of Freshwater Fish 25:340–351. doi:10.1111/eff.12214.

Coelho, M., L. Juen, and A. C. Mendes-Oliveira. 2014. The role of remnants of Amazon savanna for the conservation of Neotropical mammal communities in eucalyptus plantations. Biodiversity and Conservation 23:3171–3184.

Colwell, R. K., A. Chao, N. J. Gotelli, S.-Y. Lin, C. X. Mao, R. L. Chazdon, and J. T. Longino. 2012. Models and estimators linking individual-based and sample-based rarefaction, extrapolation and comparison of assemblages. Journal of Plant Ecology 5:3–21.

Dolan, J. R., M. E. Ritchie, A. Tunin-Ley, and M.-D. Pizay. 2009. Dynamics of core and occasional species in the marine plankton: tintinnid ciliates in the north-west Mediterranean Sea. Journal of Biogeography 35:887–895.

Douglas, M. E., and W. J. Matthews. 1992. Does morphology predict ecology? hypothesis testing within a freshwater stream fish assemblage. Oikos 65:213–224.

Echelle, A. A., and G. D. Schnell. 1976. Factor analysis of species associations among fishes of the Kiamichi River, Oklahoma. Transactions of the American Fisheries Society 105: 17–31.

Edds, D. R., W. J. Matthews, and F. P. Gelwick. 2002. Resource use by large catfishes in a reservoir: is there evidence for interactive segregation and innate differences? Journal of Fish Biology 60:739–750.

Eyre, F. H., ed. 1980. Forest cover types of the United States and Canada. Society of American Foresters, Washington, DC.

Federal Geographic Data Committee. 2008. Vegetation classification standard, FGDC-STD-005, Version 2. Federal Geographic Data Committee, Washington, DC.

Garcia, L., I. Pardo, and J. S. Richardson. 2014. A cross-continental comparison of stream invertebrate community assembly to assess convergence in forested headwater streams. Aquatic Sciences 76:29–40.

Gatz, A. J., Jr. 1979a. Community organization of fishes as indicated by morphological features. Ecology 60:711–718.

Gatz, A. J., Jr. 1979b. Ecological morphology of freshwater stream fishes. Tulane Studies in Zoology and Botany 21:91–124.

Gido, K. B., W. K. Dodds, and M. E. Eberle. 2010. Retrospective analysis of fish community change during a half-century of land use and streamflow changes. Journal of the North American Benthological Society 29:970–987.

Gido, K. B., J. F. Schaefer, and J. Pigg. 2004. Patterns of fish invasions in the Great Plains of North America. Biological Conservation 118:121–131.

Gillette, D. P., A. M. Fortner, N. R. Franssen, et al. 2012. Patterns of change over time in darter (Teleostei: Percidae) assemblages of the Arkansas River basin, northeastern Oklahoma, USA. Ecography 35:855–864.

Gotelli, N. J., and R. K. Colwell. 2001. Quantifying biodiversity: procedures and pitfalls in the measurement and comparison of species richness. Ecology Letters 4:379–391.

Gotelli, N. J., and R. K. Colwell. 2011. Estimating species richness. Pages 39–54 in A. E. Magurran and B. J. McGill, eds. Biological diversity: frontiers in measurement and assessment. Oxford University Press, Oxford, UK.

Harding, J. S., E. F. Benfield, P. V. Bolstad, G. S. Helfman, and E. B. D. Jones III. 1998. Stream biodiversity: the ghost of land use past. Proceedings of the National Academy of Sciences USA. 95:14,843–14,847.

Hocutt, C. H., and E. O. Wiley. 1986. The zoogeography of North American freshwater fishes. John Wiley, New York, NY.

Horwitz, R. J. 1978. Temporal variability patterns and the distributional patterns of stream fishes. Ecological Monographs 48:307–321.

Huet, M. 1959. Profiles and biology of western European streams as related to fish management. Transactions of the American Fisheries Society 88:155–163.

Hurlbert, S. H. 1971. The nonconcept of species diversity: a critique and alternative parameters. Ecology 52:577–586.

Hynes, N. H. B. 1972. The ecology of running waters. University of Toronto Press, Toronto, ONT.

Ibarra, M., and D. J. Stewart. 1989. Longitudinal zonation of sandy beach fishes in the Napo River basin, eastern Ecuador. Copeia 1989:364–381.

Jennings, M. D., D. Faber-Langendoen, O. L. Loucks, R. K. Peet, and D. Roberts. 2009. Standards for associations and alliances of the US national vegetation classification. Ecological Monographs 79:173–199.

Karr, J. R., K. D. Fausch, P. L. Angermeier, P. R. Yant, and I. J. Schlosser. 1986. Assessing biological integrity in running waters—a method and its rationale. Special Publication 5. Illinois Natural History Survey, Champaign, IL.

Kobayashi, S. 1982. The rarefaction diversity measurement and the spatial distribution of individuals. Japanese Journal of Ecology 32:255–258. [Cited in Gotelli and Colwell 2001.]

Kwak, T. J., and J. T. Peterson. 2007. Community indices, parameters, and comparisons. Pages 677–763 in C. S. Guy and M. L. Brown, eds. Analysis and interpretation of freshwater fisheries data. American Fisheries Society, Bethesda, MD.

Magurran, A. E. 2011. Measuring biological diversity in time (and space). Pages 85–94 in A. E. Magurran and B. J. McGill, eds. Biological diversity: frontiers in measurement and assessment. Oxford University Press, Oxford, UK.

Marsh-Matthews, E., and W. J. Matthews. 2000. Geographic, terrestrial and aquatic factors: which most influence the structure of stream fish assemblages in the midwestern United States? Ecology of Freshwater Fish 9:9–21.

Marsh-Matthews, E., and W. J. Matthews. 2010. Proximate and residual effects of exposure to simulated drought on prairie stream fishes. Pages 461–486 in K. B. Gido and D. A. Jackson, eds. Community ecology of stream fishes: concepts, approaches, and techniques. American Fisheries Society Symposium 73. American Fisheries Society, Bethesda, MD.

Matthews, W. J. 1998. Patterns in freshwater fish ecology. Chapman and Hall, New York, NY.

Matthews, W. J., and E. Marsh-Matthews. 2011. An invasive fish species within its native range: community effects and population dynamics of *Gambusia affinis* in the central United States. Freshwater Biology 56:2609–2619.

Matthews, W. J., and E. Marsh-Matthews. 2016. Dynamics of an upland stream fish community over 40 years: temporal trajectories and support for the "loose equilibrium" concept. Ecology 97:706–719.

Matthews, W. J., and H. W. Robison. 1988. The distribution of the fishes of Arkansas: a multivariate analysis. Copeia 1988:358–374.

Matthews, W. J., and H. W. Robison. 1998. Influence of drainage connectivity, drainage area and regional species richness on fishes of the Interior Highlands in Arkansas. American Midland Naturalist 139:1–19.

Matthews, W. J., A. M. Miller-Lemke, M. L. Warren, D. Cobb, J. G. Stewart, B. Crump, and F. P. Gelwick. 2004. Context-specific trophic and functional ecology of fishes of small stream ecosystems in the Ouachita National Forest. Pages 221–230 in J. M. Guildin, ed. Ouachita and Ozark Mountains Symposium: ecosystem management research. General Technical Report SRS-74. US Department of Agriculture, US Forest Service, Southern Research Station, Asheville, NC.

Maurer, B. A., and B. J. McGill. 2011. Measurement of species diversity. Pages 55–65 in A. E. Magurran and B. J. McGill, eds. Biological diversity: frontiers in measurement and assessment. Oxford University Press, Oxford, UK.

McCune, B., and J. B. Grace. 2002. Analysis of ecological communities. MjM Software Design, Gleneden Beach, OR.

Meador, M. R., and R. M. Goldstein. 2003. Assessing water quality at large geographic scales: relations among land use, water physicochemistry, riparian condition, and fish community structure. Environmental Management 31:504–517.

Miyazono, S., R. Patino, and C. M. Taylor. 2015. Desertification, salinization, and biotic homogenization in a dryland river ecosystem. Science of the Total Environment 511: 444–453.

Moyle, P. B., and M. P. Marchetti. 2006. Predicting invasion success: freshwater fishes in California as a model. BioScience 56:515–524.

Oberdorff, T., E. Guilbert, and J. C. Lucchetta. 1993. Patterns of fish species richness in the Seine River basin, France. Hydrobiologia 259:157–167.

Petanidou, T., A. S. Kallimanis, J. Tzanopoulos, S. P. Sgardelis, and J. D. Pantis. 2008. Long-term observations of a pollination network: fluctuation in species and interactions, relative invariance of network structure and implications for estimates of specialization. Ecology Letters 11:564–575.

Pires, A. M., I. G. Cowx, and M. M. Coelho. 1999. Seasonal changes in fish community structure of intermittent streams in the middle reaches of the Guadiana basin, Portugal. Journal of Fish Biology 54:235–249.

Poff, N. L., and J. V. Ward. 1989. Implications of streamflow variability and predictability for lotic community structure: a regional analysis of streamflow patterns. Canadian Journal of Fisheries and Aquatic Sciences 46:1805–1818.

Pyke, G. H. 2008. Plague minnow or mosquito fish? A review of the biology and impacts of introduced Gambusia species. Annual Review of Ecology, Evolution, and Systematics 39: 171–191.

Rahel, F. J. 1984. Factors structuring fish assemblages along a bog lake successional gradient. Ecology 65:1276–1289.

Rahel, F. J., and W. A. Hubert. 1991. Fish assemblages and habitat gradients in a Rocky Mountain–Great Plains stream: biotic zonation and additive patterns of community change. Transactions of the American Fisheries Society 120:319–332.

Roset, N., G. Grenouillet, D. Goffaux, D. Pont, and P. Kestemont. 2007. A review of existing fish assemblage indicators and methodologies. Fisheries Management and Ecology 14: 393–405.

Sanders, H. L. 1968. Marine benthic diversity: a comparative study. American Naturalist 102:243–282.

Schlosser, I. J. 1987. A conceptual framework for fish communities in small headwater streams. Pages 17–24 in W. J. Matthews and D. C. Heins, eds. Community and evolutionary ecology of North American stream fishes. University of Oklahoma Press, Norman, OK.

Schlosser, I. J. 1991. Stream fish ecology: a landscape perspective. BioScience 41:704–712.

Sheldon, A. L. 1968. Species diversity and longitudinal succession in stream fishes. Ecology 49:193–198.

Simberloff, D. 1972. Properties of the rarefaction diversity measurement. American Naturalist 106:414–418.

Taylor, C. M., and P. W. Lienesch. 1996. Regional parapatry of the congeneric cyprinids *Lythrurus snelsoni* and *L. umbratilis*: species replacement along a complex environmental gradient. Copeia 1996:493–497.

Taylor, C. M., D. S. Millican, M. E. Roberts, and W. T. Slack. 2008. Long-term change to fish assemblages and the flow regime in a southeastern US river system after extensive aquatic ecosystem fragmentation. Ecography 31:787–797.

Vaughn, C. C. 2010. Biodiversity losses and ecosystem function in freshwaters: emerging conclusions and research directions. BioScience 60:25–35.

Wolter, C., and A. Vilcinskas. 1997. Characterization of the typical fish community of inland waterways of the north-eastern lowlands in Germany. Regulated Rivers: Research and Management 13:335–343.

Worm, B., E. B. Barbier, N. Beaumont, J. E. Duffy, et al. 2006. Impacts of biodiversity loss on ocean ecosystem services. Science 314:787–790.

CHAPTER 4

Traits of Species That Influence Community Dynamics

Historical Perspective

One of the oldest questions in ecology (Elton 1927; Brown 1995) is whether abiotic factors (Andrewartha and Birch 1954) or biotic interactions (Elton 1946) control community structure and dynamics. The answer, of course, is that both do (Jackson et al. 2001). In chapter 3 we examined variation in community characteristics as a function of several abiotic factors. Here we explore traits of individual species that determine how abiotic factors act at the species level and specifically, how such traits may equip those species for life under various levels of environmental stress or suit them to particular habitats. Ecological traits also affect the ways in which species interact with others in competitive, predator–prey, or facilitative networks, but we save interspecific interactions for chapter 5 and focus in this chapter on traits of individual species.

In recent years, researchers in community ecology have rekindled the interest in biological traits of individual species (which have been studied in freshwater fish for a long time, e.g., Forbes 1878, 1880; Jordan 1884) with studies that have addressed species' traits as they relate to species' distributions (Lamouroux et al. 2002; Peterson 2003; Goldstein and Meador 2004), structure and assembly of communities (McGill et al. 2006; Ackerly and Cornwell 2007; Frimpong and Angermeier 2010; Jones et al. 2010; Sharma et al. 2011; Strecker et al. 2011; Troia and Gido 2015), or ecosystem function (Dray and Legendre 2008; Spooner and Vaughn 2008; Vaughn 2010). Traits of individual species for taxa of all kinds are being used in many predictive studies of species' potential distributions by ecological niche modeling or species distribution models (Peterson 2003; Kearney and Porter 2009). We applaud this resurgence of interest in basic biology of many species but temper that enthusiasm with a cautionary note that one size may not fit all or may not apply broadly across different populations of any species. For example, interpopulation differences in thermal or oxygen tolerances, or in life history traits, can be substantial for some minnow or darter species (Matthews and Styron 1981; Feminella and Matthews 1984; Marsh 1984).

Our studies of stream fishes have addressed a number of species-level traits that potentially affect community membership (thereby contributing to community assembly and structure), including species-specific tolerances for harsh environments; habitat or microhabitat selection or preferences; effects of habitat loss on individual species; propensity for movement, migration, or colonization; behavioral responses to disturbance or stressors; food use or trophic group membership; ecological morphology; and life-history traits. We explore these traits in the context of interactions with abiotic environmental factors that predilect the composition or dynamics of a fish community by focusing on individual species and the ways their ecological and physiological traits or behaviors interact with the habitat template at scales from very broad (e.g., east to west across the Great Plains) to microhabitats within a single watershed. We synthesize the ways that the environmental tolerances of individual species, from our own trials and those of Smale and Rabeni (1995a), help explain their distribution across geographic landscapes, from harsh to benign systems or distributions within a single stream (Matthews 1987). Tolerances of individual species also underlie the dynamics of whole communities when conditions in a system become stressful (e.g., the "crunches" of Wiens 1977). We also address habitat use patterns, identify some habitat generalist versus specialist species that are common in our streams, and examine links between morphology and ecology of individual species.

Tolerance for Harsh Environments

Ecological thought about ways that abiotic factors regulate distributions of organisms date at least to Grinnell (1914, 1917), Andrewartha and Birch (1954), and the "environmental filters" of Smith and Powell (1971) and Poff (1997). But as early as 1888, David Starr Jordan wrote in *Transactions of the American Fisheries Society* an essay titled "The Distribution of Fresh-Water Fishes," in which he noted that "the distribution of [fish] species is governed very largely by the temperature of the water. Each species has its range in this respect." Vernon (1899) wrote on "The Death Temperature of Certain Marine Organisms," describing from laboratory tests the "paralysis temperature" and the "death temperature" of several species of fishes, along with various invertebrates. And Verrill (1901) reported "A Remarkable Instance of the Death of Fishes, at Bermuda in 1901" (from unusual cold). Interest in limits of tolerance to physicochemical stressors for fish or other organisms, or the role of stressors in setting distribution limits, is not new.

But the concept of environmental filtering has reemerged in the recent literature with renewed interest in its role in community assembly (Helmus et al. 2007; Emerson and Gillespie 2008; Troia and Gido 2015). With respect to abiotic filters for stream organisms, Peckarsky (1983) proposed that streams could be placed on a gradient from "harsh" to "benign" conditions, with harsher habitats having unfavorable absolute conditions or abrupt changes in environmental conditions, and the idea has been supported by numerous subsequent studies (Allan 1995; Ross and Matthews 2014).

Ross et al. (1985) emphasized trenchant differences on a harsh–benign gradient for two of our most studied streams, showing that the upland, perennial Piney Creek in the Arkansas Ozarks was benign, compared to the much harsher lowland, intermittent Brier Creek in south Oklahoma. Both systems have experienced severe flooding (see plates 19–22) or drought (see plates 23 and 25) in the last four decades (which we address in more detail in chap. 6), but during normal or base flow conditions the differences between the two systems were obvious. In Piney Creek, from 1972 to 2012 (Matthews et al. 2014; Matthews and Marsh-Matthews 2016), we rarely found environmental conditions that would be potentially harmful to fishes, e.g., pooling up at one headwater site. In Piney Creek we have never measured oxygen or temperature conditions outside acceptable values for warm-water fishes (Matthews et al. 2014). Even during one of the worst regional droughts on record (summer 2012), temperatures measured throughout the Piney Creek watershed during 25–27 July ranged from 23.7°C to 32.8°C, and dissolved oxygen ranged from 5.0 to 9.7 ppm, values that are within acceptable limits for warm-water fishes.

In contrast, Brier Creek headwaters can exhibit dramatic fluctuations in temperature of as much as 10°C in a day, and dissolved oxygen of 0.4 to 2.0 ppm has been measured in typical isolated headwater pools in the summer (Matthews 1987). Much of Brier Creek shrinks to a limited number of small, isolated pools during droughts (Marsh-Matthews and Matthews 2010; Matthews et al. 2013), with long reaches of dry streambed (see plates 23 and 25) and temperatures known to be directly lethal to fish (Matthews 1987). These two systems reflect broad general patterns of environmental differences between upland and lower-gradient streams, as well as geographic differences related to rainfall gradients from east to west in the central United States.

We consider three kinds of situations in which the ability of a species to tolerate acute temperature or oxygen stress may relate to its ability to survive in a local community in our region: (1) east–west gradients in rainfall and temperature, (2) dichotomy between upland and lowland streams, and (3) greater stress in headwaters or small tributaries that are fed by only runoff. At the broadest scale there are strong potential differences in temperature or oxygen stress on a gradient from the mesic east to the arid west across the Great Plains.

There are pronounced gradients of decreasing rainfall and increasing evaporation and maximum annual temperature from east to west across the Great Plains from Texas through Nebraska (see the Western Regional Climate Center's climate maps at http:www.wrcc.dri.edu/climate-maps/). As a result, the western portions of the Great Plains have greatly reduced stream flows, and fish in streams toward the west across the Great Plains are on average more exposed to low stream flows and more pooling-up of streams. To the west, riverbeds also are wider, typically with channels flowing over shallow sand bottoms far from the permanent banks and with little riparian shad-ing. And many streams to the west flow mostly from surface runoff and have extremely low or no flow during severe drought. During the exceptional drought (as defined by the US Drought Monitor Classification Scheme at the University of Nebraska; see http://droughtmonitor.unl.edu) of 2012–2013, for example, the Salt Fork of the Red River, near Mangum, Oklahoma (one of our long-term sampling sites with classes), ceased to flow for almost nine months, whereas in that same year under exceptional drought (defined as above) in north Arkansas, Piney Creek continued substantial flow with the exception of one headwater site (Matthews et al. 2014).

These environmental gradients also occur across the state of Oklahoma. There is a gradual decrease of average annual rainfall from 1200 mm or more in the Ozark and Ouachita Mountains in the east to 430 mm in the western panhandle of the state (Johnson and Duchon 1995). Mean number of days over 90°F (32.2°C) increases from 60 in the Ozarks to 100 in southwestern Oklahoma. Mean monthly pan evaporation in July increases from 10 in (25.4 cm) in the east to 15 in (38.1 cm) in the Oklahoma panhandle.

As a result of all the climatic and geomorphic gradients above, streams in the western part of the region tend more toward frequent periods of low or no flow, with fish surviving only in isolated pools during substantial parts of some years (Matthews 1987; Ostrand and Wilde 2001). Toward the west, rivers are typically wide, shallow, and sand bottomed, with only limited riparian tree cover (see plates 11 and 12). As a result, solar insolation in the summer can raise midchannel temperatures to extremes such as 38°C–39°C (Matthews and Zimmerman 1990), and oxygen depletion in isolated pools also can affect distribution of fish species (Matthews 1987). In several analyses that follow we thus make the assumption that fish in streams farther west across the region have a greater chance of encountering harsh, potentially harmful physical environments. The gradients above are also exacerbated by the difference between upland streams in the eastern part of the region, typically with more sustained flows, and lower-gradient streams outside the uplands that are dependent on runoff.

The east–west gradient across the Great Plains has a strong impact on fish distribution. Maps in Miller and Robison (2004) showed that out of 148 fish species that have range limits within Oklahoma (others are ubiquitous statewide), approximately half (70 species) decline from the mesic east to the arid west. And another approximately 36 Oklahoma species are restricted to the Ozark or Ouachita uplands, with range limits coinciding with the margins of these uplifts. Parham (2009) also noted that most species in Oklahoma were found only in the eastern part of the state. In Kansas (Kansas Fishes Committee 2014) there is a similar decline in species from east to west and similar restriction of some species to the uplands of the Ozarks or the Flint Hills. Clearly, with such broad suggestions that environmental gradients influence limits of distributions, the ability of individual species to tolerate the vicissitudes of harsh environments should be important in determining their capacity to occur in or be a core species in local communities.

We consider two lines of evidence. First, we examine in more detail the range limits for individual fish species that lie within the boundaries of Oklahoma, emphasizing east–west or north–south range boundaries as they relate to temperature or rainfall gradients. Then we compare the westward distribution of numerous fish species in the Great Plains for which actual temperature and oxygen tolerance data exist (Matthews 1987 and unpub. data; Smale and Rabeni 1995a).

For all fish species that have range limits in Oklahoma, we drew by hand onto a large map (reduced as fig. 4.1) the boundaries of their ranges from Miller and Robison (2004), updated as needed by additional information we have on occurrences. To quantify areas where distribution limits coincided, we followed the approach of Pflieger (1971) to determine faunal boundaries where numerous range limits of species are close together. By convention, "major boundaries" were defined as regions in Oklahoma with 10 or more species limits within approximately 0.3 degrees of latitude or longitude (actually measured as within a 2 cm span on the original large map). "Minor boundaries" were defined as regions with five or more coinciding boundaries. Collectively, these major and minor range limits (fig. 4.1) show where in the state a substantial number of species have maximum extension to the west or to the north.

The most noteworthy result is that numerous species do have range limits at or near the western edges of the Ozark (Hill et al. 1981) or Ouachita Mountains (fig. 4.1). Then a grid consisting of $0.5°$ latitude × longitude segments (not shown) was overlain on the map of range limits and used to count the number of "hits" for range limits lines on north–south versus east–west gridlines. There was an average of 11.3 range limit hits per grid section on north–south gridlines, compared to an average of 8.1 hits per grid section on east–west gridlines. Differences from north to south across the major river basin divides in Oklahoma are obviously important zoogeographically for fish distribution. A substantial number of species that are widespread outside the Ozark or Ouachita uplifts occur only in the Red River basin (such as Blacktail Shiner, *Cyprinella venusta*; Prairie Chub, *Macrhybopsis australis*; Red River Shiner, *Notropis bairdi* (before it was introduced northward in the 1970s); Chub Shiner, *Notropis potteri*; Rocky Shiner, *Notropis suttkusi*; Orangebelly Darter, *Etheostoma radiosum*; and Bigscale Logperch, *Percina macrolepida*), and a few historically occurred only in the Arkansas River basin in Oklahoma, such as Arkansas River Shiner (*Notropis girardi*). But then, after taking into account the north–south differences due to river basin boundaries, there remains a large number of species that drop out from east to west within the state. Admittedly, there can be many reasons why some species, especially those limited to smaller drainages, may have not expanded westward, such as a limited ability to move up through large river mainstreams (e.g., Echelle et al. 1975). But over the evolutionary and geologic time available for dispersal since these species have existed, we would suspect that at least one of the limiting factors could be an intolerance for the harsher conditions that prevail toward the west and that may have been even worse during the long megadroughts in North America from about 1300 to 1600 CE

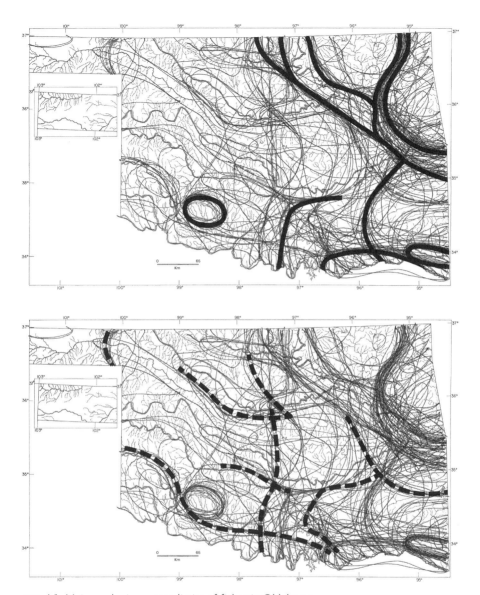

FIG. 4.1 Major and minor range limits of fishes in Oklahoma.

(Stahle et al. 2007), when there was a much drier climate across much of the United States (Cook et al. 2007).

In the harsh environments of western streams in the Great Plains, or in comparing upland to lowland-prairie species, the ability to survive periods of temperature or oxygen stress can relate directly to distribution or success of individual fish species. And tolerances can relate broadly to individual fish species distributions across other geographic regions (Smale and Rabeni 1995b). In an early pilot study, Matthews and

Hill (1977) showed that Red Shiner (*Cyprinella lutrensis*; see plate 14), which is one of the most broadly distributed and abundant minnows in low-gradient streams from Nebraska to northern Mexico, was highly tolerant of pH extremes, dissolved oxygen values as low as 1.5 ppm, and heat or cold shocks of 10°C above or 21°C below acclimation temperature. Subsequently, Matthews and Maness (1979) compared tolerances of four minnows from the South Canadian River in tests of their critical thermal maximum (CTM) and tolerance of low oxygen concentrations. Emerald Shiner (*Notropis atherinoides*) was the least tolerant for both stressors, and this species disappeared from our study reach in the South Canadian River during the hot summer of 1976 (Matthews and Hill 1980). In contrast, numbers of Red Shiner, Arkansas River Shiner, and Plains Minnow (*Hybognathus placitus*), which were more tolerant of thermal or oxygen stress or both, greatly increased at that site from spring to autumn in 1976.

In an expanded test of tolerance of minnow species typical of prairie streams versus species restricted to or more typical of upland streams, Matthews (1987) found CTMs for 5 prairie species, in summer and acclimated to 20°C, generally 1°C to 2°C higher than CTMs for 6 upland species. Within each group there was one exception, with Emerald Shiner (a prairie species) having a lower CTM than the other prairie species (consistent with the findings of Matthews and Maness 1979), and Redfin Shiner (*Lythrurus umbratilis*), considered an "upland" species, having a CTM nearly as high as the other four prairie species. But Redfin Shiner is widely distributed and is actually a more marginal species, with some populations in lowland habitats, and Emerald Shiners do penetrate into some upland streams like the lower Kiamichi River. Both are probably habitat generalists (see next section). If comparisons are limited in Matthews (1987, table 14.2) to strictly prairie and strictly upland species, the mean CTM across species was 36.09°C (standard deviation, or SD = 0.24) for prairie species and 34.32°C (SD = 0.77) for upland species. This difference of nearly 2°C between prairie and upland species could be critical at times when fish experience thermal stress in prairie rivers, likely influencing their survival and acting as a strong selective force in their evolution.

To view the potential importance of thermal tolerance a bit more broadly, Matthews (1987) reported results of CTM trials for common species that ranged in distribution from uplands to prairie or that were common in Brier Creek, Oklahoma. The CTMs for these species ranged from 31.5°C to 37.2°C. We used the data in Matthews (1987) after separating *rubellus* of Matthews (1987) into *Notropis percobromus* (Carmine Shiner) in the Ozarks and *Notropis suttkusi* in south Oklahoma and adding CTMs for Southern Redbelly Dace (*Chrosomus erythrogaster*) and Suckermouth Minnow (*Phenacobius mirabilis*) from original data sheets. For the resulting 22 species we determined from the *Atlas of North American Freshwater Fishes* (Lee et al. 1980) the maximum extent of their distribution (in whole degrees west longitude) on the Great Plains in or west of Nebraska, Kansas, Oklahoma, or north Texas, in the Platte, Kansas, Arkansas, and Red River basins (ignoring distributions in Texas coastal drainages or the uplifted Texas Hill Country). There was a significant (Spearman's rho = 0.562; $p = 0.007$)

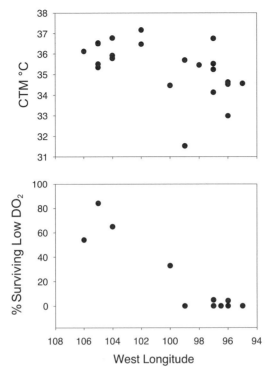

FIG. 4.2 Critical thermal maximum (CTM) of 22 species and the western extent of their ranges (*top*). Resistance to low oxygen (measured as percentage survival) for 11 species and the western extent of their ranges (*bottom*). Note that the abscissa is scaled from west to east, so that species with a more western range edge are on the left side of the axis.

positive correlation between the CTMs and the westward range limits of these species (fig. 4.2), which supports (but of course does not prove) the hypothesis that distributions of species across the gradient of climatic harshness in the Great Plains may relate to their tolerance for thermal stress.

In other trials, Matthews (1987) showed that 4 out of 5 prairie minnow species had higher percentage survival than any of the upland species in 8- to 10-hour tests in low dissolved oxygen conditions (0.2–0.9 ppm). Ability to withstand periods of low dissolved oxygen, which can accompany high temperatures in isolated pools (Ostrand and Wilde 2001), may place these prairie species at a distinct advantage in harsh environments relative to species from continuously flowing, more environmentally benign, upland streams. Across all of the minnow species ($n = 11$) tested for percentage survival in 8–10 hours at <1 ppm dissolved oxygen, there was a strong positive relationship (Spearman's rho $= 0.771$; $p = 0.004$) between the percentage surviving low dissolved oxygen and the westward extent of their ranges (fig. 4.2).

The results above were based on a relatively small number of species, selected for the trials because of their dichotomy in distribution between uplands and prairie streams. As a check on how broadly Matthews's findings might apply, we took from Smale and Rabeni (1995a) ranks for thermal tolerance (CTMs) for 34 species and ranks of tolerance for low oxygen for 36 species that occur in Missouri. For each species in Smale and Rabeni (1995a), we determined from the fish atlas the westward extent of its range on the Great Plains west of Missouri or Oklahoma. For both temperature and oxygen tolerance the species evaluated by Smale and Rabeni (1995a,b) showed weak but interesting correlations between tolerance and the maximum westward extent of their ranges on the Great Plains (CTM, Spearman's rho $=-0.331$, $n=36$, $p=0.06$; DO_2, Spearman's rho $=0.337$, $n=34$, $p=0.05$). For both variables, Spearman's rank correlation between tolerance and westward range was significant (or nearly so), but the relationship accounted for only about 9%, each, of the variance. The relationships between tolerance and distribution for the species studied by Smale and Rabeni explained only a small amount of the variance in westward range of the individual species, but amid so many other factors (below) that could influence ranges (competition, predation, zoogeographic boundaries, other?), even marginally significant findings suggest that physicochemical tolerances have potential importance for fish distribution or success in harsh habitats.

The findings from all of the tolerance tests suggest that tolerance for high temperatures or low oxygen concentrations, or both, could relate to westward distribution in harsher habitats for individual species in the central United States. But this speculation has to be tempered with alternate hypotheses. First, the patterns discussed above are influenced by the fact that Matthews (1987) intentionally included two species (Cardinal Shiner, *Luxilus cardinalis*, and Ozark Minnow, *Notropis nubilus*) that are distinctly restricted to the west by the edge of the Ozark uplift (or to the neighboring Flint Hills of east Kansas), and several of the species used by Smale and Rabeni (1995a) are also limited to the west at the edge of the Ozark uplands. This introduces a local zoogeographic question, that of the distinction between upland and lowland species, for any of numerous possible reasons. And the question likely applies to other areas with sharp boundaries between uplands and lowlands. Keep in mind that the problem we address here for upland-lowland dichotomies is all for warm-water species; that is, in this region, cold salmonid water does not exist, so we are not dealing with major cold-water to warm-water gradients involving salmonids (e.g., Rahel and Hubert 1991).

In Oklahoma, a substantial number of species from various taxonomic groups are sharply limited to the west by the border of the Ozark uplift (Hill et al. 1981; Miller and Robison 2004; Parham 2009), including Cardinal Shiner and Ozark Minnow, as well as Gravel Chub (*Erimystax x-punctatus*), Bigeye Chub (*Hybopsis amblops*), Wedgespot Shiner (*Notropis greenei*), Carmine Shiner, White Sucker (*Catostomus commersoni*), Banded Sculpin (*Cottus carolinae*), Fantail Darter (*Etheostoma flabellare*), Banded Darter (*Etheostoma zonale*), Speckled Darter (*Etheostoma stigmaeum*), and Sunburst Darter (*Etheostoma mihileze*). For some species pairs within a single genus of minnow, there are marked differences in their abundance in lowlands versus uplands. In southern Oklahoma, Red Shiner is one of the most common and abundant species in low-gradient streams, and Blacktail Shiner is common. But from lowlands to uplands in the Kiamichi and Little river systems, both are replaced by the congeneric Steelcolor Shiner (*Cyprinella whipplei*). In the Little River basin of southeast Oklahoma and western Arkansas, Taylor and Lienesch (1996) showed parapatry between the distributions of Redfin Shiner and the Ouachita Mountain Shiner (*Lythrurus snelsoni*), which replaces the former in uplands of that system. But in uplands of the Kiamichi River, where the Ouachita Mountain Shiner is lacking, Redfin Shiner occurs broadly, including large populations in headwater sites.

The reasons for the restriction of some species to uplands or, conversely, why hardy plains or prairie species do not invade highlands are unknown and in need of study at the level of individual species. We could use the tolerance hypothesis as one explanation, suggesting that inability to cope with harsher environments outside the uplifts is the factor limiting upland fish to uplands. But that cannot explain the lack of lowland species penetration into uplands. As alternatives to the tolerance hypothesis, it is possible that competition from species better adapted to harsh environments keep upland species out of the lowlands. And the reverse could be true, that in the more perennial streams of the uplands the resident species can outcompete lowland invaders, should they arrive. So, competition needs to be considered, as we do in chapter 5. Equal consideration should be given to predator control scenarios, also discussed in chapter 5. For example, Matthews and Marsh-Matthews (2007) and Marsh-Matthews et al. (2011, 2013) showed that increased numbers of intermediate-sized predators (mesopredators) like sunfish have the potential to exclude minnows from streams. And as we will see in chapter 5, lowland streams tend to have higher proportions of piscivorous fish species than upland streams. All of these biotic interactions need to be considered in more detail to provide any real explanation for the distributions of fish in our region, and most such work awaits experimental investigation.

Local adaptation to harsh conditions, or the relationship of tolerance to distribution of individual species within watersheds or drainages, can also be important, and results can differ from species to species. Feminella and Matthews (1984) showed a strong relationship between thermal tolerances of Orangethroat Darters (*Etheostoma spectabile*) (see plate 16) from different populations in close proximity in south-central Oklahoma that differed in local thermal characteristics. Populations in constant-temperature springs were least tolerant of heat stress, and the population in Brier Creek, the harshest of four habitats studied, had a markedly higher CTM than the populations from spring-fed waters. In contrast to this finding of strong local adaptation, Matthews (1986) found that populations of Red Shiner from Kansas through Texas showed no significant differences in CTM. Red Shiner is a hardy species, known to be tolerant of temperature and oxygen stress from previous testing (Matthews and Hill 1977; Matthews and Maness 1979), but it still seemed remarkable that across such a 1100 km north–south gradient there was no evidence of local adaptation. Matthews (1986) speculated that in spite of the length of the north–south gradient that was studied, all of the prairie streams where Red Shiner is common may experience very high summer temperatures, so perhaps there were no strong differences in selection pressure for thermal tolerance across all of these river basins. Regardless of the underlying factors, however, these two studies stand in sharp contrast regarding local adaptation of fish to stressful environments.

Tolerance can also be related to distribution of individual species upstream in smaller headwaters, which are potentially more stressful habitats owing to their tendency for less regular flow, pooling up (Fritz and Dodds 2005), and oxygen depletion in isolated pools (Matthews 1987). Long ago, Thompson and Hunt (1930) predicted that fish from intermittent headwaters should be more tolerant of stressful conditions than fish in larger streams, and Moyle and Li (1979) suggested that differences in distributions of warm-water stream fish could result from differences among individual species in their responses to physicochemical conditions. Matthews and Styron (1981) showed in the Roanoke River drainage of Virginia that Mountain Redbelly Dace (*Phoxinus oreas*) from a comparatively harsh, intermittent tributary were more tolerant of abrupt changes in temperature, oxygen concentrations, or pH than were three other minnow species that are typical of the more environmentally stable river mainstem. They also found that Fantail Darters from this intermittent tributary stream were more tolerant of low oxygen than two other darter species that were common in the river but not in small tributaries. Results for these minnows and darters support the idea that tolerance of harsh conditions can relate directly to the degree of penetration of species upstream into environmentally variable headwaters. Additionally, Fantail Darters from the intermittent tributary were more tolerant of low-oxygen conditions than conspecifics from the river mainstem (Matthews and Styron 1981), also suggesting the potential for fishes to adapt to local adverse conditions. Subsequently, Matthews (1987) found that in the harsh environment of Brier Creek, Oklahoma, where headwaters can change as much as 10°C daily and oxygen in isolated pools is often <1 ppm, the upstream distribution of 11 species was related not to their thermal tolerance (all were hardy) but to their relative tolerances for low-oxygen conditions.

The results for temperature or oxygen tolerances of extant species, as they occur in their present ranges, all beg the evolutionary question as to whether fish species that occur in and are more tolerant of harsh conditions across broad geographic gradients first developed an enhanced tolerance for stress (perhaps via random increases in heat shock proteins, or HSPs, or plasticity in tolerance), which allowed individuals to successfully invade harsh habitats. The results also lead us to ask, alternatively, whether fish with limited tolerance first invaded harsher streams and subsequently evolved enhanced tolerance under the strong selective pressure of episodic hard times within these harsh environments. Molecular studies might help disentangle these knotty questions but are beyond the scope of our work, and we leave those questions for consideration by future workers. With respect to the tolerance hypothesis, all of the temperature or oxygen tolerance trials done in WJM's lab simply measured responses of whole organisms. Underlying physiological mechanisms that might explain differences in tolerance, or potentially differences in performance, await detailed investigations for most of the fish species in our region. At the time WJM did most tolerance trials, for example, HSPs were just beginning to be recognized in fishes (e.g., Kothary and Candido 1982), although thermal hardening had been known in fishes for some time (Maness and Hutchison 1980). Future work might be directed toward better understanding of these or other of many possible factors in physiology that could help explain tolerances of species and their distribution patterns.

Habitat Selectivity, Habitat Specificity, and Habitat Loss
Habitat Selectivity

Physicochemical tolerances of individual species also can relate to their habitat preferences or selectivity in harsh environments. Selection of the best available microhabitats within a stream reach can be important to survival of a species during harsh conditions. The Leopard Darter (*Percina pantherina*) is a federally threatened species in southeast Oklahoma and western Arkansas that we have found (by snorkel observations) to typically swim in the water column in close proximity to boulders the size of basketballs in the Glover River, Oklahoma. In one survey to collect Leopard Darters for a mark and resight study (Schaefer et al. 2003), however, during very hot conditions when water temperature exceeded 30°C in their preferred habitat in the shallow mainstem, we found relatively few individuals. But a member of the crew, Daniel Spooner, suggested that they might be using deeper, cooler pools as refugia. Sure enough, when Dan donned SCUBA gear, he found a large number of Leopard Darters near the bottom of pools that were as much as 4°C–6°C cooler than the surrounding shallower habitats (which sometimes in midsummer reached potentially lethal temperatures of 36°C–38°C) and 4–5 m deep . . . deeper by far than anybody had previously searched for Leopard Darters! And this pattern was repeated on several dates when discharge was low and air and water temperatures were high (Schaefer et al. 2003). This finding suggests (1) the value of thermal refuges in warm-water streams by this and other species for management planning; (2) the value of "looking where they ain't" (with apologies to baseball fans); and (3) that many warm-water fish species might survive harsh periods by precisely selecting various kinds of microhabitat refugia that may not yet be known to biologists.

In the South Canadian River during intense heating in midsummer (Matthews and Hill 1980; Matthews and Zimmerman 1990), minnows were scarce in the exposed mainstream, where solar insolation raised water temperatures to near lethal, but were in large numbers in a limited amount of habitat in deeper, shaded pools near the river's edge (see plate 12). Matthews (1987) postulated that fish might survive in harsh, heterogeneous habitats like the South Canadian River by not only being highly tolerant of high temperature or low oxygen but also by being highly selective of the best microhabitats, allowing them to avoid or minimize stress. The selectivity hypothesis—that is, that minnows of the harsh prairie rivers would show more microhabitat selectivity than minnows from more benign uplands—was not supported in experimental trials in thermal gradients, but the prairie minnow species were generally more se-lective than upland species in oxygen gradients (Matthews 1987). Matthews (1987) concluded that whereas the two groups did not differ in thermal selectivity, the prai-rie species exhibited a combination of both greater tolerance for low oxygen and greater selectivity for oxygen concentrations. This combination, with substantial tolerance and the capacity to select the best habitats (or avoid the worst?) should be highly adaptive in the harsh environments of prairie rivers, where high temperatures and low oxygen concentrations (especially in some backwater pools or side channels with decaying detritus, where oxygen can be low) are common features of the environment.

At the other extreme, winter cold, there also are changes in microhabitats used by stream fishes. We have often found that water-column fish like minnows are abun-dant and broadly distributed across much of the available habitat in warm weather, but this can be quite different in winter. In summer, it is almost impossible to pull a seine any distance in pools of our study systems without catching minnows. In con-trast, in winter it is possible to seine substantial reaches of flowing channel or pools catching few minnows but find them in large numbers in deeper, slow pools. David Edds (pers. comm.) has suggested that fish near the bottom of deeper pools in gravel-bed streams in winter might actually find slightly warmer water near the substrate than at the surface because of seeps of groundwater. And David Starr Jordan long ago suggested that some fish "can reach the cool bottoms in hot weather, or the warm bot-toms in cold weather, thus keeping their own temperature more even than that of the surface of the water" (Jordan 1888).

As an example, in March 1979, with water temperature at 7°C at a mainstem site on the Roanoke River near Shawsville, Virginia, we sampled extensively in all avail-able pool, riffle, and channel habitats, but found relatively few fish, with the excep-tion of one pool that was 10 m wide, 30 m long, and 1.1 m deep and described in the field notes as "extremely stocked with minnows" (WJM413; BS 73). There we "filled a gallon jar" with these fish, taking many large adult minnows as well as juveniles. On the same day at another site farther upstream on the main Roanoke River (WJM414; BS74) with water at 9°C, we took "no fish in the channel" but found "many minnows," filling two 1-gallon collecting jars in one pool about 1.1 m deep with little current. Two days later (WJM415; BS76) at an additional mainstem site with water temperature of 5°C, we found few fish in the flowing channel and a modest number of minnows and darters on riffles but minnows "extremely numerous" in one large pool, including 424 White Shiners (*Luxilus albeolus*), 188 Crescent Shiners (*Luxilus cerasinus*) and 128 Rosefin Shiners (*Lythrurus ardens*), among others.

We found similar movement of Central Stonerollers (*Campostoma anomalum*) out of the strongly flowing mainstem of the Baron Fork of the Illinois River in northeast Oklahoma during winter, whereas they were present in the mainstem in huge schools, actively feeding on attached algae (see plates 26–30) throughout the summer. This fact, as well as potential foibles in setting up field experiments, was made clear in one comparison between summer and winter. In July 1988, WJM, with Fran Gelwick and

Marsha Stock, laboriously built arrowhead-shaped pens to admit or exclude stonerollers and other minnows, finding that the grazing effects of the stonerollers were a major factor in dynamics of attached algae and associated biotic or physical parameters in the pens (Gelwick et al. 1997). The next winter, we returned to the field site to build similar pens to repeat the summer trials. After laboriously installing the pens again in the flowing main channel of Baron Fork, we sat back to await the arrival of the stonerollers, but they never came! The experiment was a total loss. After extensive searching for the stonerollers, which had been so abundant in midchannel during the summer, we found them in large numbers "hanging out" within the water column (instead of using their typical benthic foraging habitat) with little movement and no apparent feeding, in a backwater slough off the main channel of Baron Fork (Matthews 1998, 415). The stonerollers remained in this microhabitat for days of observations, including sunny and cloudy weather conditions, never, so far as we could tell, returning to the main channel to feed. The behavior of this stoneroller population differed markedly from that of a population in a nearby smaller creek with slower currents, where they actively fed all winter (Matthews 1998, 415). David Edds (pers. comm.) also found that in cold weather in winter 1982, minnows were largely absent from water in Baron Fork shallow enough to seine and instead were shoaling in large numbers in small pools that were too deep to seine.

After realizing the marked summer-winter dichotomy in stoneroller behavior, WJM made year-round observations of the behavior of another large population of stonerollers at a site farther upstream on the Baron Fork (fig. 4.3) for several years. At this site, in northwest Arkansas, the Baron Fork flowed with substantial current in a wide, shallow pool just downstream of an abandoned bridge, from which observation conditions were ideal. During warm weather, large numbers of stonerollers regularly foraged in one or two shoals within the flowing pool, maintaining the stony substrate with a freshly grazed, bright green appearance. But in cold weather, the stonerollers abandoned the flowing channel, as they had done earlier in 1988 at the more downstream site, and took up residency in a nonflowing small backwater channel away from the main channel.

The findings with the stonerollers underscore our thoughts in chapter 1 about "knowing your fish" before you set up experiments. But that experience also underscores the propensity of fish species to use markedly different habitats or microhabitats seasonally, as Matthews and Hill (1980) also showed for minnows of the South Canadian River and as Bart (1989) showed for Ozark stream fishes in general. Our stoneroller findings also agree with the general model of Schlosser (1991), indicating that fish typically use several different microhabitats seasonally or during ontogeny.

Returning to the case of the missing stonerollers, Matthews (1987) postulated that this phenomenon could have related to a metabolic strategy that had the advantage of their not expending energy to feed in the strong currents of the midstream Baron Fork, instead avoiding currents, feeding little or perhaps none, and depending on stored reserves to survive the winter. So far nobody has tested the postulate that stonerollers can "win" the cost-benefit balance by cessation of activity and feeding in

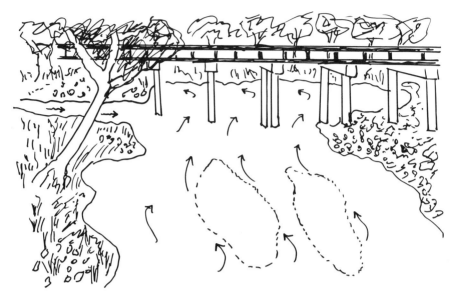

FIG. 4.3 Schematic of the distribution of Central Stonerollers (dotted outlines) in the flowing main channel of the Baron Fork of the Illinois River west of Fayetteville, Arkansas, during summer, and the small channel (*top left*) to which they retreated in winter. View is from abandoned bridge observation point, looking downstream to a newer bridge on a county road. Arrows indicate the direction of flow.

winter. But Clausen (1936) compared metabolic rates of stonerollers to seven other common species in Illinois and found that they had the second highest metabolic rate of all eight species, almost triple that of largemouth bass or bullheads. So, it appears that stonerollers have a rather high metabolic rate, and for the Common Minnow (*Phoxinus phoxinus*) in the United Kingdom, Cui and Wootton (1988) showed a strong relationship between amount of feeding and the total metabolic rate. Wootton (1990, 64) noted that Bluegill (*Lepomis macrochirus*) in Lake Opinicon, Canada, cease feeding in winter, resuming feeding in spring at 8°C–10°C. And, as we noted above, in winter, most minnows in the Roanoke River abandoned the main channel of the river and were found in pools (Surat et al. 1982). But in addition to taking up residence in deeper, slower pools, these minnows also ate much less in the winter, such that fish taken at cold temperatures in March 1979 had indices of stomach fullness approximately half that in warmer weather (Surat et al. 1982). To the extent that any of these relationships apply to our stonerollers, perhaps there is some advantage to their reduction of activity and feeding in winter. We wish somebody would test this.

Habitat Specificity
All fish species have some degree of habitat specificity (Frimpong and Angermeier 2010), as is well known to any ichthyologist, fisherman, or even casual observer at a

public aquarium. But the degree to which individual fish species are either broadly distributed across microhabitats (generalists) or specific to particular microhabitats (specialists) can be an important predictor of the potential of a fish community to resist change or, alternatively, to change substantially if requisite microhabitats for some species are lost. As a broad overview, WJM classified 126 fish species with which we are familiar from field sampling or observation, with individual species assigned subjectively by memory from years in the field, as occurring across 72 defined kinds of habitats or microhabitats. The scoring included broad categories like upland versus lowland (or both), and microhabitat categories focused on the kinds of habitats where they have been found, from head pools of springs, to other pools of various sizes and depth, location in pools (surface, water column, substrate, edge), different parts of riffles or riffle substrates, flowing channels at edges or deeper, and association with various kinds of structure such as rootwads, wood debris, boulder, long upright columns of attached algae, and substrates ranging from stony to mud-bottomed or silted backwaters. Scoring for a species was not exclusive to any one category, and most of the species were scored as associated with a substantial number of habitat or microhabitat types.

The concept of specialists versus generalists has important implications for animal communities of all kinds (McPeek 1996; Julliard et al. 2006), including fish (Taylor et al. 1993; Poff and Allan 1995). Specialization versus generalization, for example, can have implications for the ability of different reef fish to survive disturbance (Wilson et al. 2008). As a rough assessment of habitat specialists versus generalists, 41 of the 126 species were scored in 5 or fewer of the 72 habitat/microhabitat categories, and another 41 were scored as associated with 6–10 categories. We consider these 82 species to be habitat specialists. At the other extreme, 16 species were scored as associated with 16 to 45 habitat categories, comprising habitat generalists. We consider those in between to be intermediate in habitat specialization. Taxa with widest habitat use, the extreme generalists, were (number of habitats in parentheses): Largemouth Bass, *Micropterus salmoides* (44); Longear Sunfish, *Lepomis megalotis* (43); Green Sunfish, *Lepomis cyanellus* (41); Bluegill (36); Red Shiner (33); Blacktail Shiner (30); Central Stoneroller / Highland Stoneroller (26); and Western Mosquitofish, *Gambusia affinis* (26). That these species occurred across many kinds of habitats is no surprise, as some are quite tolerant of harsh conditions of temperature or oxygen (Matthews 1987). At the other extreme were species we have found only in highly specific habitats, for example, swampy habitats in southeastern Oklahoma, including Western Starhead Topminnow (*Fundulus blairae*), Pygmy Sunfish (*Elassoma zonatum*), Flier (*Centrarchus macropterus*), and Goldstripe Darter (*Etheostoma parvipinne*), in one habitat type each. In the scoring, 50 of the 126 species were classified as upland only, 28 were lowland only, and 48 were scored in both upland and lowland streams.

For more detail, a SAHN (NTSYS Version 2.2) cluster analysis based on Jaccard's index of similarity between all possible pairs of the 126 species across the 72 habitat types sharply separated species on the basis of their affinity for upland versus lowland stream sites, and there were an additional 12 subclusters based on a hierarchical series of cutpoints in the phenogram representing the clustering. In addition to the upland versus lowland dichotomy, there was a third small cluster occupied exclusively by species typical of springs or spring-influenced habitat, including only the darters *Etheostoma cragini* (Arkansas Darter), *Etheostoma microperca* (Least Darter), Sunburst Darter, and Goldstripe Darter. Beyond being an academic exercise, the propensity of individual species to occupy a wide versus narrow range of microhabitats can be related to the degree to which temporary restrictions or permanent loss of requisite habitat might result in loss of species, thus influencing the dynamics of the entire community. In fact, there are predictions of worldwide declines of specialist species across many taxa, with resulting homogenization of diversity and function in ecosystems (Clavel et al. 2011).

Habitat Loss

From our study of streams we can offer a few examples of the ways that loss of special microhabitats has altered local communities. In Brier Creek there historically were large populations of the Orangethroat Darter, but they were highly localized, limited mostly to flowing riffles with gravel and cobble. Orangethroat Darters were abundant in a local population in one short riffle just downstream from State Highway 32 (our Brier site BR-5; fig. 2.3) and in one long, shallow gravel riffle within our Brier site BR-6. As recently as a decade ago, the darter population in the riffle at BR-5 was dense, and many darters could be found in a few kicksets. But in the last decade, floods or other geomorphic processes have eroded the gravel in that riffle, and the streambed in the former riffle now consists of stony bedrock more than 25 cm lower than the original gravel streambed. This change in streambed depths was made obvious by the fortuitous circumstance that a steel pipe about 15 cm in diameter (presumably for some utility?) crosses Brier Creek exactly at that point, and whereas it was formerly buried under the gravel of the riffle, it now is suspended well above the water, owing to the loss of the gravel. Now at the site of that former riffle it is nearly impossible to find a darter, as their requisite microhabitat is gone! In this case it was not a simple downstream movement of the gravel bed, because the substrate for approximately 50 m or more downstream is largely eroded sandstone bedrock, with no good gravel riffles.

A similar situation, with geomorphic changes due to processes unknown to us but suspected to relate to flooding and erosion in the last decade (Matthews and Marsh-Matthews 2007), has occurred at Brier site BR-6. A long gravel riffle that formerly housed many Orangethroat Darters is now a long, flowing pool approximately 50–75 cm deep, with less gravel and a softer bottom than a decade ago, and here too it is now nearly impossible to find a darter. While we do not know the details of how the darter microhabitat was lost at either site, we do know that now that those microhabitats are gone, the darter populations have suffered considerably. We still do find large numbers of juvenile darters in the spring of some years at those sites, but we no longer find adults concentrated in the riffles as they formerly were.

Movement or Recolonization

Species-specific differences in ability or propensity to move can be extremely important to local fish community dynamics, particularly in small riffle-pool streams or runoff streams with headwaters that frequently dewater for substantial reaches (Schlosser 1987). In a headwater reach of Brier Creek that was rewatered in late February and early March 1983, after having been dry for the preceding six months (see plates 22 and 23), 11 species recolonized the reach by midsummer, but there were marked differences in arrival times, presumably from downstream pool refugia. Central Stonerollers were present in our first survey, about 10 days after the first of the rewatering rains, followed a month later by small numbers of Orangethroat Darters, Green Sunfish, Bigeye Shiners (*Notropis boops*), and Red Shiners. Green Sunfish had the greatest success, producing large numbers of young by midsummer. Other common Brier Creek fish, including Largemouth Bass and Longear Sunfish, did not arrive in the reach until June but also reproduced successfully (Matthews 1987). The fish species most successful at recolonizing the reach were those with greater capacity for low-oxygen environments (Matthews 1987), as described above.

Even though in the example above the timing of arrival was not strongly related to reproductive success, many studies suggest the importance of "priority effects" in community formation, so any traits enhancing movement of fish may be of interest in local community dynamics. Jacob Schaefer, a PhD student under WJM, modified the experimental mesocosms described in Matthews et al. (2006) to allow him to carry out quite creative studies of the propensity of three minnow species to cross riffles between deeper pool units as a function of riffle depth, current speed, and length, with and without predators present (Schaefer 2001). Schaefer found that current speed in the riffles was a substantial factor in rates of crossing, as was presence of (caged) predators. Of the three, Bigeye Shiners had an overall lower rate of crossing riffles than Blacktail Shiners or Central Stonerollers, and Bigeye Shiners also moved less as a result of predators being present. These results are consistent with observations in Brier Creek by snorkeling (Matthews et al. 1994) or simply searching for schools of fish among pools (Power and Matthews 1983; Power et al. 1985), during which Bigeye Shiners were consistently observed in schools in the extreme upper ends of pools, where currents from riffles made visual feeding on drift likely. Conversely, we found that presence of predators (Largemouth Bass) resulted in rapid emigration of stonerollers out of a focal pool (Power et al. 1985). Interestingly, in a mark and re-sight study, Schaefer also found that pools (with predators including black bass spp. or large sunfishes) could also differentially inhibit movement of small fishes. Movement within even a relatively short reach of a stream is complicated, but species-specific tendencies to move need to be taken into account to understand dynamics of local fish communities in small streams (Schlosser 1987).

Ecomorphology and Behavior

The relationships between morphology, ecology, and behavior have been of interest to biologists for a long time (Jordan 1884, 216): "Most of them prefer clear running water, where they lie on the bottom concealed under stones, darting when frightened or hungry with great velocity for a short distance by a powerful movement of the fan-shaped pectorals, then stopping as suddenly. They rarely use the caudal fin in swimming, and they are never seen moving or floating freely in the water like most fishes. When at rest they support themselves on their extended ventrals and anal."

Emphasis since the 1970s (Findley 1976; Gatz 1979a,b) has been placed on multi-variate analyses of a large number of physical traits of individual species, then match-ing the overall morphological traits to measured ecological or behavioral traits of fishes, as reviewed in Douglas and Matthews (1992) and Matthews (1998, 403–412). Ecomorphology evaluated across families of fishes can show broad general patterns (Gatz 1979a,b; Douglas and Matthews 1992), but it may be most useful in consider-ation of ecological relationships within families. We emphasized in chapter 3 the importance of "family" as a descriptor of fish communities, but here we emphasize variation within families that, although subtle, may have important implications for their ecology or behavior. For example, consider the minnows, all in the family Cy-prinidae, in figure 4.4. These species, all common in Oklahoma, represent a gradient of body depth from the streamlined Redfin Shiner (A) to the deep-bodied, slab-sided Striped Shiner (*Luxilus chrysocephalus*) (D). Note also that the dorsal fins on Redfin Shiner and Rocky Shiner (B) are posterior to the origin of the pelvic fins, whereas for Steelcolor Shiner (C) and Striped Shiner the dorsal origin is nearly directly above the pelvic fin origin. Finally, note the sharpness of the snout of Steelcolor Shiner, com-pared to the much more blunt snout of Striped Shiner. All such differences in body shape, fin placement, or length of snout can relate to specific differences in the ways fish move through the water column, respond to currents, or use microhabitats or feed. Below we offer several examples of the ways that differences in body or fin morphol-ogy relate to ecological or behavioral differences between species.

For 17 Roanoke River fish species across several families, Douglas and Matthews (1992) found a strong relationship between morphological differences and food use, with much of the variation consisting of differences between families. But a reduced data set limited to eight minnow species showed a strong relationship between mor-phology and microhabitat use (Douglas and Matthews 1992). The results suggested that morphology, used judiciously, can indeed help to evaluate the ecological roles of fish species in communities. The morphological evaluation by Douglas and Matthews (1992), like earlier work by Gatz (1979a,b), was based on rather classical ichthyology measurements of body proportions, largely following the metrics of Hubbs and La-gler (1949). More recently, mathematically elegant techniques of truss-based measure-ments, especially geometric morphometrics, have been used a great deal to describe or compare shape among fishes. But a recent reassessment of the morphological data in Douglas and Matthews (1992), compared to results of geometric morphometrics

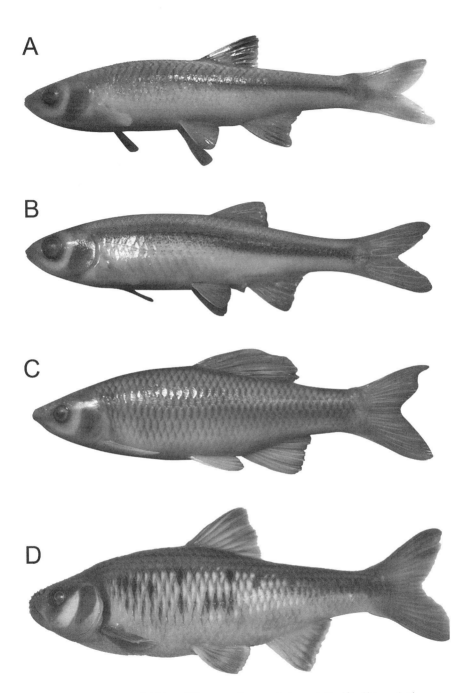

FIG. 4.4 Four minnows exhibiting differences in morphology. *A*, Redfin Shiner, *Lythrurus umbratilis*; *B*, Rocky Shiner, *Notropis suttkusi*; *C*, Steelcolor Shiner, *Cyprinella whipplei*; *D*, Striped Shiner, *Luxilus chrysocephalus*.

(Franssen et al. 2015), showed that both approaches recover ecologically important, species-specific information about habitat or diet. Both approaches can give results that can be explicitly linked to shape versus habitat use traits of individual species.

Matthews (unpub. data, from a class project in graduate school in 1976) measured a total of 25 morphological traits (fig. 4.5) for 12 species of darters that are common in northeast Oklahoma (Blair 1959). These traits included numerous standard ichthyological measurements (Hubbs and Lagler 1949) but also included traits suspected a priori to likely relate to ecology of the darters, including fine details of head structure such as size of the nares, fin structure such as thickness of the lower two rays in the pectoral fin or the first ray of the pelvic fin, fin base widths, or distance between the pelvic fins. For darters or other benthic stream fishes (Lundberg and Marsh 1976), the pectoral fin structure can be closely related to capacity for position holding, using the fin pressed to or as a prop upon the substrate or as a hydrofoil, and pelvic fins can be used to increase friction with substrates (Hynes 1972).

A simple phenogram from a correlation matrix for these darter species based on all 25 morphological characters separated the 12 species into 3 groups, with 1 species as a lone outlier (fig. 4.6). Correlations among all darters were high across the 25 morphological characters, but Least Darter separated first from all of the other species.

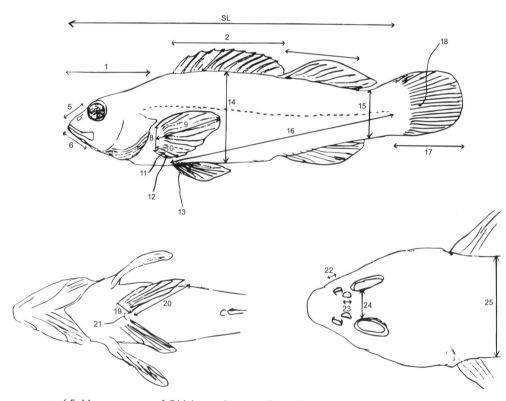

FIG. 4.5 Measurements of Oklahoma darters, selected to represent ecological traits.

This is one of the smallest of darters, with the extreme maximum or minimum values for 7 of the 25 traits, including having the relatively shortest snout, longest pectoral fin, and shortest distance from the pelvic to the caudal fin, among others. This species is scarce in the Ozarks and occupies a highly specialized backwater habitat with aquatic vegetation (Blair 1959). Then there was a separation (at about 99% correlation) of the remaining species into three groups. Darters typical of slower waters, including Slough Darter (*Etheostoma gracile*), Cypress Darter (*Etheostoma proeliare*) and Bluntnose Darter (*Etheostoma chlorosoma*), formed one group. Another group included three species typical of very swift habitats (Banded Darter; Greenside Darter, *Etheostoma blennioides*, and Speckled Darter). A third group included five species—Orangethroat Darter, Sunburst Darter, Arkansas Darter, Redfin Darter (*Etheostoma whipplei*), and Fantail Darter—that are widespread across numerous kinds of habitats or that occur in more or less intermediate current speeds. This early analysis suggested that multivariate approaches based on measurement of a variety of morphological features can relate to differences in habitat use among darters within a general geographic region.

In the Roanoke River, Virginia (see plate 4), the most common darters differ substantially in their use of microhabitats (Matthews et al. 1982). Fantail Darters are typically in the shallower, slower-flowing parts of riffles, while Roanoke Darters (*Percina roanoka*) are more abundant in the faster or deeper riffle microhabitats (Matthews et al. 1982; Roberts and Angermeier 2007). Adults of both species were tested in a glass flow tube for their ability to resist displacement as current speeds increased (measuring critical current speed, or CCS) through a combination of morphological and

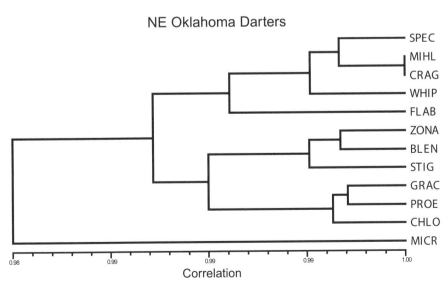

FIG. 4.6 Phenogram showing morphological relationships among 12 species of northeast Oklahoma darters.

behavioral traits (Matthews 1985). Roanoke Darters had a significantly (p < 0.001) higher mean CCS (30.2 cm/s; SD = 5.32) than Fantail Darters (24.0 cm/s; SD = 5.65), consistent in the direction of difference with the observed differences between the two species in microhabitat use in the field (Matthews et al. 1982; Roberts and An-germeier 2007). Morphologically, Roanoke Darters have traits considered by Page (1983) to be adaptive to life in flowing riffles, including a deeper body, wider head, and shorter snout length than in Fantail Darters. Behaviorally, however, Fantail Darters and Roanoke Darters differed markedly in response to increasing current speed. As current speed was initially increased, both species held the pectoral fins expanded and in contact with the glass tube substrate. But as current speeds increased, Fantail Darters increased contact between pectoral fin and substrate while moving the up-per rays posteriorly, giving the fin an angle of approximately 45° to the direction of flow. At higher current speeds, Fantail Darters exhibited an unexpected behavior that consisted of lowering the head so that the lower jaw was on the surface while sharply arching the body, so that the dorsal surface of the head and nape were at an angle of about 30° to 45° with the direction of flow, with arching of the body so pronounced that the midbody often lost contact with the substrate. In contrast, Roanoke Darters simply perched on the substrate as current speeds increased, with the head raised in normal position and the body held straight. While the behavior by Fantail Darters did not allow them to match Roanoke Darters in CCS, other trials with preserved indi-viduals did show that the behavior of Fantail Darters provided them with greater abil-ity to hold position as current speeds increased (Matthews 1985) (fig. 4.7).

Darters in riffles do not in general remain with the body directly exposed to strong currents, as they use sheltered spaces behind or beneath stones in a variety of ways (Schlosser and Toth 1984). But the ability to hold position when exposed to direct flow (rather than being swept away or out of preferred positions) seems likely to be advanta-geous to darters in larger, swifter riffles and rapids like many in the Roanoke River. That some species may be favored over others in swift habitats by morphology or behavior, or a combination of both, probably has a direct effect on the ways that darter communities are assembled in complex habitats or when multiple species coexist in riffles.

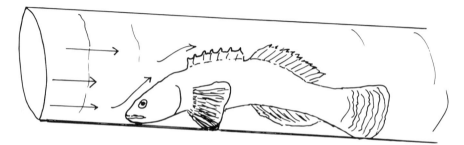

FIG. 4.7 Fantail darter (*Etheostoma flabellare*) behavior in flow tube: chin on substrate, body arched so venter is off substrate.

The studies above represent only a few of the many studies of the relationship between morphology and ecology, or morphology and behavior, for fishes, but they do suggest that for minnow, darter, or other common species, an ecomorphological or morphological-behavioral approach could be profitable for consideration of structure and dynamics of fish communities in general. We also note that in the studies above the assumption was tacitly made that morphology of a species was relatively fixed, although individuals of various sizes were included. But another potentially important consideration in ecomorphology is that body shape, which we know to be important ecologically, can also be malleable within a species, depending on the environmental conditions where populations occur. Nathan Franssen, a PhD student under EMM's direction, showed distinct differences in morphology of Red Shiners from populations that had been in artificially impounded reservoirs for multiple generations, relative to those from unimpounded reaches of the tributary rivers (Franssen 2011), pointing out the importance of considering rapid evolution of morphology as habitat conditions change or the equally plausible consideration that ecomorphology could be plastic within a generation.

Other Traits Important to Individual Species

Here we briefly consider our observations that provide some insight into things such as differences in food use or trophic group membership, schooling or shoaling or group sizes, life history traits, or the ways that seasonality may affect individual species.

Research on food use has often been done on the ways species differ in a community in the context of comparative resource use (Matthews et al. 1982; Surat et al. 1982) and as a mechanism of avoiding interspecific competition, which we consider in chapter 5. But the foods used by individual species, or their feeding behaviors, are equally of long-standing ecological interest (e.g., Forbes 1878). And particularly in light of global warming, knowing the foods typically used by individual species could help predict matches or mismatches in the future, as their phenology and that of their food organisms may respond differently to a new thermal milieu in streams. If a minnow like Dusky-stripe Shiner (*Luxilus pilsbryi*) has a broad general diet and feeds opportunistically on whatever aquatic insects are available at a particular time (Matthews et al. 1978), then changes in the timing of availability of different foods may matter little. But if different darter species have different breadths of diet (Matthews et al. 1982), those with the narrowest range of diets might fare less well if timing of availability of food items changes. In the Roanoke River, for example, the Riverweed Darter (*Etheostoma podostemone*) has the smallest mouth of the three darters whose diets were studied and a correspondingly smallest range in food sizes. In contrast, the Roanoke Darter has a comparatively larger mouth, a larger range in sizes of foods used, and a wider diet breadth, suggesting that it might have a greater capacity to accommodate any changes in the aquatic insect food resource base in the future. This is all speculative, but we suggest that knowledge of the food habits or breadth of foods used by individual species could help with predictions about their future if streams warm, flood more, and so on.

Knowing food habitats or modes of feeding of individual species can also help explain their roles in ecosystems (see chap. 9). At the simplest level, a variety of schemes have been developed to classify fishes on the basis of the foods typically used by adults of each species, with the most widely used scheme probably being that of Karr et al. (1986). But the ways individual species affect ecosystems can vary within broad feeding groups like insectivores because of differences in the ways species obtain items such as aquatic insects. Many minnows like Bigeye Shiner (WJM, pers. obs.), Dusky-stripe Shiner (Matthews et al. 1978), or Rosefin Shiner (Surat et al. 1982) eat aquatic insects drifting in the water column or hanging near the water's surface. The same kinds of aquatic insects are picked from stony surfaces by many species of darters (Matthews et al. 1982) or are winnowed from the gravels of the streambed by suckers such as various redhorses (*Moxostoma* spp.) or Northern Hog Sucker (*Hypentelium nigricans*). While taking similar foods, the species in the groups above can have different impacts in an ecosystem, in that some disturb streambed substrates, suspending biofilms and organic matter into the water column, whereas others simply remove the prey organism without any other disturbance to the stream. Matthews (1998, 79) suggested expansion of trophic schemes to include as many as 17 distinct modes of feeding, with a major dichotomy between species that do versus those that do not disturb substrates to obtain food items. We will return to the implications of feeding mode in chapter 9, but for now the point is to encourage studies of detailed differences among individual species in what they eat and how they obtain the food items. Such studies are not easy to do. In a class taught by Tony Echelle in summer 1976, we all took turns seining fish, sampling benthos, and running drift nets in a stream in south Oklahoma, every few hours for a 24-hour cycle, so we could compare stomach contents of Blacktail Shiners with the composition of the benthos versus the drift. I don't remember the results but do clearly remember finding two water moccasins in one of the drift nets at about 2:00 a.m.!

Another aspect of understanding differences among individual species with respect to their behavior in a community is to consider the extent to which they range from being solitary to forming large groups and what benefits might accrue from solitary existence or from various group sizes. Most freshwater species probably do not "school" in the formal sense that members of a school swim in coordination and turn as a group (Pitcher 1986), as many marine species do. The terminology most used that would apply to our species that form large groups is probably a "shoal," requiring no closely coordinated movements. Even without all swimming in parallel or turning simultaneously, however, many of the stream species we study do move together in groups.

For example, Central Stonerollers (see plate 27) form large groups (sometimes thousands of individuals) in clear, gravel-bottomed streams like the Baron Fork (Gelwick et al. 1997). There, stonerollers in a shoal covering an area of several square meters of substrate within a long pool will actively feed on attached algae in the substantial currents of midchannel, all oriented upstream and swimming to hold position. As fish in the shoal feed, they gradually drift downstream (i.e., backward) as a group because of the currents, until they near the downstream end of the pool. Then, as a group, they stop feeding and swim rapidly toward the upper part of the pool, where they resume feeding as a group. The advantages of this group behavior may be at least twofold. First, the coordinated feeding of the shoal on specific stony areas can help keep patches of algae open, with the most intensive feeding by individuals biting or scraping algae at the edges of the patch (Matthews et al. 1986). Alternatively, or perhaps in addition, fish in the shoal may find some of the well-known protection from predators by swimming with the "herd." For the stonerollers, the most active piscivores in Baron Fork are large Smallmouth Bass and various wading birds like Great Blue Herons. When startled, a shoal of stonerollers tends to act as a group, as an observer can readily see while walking the bank. But if a Smallmouth Bass approaches or swims through a shoal of feeding stonerollers, the minnows typically exhibit a "fountain effect" (Pitcher 1986, 307; Magurran and Pitcher 1987), with individuals parting to make a pathway for the bass, then circling around and taking up new positions behind the bass.

In general, based on underwater and bankside observations, as well as observations by seining, we find minnows in groups of a dozen to hundreds of individuals, the same as Western Mosquitofish. In contrast, catfish such as Channel Catfish (*Ictalurus punctatus*) are often solitary as adults, and adult bullhead catfishes are often in pairs. Suckers like *Hypentelium* and *Moxostoma* usually are either in groups of twos or threes, occasionally swimming in larger groups. Between these extremes are species such as Blackstripe Topminnows (*Fundulus notatus*) and Blackspotted Topminnows (*Fundulus olivaceus*) that are typically seen swimming just below the surface of pools in groups of three or four to perhaps a dozen individuals. Centrarchids are interesting in that group size differs markedly among species or by size of individuals within species. Of those that we see most often, Longear Sunfish tend to occur in groups of five or six adults to sometimes dozens of juveniles. They will often approach a snorkeling observer as a group, coming near enough to inspect the intruder and then backing away while continuing to observe, by sculling backward with the pectoral fins. Green Sunfish are more solitary as adults and less likely to form groups, with the exception that small juveniles sometimes are seen in groups of a dozen or more. Redear Sunfish (*Lepomis microlophus*) and Orangespotted Sunfish (*Lepomis humilus*) are often found in our lower-gradient streams but rarely in groups, occurring most often as singletons. Very small Largemouth Bass (just off the nest) are often found in groups of a hundred or more, but for juveniles and adults we rarely see more than two or three individuals swimming together. Group sizes are quite characteristic of individual species, and all such traits should have implications for community dynamics, some of which we explore in chapter 5. In general, however, "our" species run the spectrum from solitary to forming large groups, all with implications for ability to find food and potentially mates and to potentially avoid predators.

Reproductive or life history traits of individual species can also have huge implications for their ability to successfully maintain populations, repopulate after disturbances, or potentially invade new habitats or ranges. Schlosser (1987) provided a robust general model for the distribution of different species in small, warm-water streams, blending their reproductive potential with the differing kinds of habitat (stability, predator pressure) from headwaters to lower mainstem. The life history traits of some species we know from our surveys or experiments help make it possible for them to live where they do. Many of the streams we work in are highly seasonal, with spring flooding and late summer–autumn drying, and the entire region varies greatly in rainfall or temperatures from year to year (Matthews and Marsh-Matthews 2011). Some of the hardiest, most resilient, and widely distributed fishes we work with (Western Mosquitofish, Red Shiner, Green Sunfish, or Longear Sunfish) have reproductive traits that seem related to life in these highly variable streams. Western Mosquitofish are live-bearing poeciliids, capable of rapidly repopulating a system, sometimes in a boom-and-bust population pattern, but rarely leaving a community entirely (see plate 15). This species has an interbrood interval of approximately a month, and females often have several broods a year. We found (Matthews and Marsh-Matthews 2011) that in years with early springtime warming, mosquitofish populations in late summer were larger than in other years. Because onset of reproduction by mosquitofish is temperature dependent, starting at about 18°C in our area (EMM, pers. obs.), an early spring with warming several weeks earlier than usual can essentially provide mosquitofish with an additional brood at the beginning of a year, and the offspring from that brood can themselves be reproductive adults within five to six weeks of birth (Campton and Gall 1988). The adults from this early brood, by also reproducing earlier in the year, put "money in the bank" on which to draw interest (i.e., more kids), and the population by late summer builds up to a larger size than is the case in years with colder, later springs (Matthews and Marsh-Matthews 2011).

As an initial experimental test of this "warmer = more babies" hypothesis, we stocked 16 adult, pregnant female Western Mosquitofish from a local creek in each of four outdoor mesocosm units open to direct sunlight and in four units covered with shade cloth, in early May 2014. From then until we ended the experiment on 20–21 July, the "sun" units averaged about 3°C to 5°C warmer, and all units grew modest to dense crops of attached algae (which could protect neonate mosquitofish from cannibalism). At the end of the experiment, the warmer sun units averaged 96 more total mosquitofish than the cooler "shade" units, different at $P = 0.05$ (one-way ANOVA) (Marsh-Matthews and Matthews, unpub. data). While this was only a pilot study and did not include the earlier period of the year when spring warming had apparently influenced Western Mosquitofish population sizes in the wild, it suggests that warmer temperatures can result in more output of young.

Red Shiners also have reproductive capacity that makes them capable of rapid population expansion, relative to most other minnows. Minnows typically do not reproduce until at least their second year of life (Year 1 in fisheries jargon), but Red Shiners that are spawned in May can reproduce by August (Marsh-Matthews et al. 2002). Red shiners also continue producing broods until well into autumn, as late as October or early November (Marsh-Matthews et al. 2002), and we have sometimes found large numbers of extremely small Red Shiners (e.g., 14 mm total length) in streams during the coldest parts of winter. One might wonder whether small, newly hatched Red Shiners entering winter would be good for anything, as winter mortality of small individuals is well known for many fish species. But we overwintered three size classes of small Red Shiners in outdoor experimental streams, including extremely small individuals, and found that all had very high survival rates, remaining over winter at a small size with little growth but putting on a burst of growth and becoming reproductive by May (Marsh-Matthews, Matthews, Gido, unpub. data).

Timing of reproduction in seasonally variable streams can be a key to successful reproduction, and midwestern species differ greatly in this trait. Common and abundant *Lepomis* species like Green Sunfish and Longear Sunfish (see plate 18) nest actively throughout much of the summer in streams like Brier Creek. If flooding earlier in a year washes away sunfish larvae, the sunfish resume nesting and can produce young for the rest of the summer (Harvey 1987). And we observed Central Stonerollers and *Moxostoma* (probably *M. erythrurum*) to rapidly resume reproductive activities in Piney Creek after an extreme flood event (Matthews et al. 2014). In contrast, some of our most common darters (Percidae), like Orangebelly and Orangethroat Darters, reproduce in late winter or early spring and are unlikely in summer to replace any young-of-year that might be killed or washed away by flood. Young Orangebelly and Orangethroat Darters are most often found in our south Oklahoma creeks in sloping, coarse sand or cobble substrate at the edge of quiet pools, and that microhabitat would seem especially vulnerable to being moved or rearranged by erosive flooding, with potential for mortality of the younger individuals (whereas we have found that adults can move during floods to protected habitats and resume residency on riffles once floodwaters subside; Matthews et al. 2013).

Some Additional Thoughts

Traits of individual species can strongly influence community structure or dynamics, as they can influence interspecific interactions (chap. 5), temporal variation in community structure (chaps. 6–8), or the effects of species in ecosystems (chap. 9). Thus traits of individual species are important. But in considering life history or other traits, it is questionable to classify any species as though one size fits all throughout its range or across environmental vagaries. Some traits—like thermal tolerance of darters (Feminella and Matthews 1984), reproductive traits of darters (Marsh 1984), or microhabitat use by minnows (Matthews and Hill 1980)—can vary substantially among populations with time of year or among years (Matthews and Marsh-Matthews 2011). Additionally, many traits likely have a strong component of plasticity (Marsh 1984), all of which makes a clear definition of the traits of a given species challenging. This is not to say that studies broadly based on traits assigned at the species level are to be avoided, but it does suggest that users should be aware that there can be extremes of variation within a species for important traits. The truth is that we, and biologists in general, have barely begun to scratch the surface with detailed studies about the individual traits of many species that we work with (Matthews 2015). This varies from region to region, but a review of available information in numerous recent "state fish books" for the United States (Matthews 2015) showed that a substantial proportion of species still lack much basic life history or general biological information and are in need of study. This is true in spite of much emphasis in recent years on studies or summaries based on traits of individual species. One problem is that broad classifications of species traits are sometimes done with substantial guesswork, placing species about which little is really known into trait categories based on their relatedness to other better-known species ("it's a water-column minnow, so I guess it's an insectivore like other water-column minnows, even though we don't really have data"). The other problem is that sometimes classifications of species' traits have been done by what has recently been defined as GOBSAT (Good Old Boys Sitting Around a Table) biology (Miller and Petrie 2000), with "experts" deciding on traits of species by consensus. WJM took part in one of these exercises (Jester et al. 1992) to define tolerances of fishes in Oklahoma. This is not to say that expert judgment or assigning little-known species to trait groups on the basis of similarity to other species produces information that is necessarily wrong. It may be the best that can be done with the current state of knowledge about some species. But the entire cottage industry of trait-based biology would be better grounded if detailed, basic biological information were available for more species. A lot of good master's theses could still be done on life history investigations of little-known species, and they would be valuable contributions to the knowledge base that is increasingly being used for "big picture" meta-analyses or similar studies.

Summary

Details of the biology of an individual species (i.e., species traits) have long been recognized as important determinants of the species' distribution. Traits related to physiology, ecology, anatomy, and behavior can all influence not only where a species occurs but also its co-occurrence with other species. In this chapter we have focused on the factors that affect distributions of species across environmental gradients and hence potentially influence membership in particular communities.

Traits related to physiological tolerances of extreme conditions (such as temperature and dissolved oxygen) may interact with environmental conditions to "filter" potential community members, particularly in harsh or extreme environments. Across the Great Plains, there is a general east-to-west gradient of increasing harshness, with western sites having lower rainfall, higher maximum temperatures, and higher evaporation. Western stream fish communities are therefore exposed to periods of greater intermittency, higher temperatures, and lower dissolved oxygen than those in the eastern Great Plains. An analysis of western limits of distribution for fishes in Oklahoma showed a progression of western range limits across the state, with about half of the species in the state "dropping out" of the species pool toward the west. Studies of tolerance for high temperatures and low oxygen support environmental filtering as the mechanism for this westward decline in species because, of the species tested, there was a strong relationship between tolerance and western range limit.

Tolerance limits may not limit species absolutely because behaviors related to habitat use and selectivity may act to mitigate periods of harsh conditions and allow species to survive the most extreme environments. Examples are use of thermal refugia during high and low temperatures and avoidance of fast currents (with increased metabolic demands for swimming in the current) during periods of low productivity.

Also among species' traits that may affect community membership is the ability to use a wide variety of habitats or resources (i.e., being a generalist) versus a more specialized strategy. An analysis of 126 species based on habitat use showed that 65% were extreme habitat specialists whereas less than 10% were habitat generalists, with the remaining species intermediate. This result emphasizes the role of habitat availability in determining community structure and the potential for habitat loss (due to natural or anthropogenic causes) to alter community structure. And as habitats are altered, ability or propensity to move within or between streams may be crucial for species' persistence in a system.

Ecomorphological studies across a variety of stream fish species have repeatedly demonstrated the role of basic anatomy in a species' ability to occupy and persist in particular habitats or to use particular food resources. Morphology of 12 species of darters in Oklahoma was a strong predictor of the type of habitat occupied. Ability to resist displacement in increasing current speeds (tested in the laboratory) of two darter species in the Roanoke River matched the relative current speeds of microhabitats used in the field. But the species with the lower critical current speed exhibited a behavior in the laboratory that suggested it could hold position at currents faster than expected based on its morphology alone. Thus there may be complex combinations of species' traits that determine the habitats they can occupy.

Other important traits include those related to feeding, shoaling, and reproduction. Traits related to feeding will determine not only what a species can consume but also how its foraging efficiency relates to that of similar members of the community. Diet breadth may be an important predictor of a species' ability to use alternate or modified habitats or withstand changes in food availability. The propensity to form shoals or groups of various sizes varies among stream fishes and affects a number of other traits, such as foraging and predator avoidance. Reproductive traits of individual species can have enormous impacts on population and community dynamics, as species differ in age at maturity, timing of reproduction, and number of reproductive bouts in a season.

Numerous species traits (and surely more than have been studied to date) affect distribution and community membership, but many traits that have been studied at the species level are in fact plastic. Although phenotypic plasticity adds "noise" to trait-based studies of multiple species, it is precisely the trait of plasticity itself that may be the most important determinant of where, when, and with whom species are found.

References

Ackerly, D. D., and K. W. Cornwell. 2007. A trait-based approach to community assembly: partitioning of species trait values into within- and among-community components. Ecology Letters 10:135–145.

Allan, J. D. 1995. Stream ecology: structure and function of running waters. Chapman and Hall, London, UK.

Andrewartha, H. G., and L. C. Birch. 1954. The distribution and abundance of animals. University of Chicago Press, Chicago, IL.

Bart, H. L., Jr. 1989. Fish habitat association in an Ozark stream. Environmental Biology of Fishes 24:173–186.

Blair, A. P. 1959. Distribution of the darters (Percidae, Etheostomatinae) of northeastern Oklahoma. Southwestern Association of Naturalists 4:1–13.

Brown, J. H. 1995. Macroecology. University of Chicago Press, Chicago, IL.

Campton, D. E., and G. A. E. Gall. 1988. Responses to selection for body size and age at sexual maturity in the mosquitofish, *Gambusia affinis*. Aquaculture 68:221–241.

Clausen, R. G. 1936. Oxygen consumption in fresh water fishes. Ecology 17:216–226.

Clavel, J., R. Julliard, and V. Devictor. 2011. Worldwide decline of specialist species: toward a global functional homogenization? Frontiers in Ecology and the Environment 9:222–228.

Cook, E. R., R. Seager, M. A. Cane, and D. W. Stahle. 2007. North American drought: reconstructions, causes, and consequences. Earth-Science Reviews 81:93–134.

Cui, Y., and R. J. Wootton. 1988. Effects of ration, temperature and body size on the body composition, energy content and condition of the minnow, *Phoxinus phoxinus* (L.). Journal of Fish Biology 32:749–764.

Douglas, M. E., and W. J. Matthews. 1992. Does morphology predict ecology? hypothesis testing within a freshwater stream fish assemblage. Oikos 65:213–224.

Dray, S., and P. Legendre. 2008. Testing the species traits-environmental relationships: the fourth corner problem revisited. Ecology 89:3400–3412.

Echelle, A. A., A. F. Echelle, M. H. Smith, and L. G. Hill. 1975. Analysis of genic continuity in a headwater fish, *Etheostoma radiosum* (Percidae). Copeia 1975:197–204.

Elton, C. 1927. Animal ecology. Sidgewick & Jackson, London, UK.

Elton, C. 1946. Competition and the structure of ecological communities. Journal of Animal Ecology 15:54–68.

Emerson, B. C., and R. G. Gillespie. 2008. Phylogenetic analysis of community assembly and structure over space and time. Trends in Ecology and Evolution 23:619–630.

Feminella, J. W., and W. J. Matthews. 1984. Intraspecific differences in thermal tolerance of *Etheostoma spectabile* (Agassiz) in constant versus fluctuating environments. Journal of Fish Biology 25:455–461.

Findley, J. S. 1976. The structure of bat communities. American Naturalist 110:129–139.

Forbes, S. A. 1878. The food of Illinois fishes. Bulletin of the Illinois State Laboratory of Natural History 1:71–89.

Forbes, S. A. 1880. The food of the darters. American Naturalist 14:697–703.

Franssen, N. R. 2011. Anthropogenic habitat alteration induces rapid morphological divergence in a native stream fish. Evolutionary Application 4:791–804.

Franssen, N. R., C. G. Goodchild, and D. B. Shepard. 2015. Morphology predicting ecology: incorporating new methodological and analytical approaches. Environmental Biology of Fishes 98:713–724.

Frimpong, E. A., and P. L. Angermeier. 2010. Comparative utility of selected frameworks for regionalizing fish-based bioassessments across the United States. Transactions of the American Fisheries Society 139:1872–1895.

Fritz, K. M., and W. K. Dodds. 2005. Harshness: characterization of intermittent stream habitat over space and time. Marine and Freshwater Research 56:13–23.

Gatz, A. J., Jr. 1979a. Community organization of fishes as indicated by morphological features. Ecology 60:711–718.

Gatz, A. J., Jr. 1979b. Ecological morphology of freshwater stream fishes. Tulane Studies in Zoology and Botany 21:91–124.

Gelwick, F. P., M. S. Stock, and W. J. Matthews. 1997. Effects of fish, water depth, and predation risk on patch dynamics in a north-temperate river ecosystem. Oikos 80:382–398.

Goldstein, R. M., and M. R. Meador. 2004. Comparisons of fish species traits from small streams to large rivers. Transactions of the American Fisheries Society 133:971–983.

Grinnell, J. 1914. An account of the mammals and birds of the lower Colorado Valley, with especial reference to the distributional problems presented. University of California Publications in Zoology 12:51–294.

Grinnell, J. 1917. The niche-relationships of the California thrasher. Auk 34:427–433.

Harvey, B. C. 1987. Susceptibility of young-of-the-year fishes to downstream displacement by flooding. Transactions of the American Fisheries Society 116:851–855.

Helmus, M. R., K. Savaage, M. W. Diebel, J. T. Maxted, and A. R. Ives. 2007. Separating the determinants of phylogenetic community structure. Ecology Letters 10:917–925.

Hill, L. G., W. J. Matthews, T. Schene, and K. Asbury. 1981. Notes on fishes of Grand River, Chouteau Creek, and Pryor Creek, Mayes County, Oklahoma. Proceedings of the Oklahoma Academy of Science 61:76–77.

Hubbs, C. L., and K. F. Lagler. 1949. Fishes of the Great Lakes region. Bulletin No. 26. Cranbrook Institute of Science, Bloomfield Hills, MI.

Hynes, H. B. N. 1972. The ecology of running waters (second impression). University of Toronto Press, Toronto, ONT.

Jackson, D. A., P. R. Peres-Neto, and J. D. Olden. 2001 What controls who is where in freshwater fish communities—the roles of biotic, abiotic, and spatial factors. Canadian Journal of Fisheries and Aquatic Sciences 58:157–170.

Jester, D. B., A. A. Echelle, W. J. Matthews, J. Pigg, C. M. Scott, and K. D. Collins. 1992. The fishes of Oklahoma, their gross habitats, and their tolerance of degradation in water quality and habitat. Proceedings of the Oklahoma Academy of Science 72:7–19.

Johnson, H. L., and C. E. Duchon. 1995. Atlas of Oklahoma climate. University of Oklahoma Press, Norman, OK.

Jones, N. E., G. J. Scrimgeour, and W. M. Tonn. 2010. Fish species traits and communities in relation to a habitat template for Arctic rivers and streams. Pages 147–156 in K. B. Gido and D. A. Jackson, eds. Community ecology of stream fishes: concepts, approaches, and techniques. American Fisheries Society Symposium 73. American Fisheries Society, Bethesda, MD.

Jordan, D. S. 1884. Manual of the vertebrate animals of the northern United States, 4th ed. Jansen, McClurg & Company, Chicago, IL.

Jordan, D. S. 1888. The distribution of fresh-water fishes. Transactions of the American Fisheries Society 17:4–29.

Julliard, R., J. Clavel, V. Devictor, F. Jiguet, and D. Couvet. 2006. Spatial segregation of specialists and generalists in bird communities. Ecology Letters 9:1237–1244.

Kansas Fishes Committee. 2014. Kansas fishes. University of Kansas Press, Lawrence, KS.

Karr, J. R., K. D. Fausch, P. L. Angermeier, P. R. Yant, and I. J. Schlosser. 1986. Assessing biological integrity in running waters—a method and its rationale. Special Publication 5. Illinois Natural History Survey, Champaign, IL.

Kearney, M., and W. Porter. 2009. Mechanistic niche modelling: combining physiological and spatial data to predict species' ranges. Ecology Letters 12:334–350.

Kothary, R. K., and E. P. M. Candido. 1982. Induction of a novel set of polypeptides by head shock or sodium arsenite in cultured cells of rainbow trout, *Salmo gairdnerii*. Canadian Journal of Biochemistry 60:347–355.

Lamouroux, N., N. L. Poff, and P. L. Angermeier. 2002. Intercontinental convergence of stream fish community traits along geomorphic and hydraulic gradients. Ecology 83: 1792–1807.

Lee, D. S., C. R. Gilbert, C. H. Hocutt, R. E. Jenkins, D. E. McAllister, and J. R. Stauffer Jr. 1980. Atlas of North American freshwater fishes. North Carolina State Museum of Natural History, Raleigh, NC.

Lundberg, J. G., and E. Marsh. 1976. Evolution and functional anatomy of the pectoral fin rays in cyprinoid fishes, with emphasis on the suckers (Family Catostomidae). American Midland Naturalist 96:332–349.

Magurran, A. E., and T. J. Pitcher. 1987. Provenance, shoal size and the sociobiology of predator-evasion behavior in minnow shoals. Proceedings of the Royal Society of London Part B 229:439–465.

Maness, J. D., and V. H. Hutchison. 1980. Acute adjustment of thermal tolerance in vertebrate ectotherms following exposure to critical thermal maxima. Journal of Thermal Biology 5: 225–233.

Marsh, E. 1984. Egg size variation in central Texas populations of *Etheostoma spectabile* (Pisces: Percidae). Copeia 1984:291–301.

Marsh-Matthews, E., and W. J. Matthews. 2010. Proximate and residual effects of exposure to simulated drought on prairie stream fishes. Pages 461–486 in K. B. Gido and D. A. Jackson, eds. Community ecology of stream fishes: concepts, approaches, and techniques. American Fisheries Society Symposium 73. American Fisheries Society, Bethesda, MD.

Marsh-Matthews, E., W. J. Matthews, and N. R. Franssen. 2011. Can a highly invasive species re-invade its native community? the paradox of the Red Shiner. Biological Invasions 13: 2911–2924.

Marsh-Matthews, E., W. J. Matthews, K. B. Gido, and R. L. Marsh. 2002. Reproduction by young-of-year Red Shiner (*Cyprinella lutrensis*) and its implications for invasion success. Southwestern Naturalist 47:605–610.

Marsh-Matthews, E., J. Thompson, W. J. Matthews, A. Geheber, N. R. Franssen, and J. Barkstedt. 2013. Differential survival of two minnow species under experimental sunfish predation: implications for re-invasion of a species into its native range. Freshwater Biology 58:1745–1754.

Matthews, W. J. 1985. Critical current speeds and microhabitats of the benthic fishes *Percina roanoka* and *Etheostoma flabellare*. Environmental Biology of Fishes 12:303–308.

Matthews, W. J. 1986. Geographic variation in thermal tolerance of a widespread minnow *Notropis lutrensis* of the North American mid-west. Journal of Fish Biology 28:404–417.

Matthews, W. J. 1987. Physicochemical tolerance and selectivity of stream fishes as related to their geographic ranges and local distributions. Pages 111–120 in W. J. Matthews and D. C. Heins, eds. Community and evolutionary ecology of North American stream fishes. University of Oklahoma Press, Norman, OK.

Matthews, W. J. 1998. Patterns in freshwater fish ecology. Chapman and Hall, New York, NY.

Matthews, W. J. 2015. Basic biology, good field notes, and synthesizing across your career. Copeia 103:495–501.

Matthews, W. J., and L. G. Hill. 1977. Tolerance of the Red Shiner, *Notropis lutrensis* (Cyprinidae) to environmental parameters. Southwestern Naturalist 22:89–98.

Matthews, W. J., and L. G. Hill. 1980. Habitat partitioning in the fish community of a southwestern river. Southwestern Naturalist 25:51–66.

Matthews, W. J., and J. D. Maness. 1979. Critical thermal maxima, oxygen tolerances and success of cyprinid fishes in a southwestern river. American Midland Naturalist 102: 374–377.

Matthews, W. J., and E. Marsh-Matthews. 2007. Extirpation of Red Shiner in direct tributaries of Lake Texoma (Oklahoma-Texas): a cautionary case history from a fragmented river-reservoir system. Transactions of the American Fisheries Society 136:1041–1062.

Matthews, W. J., and E. Marsh-Matthews. 2011. An invasive fish species within its native range: community effects and population dynamics of *Gambusia affinis* in the central United States. Freshwater Biology 56:2609–2619.

Matthews, W. J., and E. Marsh-Matthews. 2016. Dynamics of an upland stream fish community over 40 years: trajectories and support for the loose equilibrium concept. Ecology 97:706–719.

Matthews, W. J., and J. T. Styron Jr. 1981. Tolerance of headwater vs. mainstream fishes for abrupt physicochemical changes. American Midland Naturalist 105:149–158.

Matthews, W. J., and E. G. Zimmerman. 1990. Potential effects of global warming on native fishes of the southern Great Plains and the Southwest. Fisheries 15:26–32.

Matthews, W. J., J. R. Bek, and E. Surat. 1982. Comparative ecology of the darters *Etheostoma podostemone*, *E. flabellare* and *Percina roanoka* in the upper Roanoke River drainage, Virginia. Copeia 1982:805–814.

Matthews, W. J., K. B. Gido, G. P. Garrett, F. P. Gelwick, J. G. Stewart, and J. Schaefer. 2006. Modular experimental riffle-pool stream system. Transactions of the American Fisheries Society 135:1559–1566.

Matthews, W. J., B. C. Harvey, and M. E. Power. 1994. Spatial and temporal patterns in the fish assemblages of individual pools in a midwestern stream (U.S.A.). Environmental Biology of Fishes 39:381–397.

Matthews, W. J., E. Marsh-Matthews, G. L. Adams, and S. R. Adams. 2014. Two catastrophic floods: similarities and differences in effects on an Ozark stream fish community. Copeia 2014:682–693.

Matthews, W. J., E. Marsh-Matthews, R. C. Cashner, and F. Gelwick. 2013. Disturbance and trajectory of change in a stream fish community over four decades. Oecologia 173:955–969.

Matthews, W. J., M. E. Power, and A. J. Stewart. 1986. Depth distribution of *Campostoma* grazing scars in an Ozark stream. Environmental Biology of Fishes 17:291–297.

Matthews, W. J., W. D. Shepard, and L. G. Hill. 1978. Aspects of the ecology of the Dusky-stripe Shiner, *Notropis pilsbryi* (Cypriniformes: Cyprinidae) in an Ozark stream. American Midland Naturalist 100:247–252.

McGill, B. J., B. J. Enquist, E. Weiher, and M. Westoby. 2006. Rebuilding community ecology from functional traits. Trends in Ecology and Evolution 21:178–185.

McPeek, M. A. 1996. Trade-offs, food web structure, and the coexistence of habitat specialists and generalists. American Naturalist 148:S124–S138.

Miller, J., and J. Petrie. 2000. Development of practice guidelines. The Lancet 355:82–83.

Miller, R. L., and H. W. Robison. 2004. Fishes of Oklahoma. University of Oklahoma Press, Norman, OK.

Moyle, P. B., and H. W. Li. 1979. Community ecology and predator-prey relations in warmwater streams. Pages 171–180 in H. Clepper, ed. Predator-prey systems in fisheries management. Sport Fishing Institute, Washington, DC.

Ostrand, K. G., and G. R. Wilde. 2001. Temperature, dissolved oxygen, and salinity tolerances of five prairie stream fishes and their role in explaining fish assemblage patterns. Transactions of the American Fisheries Society 130:742–749.

Page, L. M. 1983. Handbook of darters. TFH, Neptune City, NJ.

Parham, R. W. 2009. Structure of assemblages and recent distribution of riverine fishes in Oklahoma. Southwestern Naturalist 54:382–399.

Peckarsky, B. L. 1983. Biotic interactions or abiotic limitations? a model of lotic community structure. Pages 303–323 in T. D. Fontaine III and S. M. Bartell, eds. Dynamics of lotic ecosystems. Ann Arbor Science, Ann Arbor, MI.

Peterson, A. T. 2003. Predicting the geography of species' invasions via ecological niche modeling. Quarterly Review of Biology 78:419–433.

Pflieger, W. L. 1971. A distributional study of Missouri fishes. University of Kansas Publications Museum of Natural History 20:225–570.

Pitcher, T. J. 1986. Functions of shoaling behavior in teleosts. Pages 294–337 in T. J. Pitcher, ed. The behavior of teleost fishes. Johns Hopkins University Press, Baltimore, MD.

Poff, N. L. 1997. Landscape filters and species traits: towards mechanistic understanding and prediction in stream ecology. Journal of the North American Benthological Society 16: 391–409.

Poff, N. L., and J. D. Allan. 1995. Functional organization of stream fish assemblages in relation to hydrological variability. Ecology 76:606–627.

Power, M. E., and W. J. Matthews. 1983. Algae-grazing minnows (*Campostoma anomalum*), piscivorous bass (*Micropterus* spp.), and the distribution of attached algae in a small prairie-margin stream. Oecologia 60:328–332.

Power, M. E., W. J. Matthews, and A. J. Stewart. 1985. Grazing minnows, piscivorous bass and stream algae: dynamics of a strong interaction. Ecology 66:1448–1456.

Rahel, F. J., and W. A. Hubert. 1991. Fish assemblages and habitat gradients in a Rocky Mountain–Great Plains stream: biotic zonation and additive patterns of community change. Transactions of the American Fisheries Society 120:319–332.

Roberts, J. H., and P. L. Angermeier. 2007. Spatiotemporal variability of stream habitat and movement of three species of fish. Oecologia 151:417–430.

Ross, S. T., and W. J. Matthews. 2014. Evolution and ecology of North American freshwater fish assemblages. Pages 1–49 in M. L. Warren Jr. and B. M. Burr, eds. Freshwater fishes of North America. Vol. 1. Petromyzontidae to Catostomidae. Johns Hopkins University Press, Baltimore, MD.

Ross, S. T., W. J. Matthews, and A. A. Echelle. 1985. Persistence of stream fish assemblages: effects of environmental change. American Naturalist 126:24–40.

Schaefer, J. 2001. Riffles as barriers to interpool movement by three cyprinids (*Notropis boops*, *Campostoma anomalum*, and *Cyprinella venusta*). Freshwater Biology 46:379–388.

Schaefer, J. F., E. Marsh-Matthews, D. E. Spooner, K. B. Gido, and W. J. Matthews. 2003. Effects of barriers and thermal refugia on local movement of the threatened leopard darter, *Percina pantherina*. Environmental Biology of Fishes 66:391–400.

Schlosser, I. J. 1987. A conceptual framework for fish communities in small headwater streams. Pages 17–24 in W. J. Matthews and D. C. Heins, eds. Community and evolutionary ecology of North American stream fishes. University of Oklahoma Press, Norman, OK.

Schlosser, I. J. 1991. Stream fish ecology: a landscape perspective. BioScience 41:704–712.

Schlosser, I. J., and L. A. Toth. 1984. Niche relationships and population ecology of rainbow (*Etheostoma caeruleum*) and fantail (*E. flabellare*) darters in a temporally variable environment. Oikos 42:229–238.

Sharma, S., P. Legendre, M. De Cáceres, and D. Boisclair. 2011. The role of environmental and spatial processes in structuring native and non-native fish communities across thousands of lakes. Ecography 34:762–771.

Smale, M. A., and C. F. Rabeni. 1995a. Hypoxia and hyperthermia tolerances of headwater stream fishes. Transactions of the American Fisheries Society 124:698–710.

Smale, M. A., and C. F. Rabeni. 1995b. Influences of hypoxia and hyperthermia on fish species composition in headwater streams. Transactions of the American Fisheries Society 124: 711–725.

Smith, C. L., and C. R. Powell. 1971. The summer fish communities of Brier Creek, Marshall County, Oklahoma. American Museum Novitates 2458:1–30.

Spooner, D. E., and C. C. Vaughn. 2008. A trait-based approach to species' roles in stream ecosystems: climate change, community structure and material cycling. Oecologia 158: 307–317.

Stahle, D. W., F. K. Fye, E. R. Cook, and R. D. Griffin. 2007. Tree-ring reconstructed mega-droughts over North America since A.D. 1300. Climatic Change 83:133–149.

Strecker, A. L., J. M. Casselman, M. J. Fortin, D. A. Jackson, M. S. Ridgeway, P. A. Abrams, and B. J. Shuter. 2011. A multi-scale comparison of trait linages to environmental and spatial variables in fish communities across a large freshwater lake. Oecologia 166:819–831.

Surat, E. M., W. J. Matthews, and J. R. Bek. 1982. Comparative ecology of *Notropis albeolus*, *N. ardens* and *N. cerasinus* (Cyprinidae) in the upper Roanoke River Drainage, Virginia. American Midland Naturalist 107:13–24.

Taylor, C. M., and P. W. Lienesch. 1996. Regional parapatry of the congeneric cyprinids *Lythrurus snelsoni* and *L. umbratilis*: species replacement along a complex environmental gradient. Copeia 1996:493–497.

Taylor, C. M., M. R. Winston, and W. J. Matthews. 1993. Fish species-environment and abundance relationships in a Great Plains river system. Ecography 16:16–23.

Thompson, D. H., and F. D. Hunt. 1930. The fishes of Champaign County: a study of the distribution and abundance of fishes in small streams. Illinois Natural History Survey Bulletin 19:5–101.

Troia, M. J., and K. B. Gido. 2015. Functional strategies drive community assembly of stream fishes along environmental gradients and across spatial scales. Oecologia 177:545–559.

Vaughn, C. C. 2010. Biodiversity losses and ecosystem function in freshwaters: emerging conclusions and research directions. BioScience 60:25–35.

Vernon, H. M. 1899. The death temperature of certain marine organisms. Journal of Physiology 25:131–136.

Verrill, A. E. 1901. A remarkable instance of the death of fishes, at Bermuda in 1901. American Journal of Science (Series 4) 12:88–95.

Wiens, J. A. 1977. On competition and variable environments. American Scientist 65:590–597.

Wilson, S. K., S. C. Burgess, A. J. Cheal, M. Emslie, R. Fisher, I. Miller, N. V. Polunin, and H. P. Sweatman. 2008. Habitat utilization by coral reef fish: implications for specialists vs. generalists in a changing environment. Journal of Animal Ecology 77:220–228.

Wootton, R. J. 1990. Ecology of teleost fishes. Chapman and Hall, London, UK.

CHAPTER 5

Interactions among Species

Historical Perspective

Of the species that could occur in stream fish communities by virtue of traits that make them suited for the local environment, many may be excluded or diminished in local abundance by interactions such as interspecific competition or encounters with predators. Conversely, facilitation may enhance local occurrence or abundance. In this chapter we briefly review the history of studies of biotic interactions, examine our own work and that of some of our students on interactions, with a focus on competition and predation, and then end the chapter with brief consideration of facilitation and cases in which we failed to observe effects of expected interactions.

Effects of biotic interactions on community structure have been controversial. Most biologists who are now in their 60s, like us, grew up professionally near the end of the "competition era," when the "idea that competition is the sole or even the principal mechanism" affecting community structure was being challenged (Connell 1975, 461). The idea that interspecific competition was so important to community structure had its roots in the 1920s and 1930s. Mathematical theory (Volterra 1928), experiments with microorganisms (Gause 1932, 1934), "MacArthur's warblers" (MacArthur 1958), and the competitive exclusion principle (Hardin 1960) all emphasized interspecific competition as the dominant regulator of community structure. By 1975, there was still much support for the role of competition in community structure, as evidenced by numerous papers in the MacArthur volume (Cody and Diamond 1975) and the *Community Ecology* volume edited by Diamond and Case (1976). For example, J. H. Brown (1975) was unequivocal, stating that "competition is a major force in the structuring of rodent communities," and Diamond's (1975) famous "checkerboard pattern" was based largely on competitive exclusion. But even in those influential compilations, some authors questioned the primacy of competition, including Moulton and Pimm (1976), who noted, "Too many results once explained by competition can be explained by other hypotheses."

Readers who went to graduate school in the 1980s probably remember the backlash that made "competition" almost a dirty word with reviewers and editors. Even before 1975, some authors had questioned the importance of competition, beginning with Andrewartha and Birch (1954). This skepticism was fueled by Wiens's (1977) essay on variable environments and "crunches" and Connell's (1980) "ghost of competition past," followed by the landmark Grossman et al. (1982) paper that argued for dominance of stochastic control in stream fish communities. Additionally, the seminal paper by Paine (1969) and subsequent contributions showed the importance of keystone predators (Power et al. 1996), and Connell (1975) concluded that predation can trump competition under a variety of environmental conditions. And it has long been recognized by others that the role of competition can vary with time and space (Roughgarden and Diamond 1976) or that factors regulating communities can change with changes in environmental harshness or disturbance (Chase 2007; Lepori and Malmqvist 2009). All these works have refocused community ecology toward the importance of abiotic regulating factors, temporal variation, and the importance of predation and away from the preponderance of highly stable communities dominated by competitive interactions.

After Grossman et al. (1982) and in light of growing emphasis on abiotic factors (Peckarsky's (1983) harsh-benign hypothesis), there was movement toward recognizing "stochastic events" or "abiotic control" to be of primary importance in stream communities (Death 2010; Grossman and Sabo 2010). The review by Jackson et al. (2001) indicated that predation and abiotic factors are both highly important in the structure of many lake and stream fish communities. It is clear that under some circumstances, inter- or intraspecific competition can be important (Calsbeek and Cox 2010) or that competition and predation substantially interact (Gurevitch et al. 2000; Chase et al. 2002). Power and Matthews (1983) and Power et al. (1985) showed that predators could have strong top-down effects in trophic cascades. Most ecologists now view stream communities on a gradient, from ones controlled by harsh abiotic factors, disturbance, or stress to those dominated by biotic interactions. But the ways that harshness affects streams continue to be debated conceptually (Death 2010), and questions remain about empirical proof for the role of environmental harshness in animal communities (Barrio et al. 2013).

Competition

For reviews of interspecific competition among fishes, see Matthews (1998, 455–497) and Ross (2013, 227–246). Connell (1975) argued that evidence for competition in communities existed at three increasingly convincing levels, including (1) complementarity in distribution, microhabitats, or food habits, that is, "segregation"; (2) natural experiments, including resource use patterns at sites where a species does or does not occur in contact with a suspected competitor; and (3) evidence from manipulative experiments (best). With our data or some from our students we first consider complementary distributions of similar species at geographic scales and within watersheds; then examine patterns in comparative resource use or "resource partitioning"; and finally consider experimental evidence for competition in stream fish communities. Most of our research provides evidence in the first two categories. But at the end of this section we will look for a consensus about the evidence for potential importance of competition in the various communities that we, or our students, have examined. For many of the examples below it is not possible to determine conclusively that competition was actually happening, but we can ask whether a pattern is consistent with expectations from a competition-based explanation.

Complementarity or Species Repulsions at a Geographic Scale

The first (but weakest) clue that competition may be affecting community structure comes from complementary distributions of similar species or repulsion in distributions of related species. When two closely related species, within their ranges and in close proximity, have complementary occurrence or abundances, competition could be the cause. But any lack of overlap could also be caused by simple differences in habitat selection as a result of morphology, phylogeny, or other noncompetitive factors or environmental filtering (Troia and Gido 2014). In tests of distribution patterns for evidence of competition, repulsion among similar species suggests that competition is important, whereas clustering of similar species in space can suggest that convergence to similar habitats overrides any potential effect of competition.

Matt Winston's PhD dissertation with WJM provided evidence at a geographic scale for the potential importance of competition in minnow distribution. Winston (1995) sampled 219 sites in streams of all sizes (see, e.g., plate 11) throughout the Red River basin in Arkansas and Oklahoma and calculated an index of co-occurrence for all possible pairs that were in potential contact. Species pairs that were morphologically most similar (based on 56 measurements of head, body, and fins) occurred together less often than predicted by a random model. Winston (1995) tested his results against several competing explanatory models, including the potential influence of phylogenetic distance, to explain the significantly lower co-occurrence of morphologically similar species, and concluded that interspecific competition was the best explanation for the repulsed distributions. (Consistent with competition? Yes.)

As part of his PhD dissertation under WJM's direction, Chris Taylor surveyed benthic fish in riffles in 42 different stream reaches in the Ozark uplift in Oklahoma (Taylor 1996). He found negative correlations in occurrence for three pairs of common and widespread species, including Banded Sculpin (*Cottus carolinae*) versus Orangethroat Darter (*Etheostoma spectabile*), Banded Sculpin versus Slender Madtom (*Noturus exilis*), and Orangethroat Darter versus Fantail Darter (*Etheostoma flabellare*). Darters, madtoms, and sculpins are distantly related phylogenetically, but they share many common ecological features, often using the same kinds of riffle habitat. We return later in this chapter to the experiment stimulated by his survey, but for now

Table 5.1. Abundance of small fishes across six Ozark watersheds

Species	Big Creek	Cane Creek	Jane's Creek	Fourche River	Strawberry River	Piney Creek
Bigeye Shiner	abundant	abundant	abundant	abundant	rare	occasional
Carmine Shiner	absent	absent	rare	absent	common	common
Brook Silverside	abundant	common	common	common	rare	rare

Source: Matthews (1982).

let us state that Taylor (1996) provides another example of complementary distributions of ecologically similar species at a geographic scale. (Consistent with competition? Yes.)

Matthews (1982) examined abundances of 13 common minnows or small "minnow-like" water-column insectivores across 8 Ozark streams, to ask whether there was more complementarity among species than predicted by chance (table 5.1). Using several master's theses that were completed at Arkansas State University, Matthews (1982) scored abundances of species from "absent" to "abundant" on the basis of designations by the original authors and examination of the raw data. Nine species pairs were detected across the six smallest streams that were never mutually common or abundant. Two were particularly interesting because all of those species were widespread in the region, including Carmine Shiner (*Notropis percobromus*; as *Notropis rubellus* in Matthews 1982) versus Brook Silverside (*Labidesthes sicculus*) and Carmine Shiner versus Bigeye Shiner (*Notropis boops*). The mutual exclusion in abundances of species in these two pairs could not be explained by any known biological traits of the species. Mutual exclusion of Bigeye and Carmine shiners across these Ozark streams is particularly interesting, as Bigeye Shiner and Rocky Shiner (*Notropis suttkusi*) are also complementary in abundance in the Kiamichi River (see below). The Rocky Shiner and Carmine Shiner are closely related in the *Notropis rubellus* species group, only recognized as distinct from Rosyface Shiner (*Notropis rubellus*) in recent decades (Humphries and Cashner 1994; Wood et al. 2002; Berendzen et al. 2008). But a random model (created by dealing playing cards into piles, in the days before any computers were available locally) indicated that chance alone could account for the numbers of mutually exclusive species pairs observed across these watersheds. Matthews (1982) thus concluded, "Overall, mutual exclusion among species pairs seems unlikely to have a significant influence on the assembly of 'minnow' species . . . in these small Ozark streams." And a subsequent reanalysis by Biehl and Matthews (1984) to correct an error in the original paper retained this conclusion. In these analyses two mutually nonabundant species pairs were interesting from a competition perspective but could hypothetically be accounted for by random distributions of species among watersheds. (Consistent with competition? No.)

Complementarity or Species Repulsions within Watersheds

Two of the three examples above of geographic complementarity of ecologically similar species suggested that competition could be important, but in one the repulsion within species pairs was no greater than could be explained by random distributions. Complementary distribution or abundances of similar species within a watershed may provide a more focused view than we can get from geographically broader evaluation of distributions. If two species occur in the same watershed, there is no doubt (barring the existence of barriers like large waterfalls) that the species could occur together. The simplest within-watershed pattern is the distribution of one species downstream and a similar one upstream (Taylor and Lienesch 1996; Duvernell and Schaefer 2014), for any of many possible reasons (Matthews 1998, 296–312; Troia and Gido 2014). A more compelling suggestion that competition could be important is for similar species to "flip-flop" in abundance at sites in a watershed in a pattern that is not explainable by longitudinal factors alone.

Four abundant, insectivorous minnow species have complementary distributions among our Kiamichi River sites (see plate 3). Steelcolor Shiner (*Cyprinella whipplei*) and Rocky Shiner are more abundant in downstream sites, whereas Bigeye Shiner and Redfin Shiner (*Lythrurus umbratilis*) are scarce downstream but highly abundant upstream (fig 5.1). Similarly, in Piney Creek (see plate 1) there is complementarity in abundances across sites for the two most common riffle-dwelling darter species. The Rainbow Darter (*Etheostoma caeruleum*) is most abundant in larger downstream riffles or rapids, and Orangethroat Darter dominates upstream in smaller headwater riffles (fig. 5.1).

But the complementary patterns for these Kiamichi River and Piney Creek species pairs are consistent with a more general "longitudinal zonation" and do not by themselves support competition to the exclusion of other possible factors. Dewey (1988) found that Orangethroat and Rainbow Darters used highly similar foods in Illinois streams, so it is likely that these two species might compete. And Matthews (1998, 465) noted from personal observations that where the two did co-occur in substantial numbers (P-6), they segregated within riffles, with Orangethroat Darters near shore over smaller substrates and Rainbow Darters toward midriffle with larger cobbles (Matthews 1998, 465). But we have no hard data to support those observations. (Consistent with competition? Possibly, but inconclusive.)

More complicated distributions in Piney Creek cannot be explained by simple upstream-downstream differences. Knobfin Sculpin (*Cottus immaculatus*) occurred mostly at downstream sites in large riffles or rapids (see plates 1 and 19), and Banded Sculpin mostly occurred upstream, where riffles are smaller and less turbulent. The sculpin relative abundances (fig. 5.2) were reversed at sites P-4, P-5, and P-6. Likewise, three minnow species pairs with complementarity abundance in Piney Creek cannot be explained by simple longitudinal differences. These pairs (only one shown in fig. 5.2) each included the highly abundant Duskystripe Shiner (*Luxilus pilsbryi*) versus Telescope Shiner (*Notropis telescopus*), Bigeye Shiner, and Whitetail Shiner (*Cyprinella ga-*

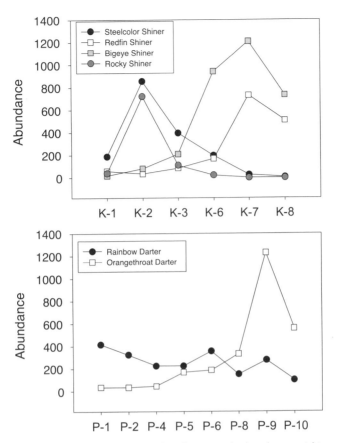

FIG. 5.1 Two examples of complementary distributions of related taxa within a watershed. *Top*, Minnows in Kiamichi River. *Bottom*, Darters in Piney Creek. These distribution patterns are also consistent with longitudinal zonation in species distributions within a stream and therefore do not provide convincing evidence for competition driving the observed pattern.

lactura). Duskystripe Shiners were more abundant than the other three species at the two most downstream sites, but relative abundances reversed at midreach sites and again in headwater sites P-9 and P-10, where Duskystripe Shiners were more abundant (fig. 5.2). The two sculpin species use similar foods, dominated by crayfish and aquatic insects (Cooper 1975 cited in Robison and Buchanan, 1988; Phillips and Kilambi 1996). The three minnow species also eat similar aquatic and terrestrial insects (Trautman 1957; Outten 1958; Matthews et al. 1978; Etnier and Starnes 1993), so the potential for them to compete seems to exist. The patterns for minnows and sculpins in Piney Creek suggest a classic checkerboard pattern (Diamond 1975) consistent with patterns expected if one species excludes or suppresses another at local sites. (Consistent with competition? Yes.)

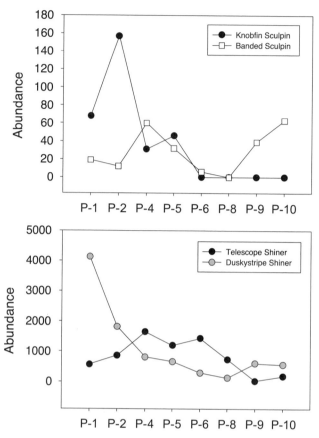

FIG. 5.2 Two examples of complementary distributions of related taxa within the Piney Creek watershed. *Top*, Sculpins. *Bottom*, Minnows. These distribution patterns are not consistent with longitudinal zonation in species distributions within a stream and therefore provide evidence for competition as a possible mechanism driving the observed pattern.

Finally, consider the similarities and differences in distribution of morphologically similar minnows within a watershed. For the 15 most common minnow (Cyprinidae) species in Piney Creek we measured 23 morphological characteristics of body, head, and fins, using traditional ichthyology measurements (Hubbs and Lagler 1949; Douglas and Matthews 1992) on one adult specimen of each species, judged by eye to be typical. Residuals from regression of measured traits on the standard length of each specimen were used to remove effects of size, and then product moment correlation was used to produce a triangular morphological similarity matrix for all possible pairs of the 15 species. Separately, a triangular similarity matrix was constructed for all possible pairs from product moment correlation of their raw abundances across all 144 of our space-time samples (12 sites × 12 surveys) in Piney Creek from headwaters to lower mainstem from 1972 to 2012.

In the abundance matrix based on all 144 samples, product moment correlation values were positive and >0.20 (as an arbitrary index, not as a measure of the statistical significance of any particular pairs) for 23 species pairs (21.9% positive out of 105 possible pairs), whereas only one pair had a negative $r < -0.20$. More positive than negative associations were indicated among these common Piney Creek minnows. The abundance and morphology matrices were congruent at $p = 0.06$ (Mantel $r = 0.168$), suggesting that, considered across the entire watershed, the morphologically more similar minnow species tended to occur together more than by chance, not suggesting any overall competition-based repulsion of similar species. As suggested by Keith Gido, we also tested congruence of correlations of minnow distribution on the basis of rank abundance versus morphology, and those matrices were congruent at $p = 0.07$ (Mantel $r = 0.164$), suggesting that either raw or ranked abundance gave similar results, that is, that morphologically similar species tended to occur together, not segregated from each other. These results for minnows in the watershed as a whole contrasted with the finding by Winston (1995) that the morphologically most similar minnow species had significantly repulsed distributions at a larger geographic scale. (Minnow morphology and distribution in Piney Creek consistent with competition? No.)

As also suggested by Keith Gido, we examined separately the abundance versus morphological relationships for these species for middle and lower mainstem sites (P-1, P-2, P-4, P-5, P-6, and M-1; see fig. 2.1) and the headwater sites (P-7, P-8, P-9, P-10, M-2, and M-3). For the mainstem sites, there were again exactly 23 species pairs (21.9% of possible pairs) with positive product moment correlation values >0.20, and only one pair with negative correlation value −0.20. The patterns in distributional associations that we found for the entire watershed held true if only the larger, mainstem sites, where minnows were generally more abundant, were considered. For the headwater sites, two species (Ozark Shiner, *Notropis ozarcanus*, and Wedgespot Shiner, *Notropis greenei*) were omitted because they did not occur, and Carmine Shiner was omitted because only five individuals were found among all the headwater sites, leaving 20 minnow species, with 66 possible pairs, in the analysis. For those 66 pairs, 15 (22.7%) had positive product moment correlation equal or greater than 0.20, but 5 (7.6%) had negative correlation values equal to or less than −0.20. From this we would conclude that while associations among minnows were strongly biased toward positive in the middle to lower mainstem sites, there were both positive and negative associations among species in the smaller headwater sites. Perhaps, as Keith Gido suggested, if competition plays any role in the distributions of minnow species in Piney Creek, it would be more intense or potentially important in headwaters, where pools tend to be smaller overall or to shrink relatively more during low-water times, compared to the wider and deeper pools downstream. This is completely speculative and in need of more detailed evaluation of food or microhabitat use by minnows in different parts of this or other systems, but it is consistent with Diamond's (1975, 423) suggestion that combinations of species that were stable on larger islands (our mainstem sites?) could be unstable on smaller islands (our headwater sites?).

Comparative Resource Use or "Resource Partitioning"

Stronger evidence about competition as a factor in fish community dynamics could be found by comparing microhabitats or foods of similar, co-occurring fish species (Matthews 1998, 488–489). Segregation of potential competitors across microhabitat gradients or by food habits suggests active "resource partitioning" (Ross 1986) that could ameliorate competition. Schoener (1974) posited the importance of niche separation along three axes, including food, habitat, and time. From a broad review of fish assemblages worldwide, Ross (1986) concluded that the strongest niche separation was typically for food use, followed by habitat, and lastly by temporal separation of activities. Matthews (1998) reviewed a large number of papers on habitat or food segregation in freshwater fish, finding at least some indication of resource segregation in most, and about half of the original authors inferred that competition was involved.

The three most common darter species in the Roanoke River mainstem (see plate 4) occurred together in similar abundance patterns (fig. 5.3) and could often be collected in the same seine haul or kickset (Matthews et al. 1982). All three species overlapped strongly in food use (>90% for all pairs), microhabitats (overlapped 78%–90%), and time of feeding, differing only modestly across current speeds and size of food items. Differences across current speeds could be explained by behavioral responses of the darters in flow tube experiments (Matthews 1985), and interspecific differences in mouth size coincided with sizes of foods they used (Matthews et al. 1982). But overall these darters were similar ecologically, providing no compelling evidence that resource partitioning facilitated their coexistence. Surber samples at mainstem locations in the Roanoke River where all three darters were often collected showed that benthic invertebrates exceeded 2100/m^2, suggesting that they coexisted where food was abundant (Matthews et al. 1982). Whitney et al. (2014) provided another example, in which higher standing crops of algae and macroinvertebrates ("basal food") facilitated production of native fishes in the Gila River, New Mexico. (Roanoke darters consistent with competition? No.)

Similarly, Surat et al. (1982) examined microhabitat and food use of the three most common water-column minnows in the Roanoke River, which also broadly co-occurred (fig. 5.3) and which, like the darters, were often found together in a single seine haul. These three minnows all overlapped strongly in food use, microhabitats, and activity (feeding) times (Surat et al. 1982). The only evident resource segregation was that the more streamlined Rosefin Shiner (*Lythrurus ardens*) tended to be higher in the water column than the slab-sided and more benthic Crescent Shiner (*Luxilus cerasinus*), with White Shiner (*Luxilus albeolus*) intermediate between the two. For both minnows and darters in the Roanoke River, the three most abundant species in two families showed more resource use overlap than was typical in many other studies of stream fishes (Matthews 1998), yet they coexisted in large numbers within the same stream reaches and even the same pools. Perhaps for both minnows and darters, their invertebrate food items in the Roanoke River, with its diverse, stony-bottomed substrates, are sufficient to ameliorate competitive effects among the species. (Consistent with competition? No.)

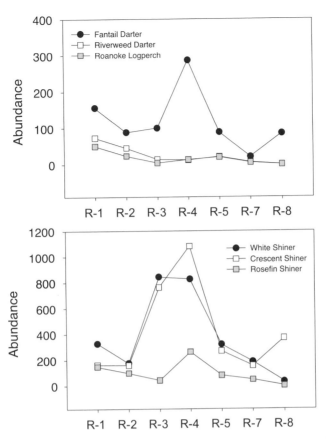

FIG. 5.3 Similarity of longitudinal distributions of three darter species (*top*) and three minnow species (*bottom*) in the Roanoke River.

In Blaylock Creek, Ouachita National Forest, Arkansas, there was a complicated situation that differed for two separate minnow pairs. In the first case, Redfin Shiners and Bigeye Shiners had almost identical distributions (fig. 5.4). Stomachs of 226 Redfin Shiners and 95 Bigeye Shiners (from several creeks in the area, not just Blaylock Creek) showed that these two species used similar foods ($r = 0.823$). Therefore, no known separation in resource use allowed coexistence of these species. Comparative use of microhabitats within sites were they co-occur has not been studied in detail, but both are pool-dwelling, water-column species. We have no quantitative measure of food availability in Blaylock Creek, but dip-netting by Marie Miller and Jeff Stewart (Matthews et al. 2004) provided a reference collection for fish stomach analyses and showed that benthic invertebrates were generally abundant in Blaylock Creek, suggesting that foods might not be limiting. Metcalf et al. (1997), however, cited unpublished data to suggest that invertebrate densities in Blaylock Creek were relatively low. Perhaps these species do compete for foods, but if so, that competition is not

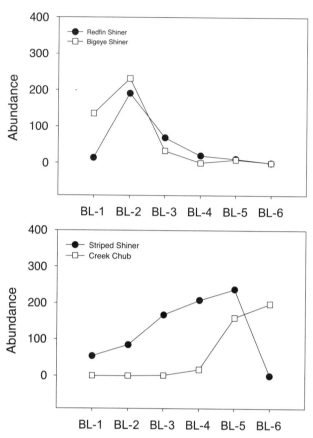

FIG. 5.4 Distributions for two pairs of minnows with similar food habits in Blaylock Creek.

sufficient to cause them to have complementary distributions. (Consistent with competition being important? No.)

Of all minnows examined in Blaylock Creek, Creek Chub (*Semotilus atromaculatus*) and Striped Shiner (*Luxilus chrysocephalus*) had the greatest similarity in food use ($r = 0.854$) (Matthews et al. 2004), and they were partly separated across sites in Blaylock Creek (fig. 5.4). Striped Shiners were moderately to highly abundant except at the most upstream site, but Creek Chubs were abundant only at the two most upstream sites (BL-5 and BL-6). Competition for similar foods could hypothetically result in the negative association for this species pair at the upstream sites, but it is equally possible that habitat availability drives their switching abundances between BL-5 and BL-6. At BL-5, a wide, waist-deep pool often contained both species, particularly with large adult Striped Shiners. But the most upstream site (BL-6) lacked large, deep pools, and we took most Creek Chubs in numerous smaller, knee-deep "step-pools" throughout the reach (Matthews, pers. obs.). The smaller pools at BL-6 were unlikely to be good habitat for the relatively large-bodied Striped Shiner. What's going on here? Competitive exclusion of Striped Shiner by Creek Chubs via use of similar food items, or a simple dichotomy in the best habitat for these two species? Again, we have an example where competition could be a factor in their distributions, but the equally plausible habitat-selection hypothesis cannot be discounted. (Consistent with competition? Unclear; maybe no.)

One of the clearest cases of ecological segregation by two closely related fishes was in Bruce Moring's MS thesis at Angelo State University, directed by EMM (Moring 1986), on comparative microhabitat and food use by Orangethroat Darters and Greenthroat Darters (*Etheostoma lepidum*). The two are not sister species, but both are in darter subgenus *Oligocephalus* (Lang and Mayden 2007) and are similar in body shape and general appearance. In the Edwards Plateau of central Texas, the endemic Greenthroat Darters are most abundant in springheads or runs, often associated with vegetation, while the more cosmopolitan Orangethroat Darters are distributed more downstream, in swifter currents and over clean gravel. Zones of sympatry are common, however (Moring 1986). At one syntopic site, seasonal samples by Moring showed consistent spatial separation of the two species, with 75% of the Greenthroat Darters in vegetated edge habitats with little current and 69% of the Orangethroat Darters in more swiftly flowing riffles lacking vegetation. The two species differed in food use, with calculated overlap values from only 0.08 to 0.35, consistent with availability of food items in their two differing microhabitats. In general, Orangethroat Darters ate more chironomids, and Greenthroat Darters fed almost exclusively on amphipods (which were more available in the marginal, vegetated habitats). In marked contrast to the findings of Matthews et al. (1982) for darters in Roanoke River, Moring (1986) provides an example of two coexisting darter species strongly separated in microhabitat and food use. This spatial separation might minimize any potential effects of competition between these two species, so long as stream flow from springs continues to provide adequate availability of the different kinds of microhabitats. Also, Moring's study was in a first-order headwater site, whereas the darters studied by Matthews et al. were in large, higher-order, mainstem riffles, suggesting that competition might be more intense in smaller habitats. The question remains whether their differences might be related to competitive interactions during the evolution of the two taxa in central Texas, with a shift of the more localized Greenthroat Darter to the marginal spring habitats, or whether their ecological differences are related to deeper phylogenetic differences between the two. (Consistent with competition? Possibly, in the past.)

Experiments

The gold standard to test for competition is experimental, either in controlled experiments in the field (Connell 1975; Ross 1986; Brown et al. 2001), laboratory (Baltz et al. 1982; Fausch 1988), or mesocosms (Matthews et al. 2006). In an experiment that was unrelated to any of the cases described above but also based on extensive field observations, Matthews et al. (1992) used small pool mesocosms to determine whether juveniles of native Largemouth Bass (*Micropterus salmoides*) and of Striped Bass (*Morone saxatilis*) introduced in a reservoir (Lake Texoma) would shift food use in the presence of the other. In 1200 L circular tanks, food use by Striped Bass juveniles was unaffected by presence of Largemouth Bass, but juvenile Largemouth Bass on average ate fewer copepods (the primary food item) when with Striped Bass. But in spite of a substantial trend (mean of 84.8 copepods per Largemouth Bass when alone vs. 57.8 when with Striped Bass), differences were not statistically significant because of low sample size ($N = 3$ for each treatment). By way of critical analysis we have to conclude that the main thing learned from Matthews et al. (1992) was that a sample size of $N = 3$, no matter how much work it may take to set up a trial, is a bad idea!

One of the most convincing experimental demonstrations of competition between darters was by Larry Greenberg, in the Little River, Tennessee. Larry briefly visited in WJM's lab for research after completing his PhD, but his ground-breaking work with darters (Greenberg 1988) was part of his earlier dissertation. Greenberg's dissertation included a straightforward but elegant field manipulation to evaluate interactions between Snubnose Darter (*Etheostoma simoterum*) and Redline Darter (*Etheostoma rufilineatum*). Greenberg first mapped densities of the two species in strips counts, finding that Snubnose Darters mostly occupied near-shore habitat in contrast to Redline Darters, which used habitat farther from shore. He then removed Redline Darters, after which Snubnose Darters became much more abundant near midstream, away from their previously preferred microhabitat. Then he replaced the Redline Darters, and the Snubnose Darters moved back into their original near-shore habitat! To this day, Greenberg (1988) remains one of the most convincing field manipulations suggesting that interactions between species can strongly affect local community structure. In a second trial in a different year and under much colder conditions, however, the response was more muted, which also underscored the influence of environmental conditions on the outcome of such interactions, that is, that interactions among fish species may be highly context dependent. (Consistent with competition? Yes, but context dependent.)

Another experiment, to test a negative association detected in a field survey, was done by Chris Taylor as part of his PhD dissertation. Taylor had surveyed benthic riffle fishes, as described previously, across 42 different stream reaches in the Ozark uplift of northeastern Oklahoma (Taylor 1996). He subsequently set up a large experiment with in-stream cages perpendicular to the direction of flow and varying sharply in current speeds across sections of the cages, to test interactions between the two species that had the strongest negative association in his surveys (Orangethroat Darter and Banded Sculpin). The experiment was fully replicated and carried out in a series of trials over a period of five weeks. When Orangethroat Darters were alone, they used the fastest and deepest microhabitat. But when with Banded Sculpins, the Orangethroat Darters selected shallow, nonflowing habitat (Taylor 1996). Although large adult sculpin can eat small fish, no young-of-year were used, and Taylor's observations indicated that no predation took place during the trials, as fish were accounted for. From this simple but elegant experiment, the most logical conclusion was that asymmetric competition (whether exploitative or interference we don't know) existed between the two species, with the sculpins negatively affecting habitat use by the darters. (Consistent with competition? Yes.)

Predation

The importance of predation in community structure is well established (Sih et al. 1985; Black and Hairston 1988). Theoretical models (Morin 2011, 120–133) suggest many ways in which predators can have strong effects in communities. Theoretical or conceptual models (e.g., Glasser 1979; Hofbauer and Sigmund 1989) and both classical (Paine 1966, 1969) and recent (Britten et al. 2014) empirical research suggest that predators can stabilize communities. For example, predators can exert "compensatory mortality" in a community by consuming a prey species that might otherwise become a community dominant at some cost to other species (Glasser 1979). Holling's (1959) models suggest that predators react to prey density in a variety of ways, from linear to asymptotic functional responses. By "switching" (Murdoch, 1969) between prey types, predators might damp population fluctuations in both prey species and stabilize communities.

Jackson et al. (2001) concluded that predation is one of the most important factors regulating fish community structure. In this chapter we address our own work related to both direct consumption and the effects of predator threat. Direct consumption (Preisser et al. 2005), removing smaller fish, is clearly important, but predator threat can result in costly behavior (Lima 1998) and changes in community dynamics.

None of our own research directly tests the elegant models of Holling, Murdoch, or other theoretical works. But a survey of predator and prey densities across our empirical databases is instructive. In high-predation environments, bursts of reproductive output by boom-and-bust species like Western Mosquitofish (Matthews and Marsh-Matthews 2011) or riverine minnow species (Matthews and Hill 1980) might be consumed by medium-sized or larger piscivores, lessening the swings in population that in turn affect community dynamics. Such a result would be consistent with Glasser's (1979) conceptual models, which predicted that by consuming disproportionally abundant prey, predators will cause species abundances to be more even within a community.

Predator:Prey Ratios

One index of the potential for predator pressure to regulate dynamics of a community is simply the number or density of piscivores relative to that of their potential prey, in ratio-dependent predator-prey models (Arditi and Ginzburg 1989). These models have been controversial (Abrams 1994; Abrams and Ginzburg 2000), but the issue now seems resolved in their favor (Sen et al. 2012). We will compare the predator:prey ratios for our local and global fish communities to ask whether there is greater predator pressure in certain types of stream environments.

Across our 9 global streams and 31 local sites, we used raw abundances of prey (fish species that are relatively small as adults) and species that are at least partly piscivorous as adults to produce predator:prey ratios. We make the assumption that the abundance of juvenile predatory species like some sunfish (*Lepomis* spp.) and all black bass (*Micropterus* spp.) suggests potential predator pressure because some grow to adult size and become piscivores. We have found in experiments that sunfish stocked in the spring as small young-of-year can by autumn grow to a size large enough to eat minnows and darters (Marsh-Matthews et al. 2011). To calculate predator:prey ratios, we included as piscivores all gars (*Lepisosteus* spp.), large-bodied catfish (*Ictalurus* and *Pylodictis*), black basses, crappies (*Pomoxis* spp.), Fliers (*Centrarchus macropterus*), and three sunfish including Green Sunfish (*Lepomis cyanellus*), Warmouth (*Lepomis gulosus*), and Longear Sunfish (*Lepomis megalotis*). Longear Sunfish were considered to be "insectivores" by Karr et al. (1986) but have been directly observed to eat adult minnows in our experiments (Marsh-Matthews et al. 2013). As prey, we included all native minnows, madtom catfishes (*Noturus* sp.), topminnows (*Fundulus* spp.), Western Mosquitofish (*Gambusia affinis*), Pirate Perch (*Aphredoderus sayanus*), Pygmy Sunfish (*Elassoma zonatum*), sculpins (*Cottus* spp.), and all darters (*Etheostoma* and *Percina*). Darters, madtom catfish, and sculpins typically live in riffles where deep-bodied centrarchids rarely hunt, but they can be forced into pools with piscivores during low-water conditions.

From predator:prey ratios for our nine global streams (table 5.2), we can form general conclusions about how we can expect predation to be important. For example, Piney Creek has relatively few piscivores and large numbers of minnows and a predator:prey ratio of only 0.033. In contrast, Brier Creek is awash in potential predators, with numerous Green Sunfish, Longear Sunfish, and Largemouth Bass and a predator:prey ratio of 0.361, more than 10 times the relative density of predators in Brier Creek than in Piney Creek. In the 9 global streams, predator:prey ratios ranged from a high of 0.361 in Brier Creek to a low of 0.004 in the Roanoke River.

For these nine streams with multiple sample sites, there was a suggestion of higher predator:prey ratios in streams that were intermittent and in lower-gradient systems (fig. 5.5). But only two points—Brier and Crutcho Creeks—drove the trend for higher predator:prey ratios in streams with lower gradients. During much of the year, Brier and Crutcho Creeks are relatively slow-flowing streams, dominated by pool or channel habitat, with some riffles but a preponderance of slow habitats, with suitable substrates for nesting, favoring centrarchids. Six of the other systems are in the uplands of the Ozark or Ouachita Mountains of Arkansas and Oklahoma, and the seventh is the Roanoke River, in the Blue Ridge–Appalachian Mountains of western Virginia. The upland streams were characterized more by large numbers of minnows or other small-bodied fishes and lower abundances of piscivores. The difference may be that

Table 5.2. Predator:prey ratios for 9 global and 31 local stream fish communities

Community Type	Stream	Predator:Prey Ratio
Global	Brier Creek	0.3610
	Crutcho Creek	0.3404
	Bread Creek	0.0837
	Kiamichi River	0.0595
	Alum Creek	0.0389
	Piney Creek	0.0333
	Blaylock Creek	0.0076
	Crooked Creek	0.0066
	Roanoke River	0.0043
Local	Lukfata slough	3.3659
	Choctaw2	0.6170
	Borrows, US 70	0.5746
	Little Glasses Creek	0.5440
	Hauani Creek	0.2255
	Brier5	0.1783
	Bread4	0.1538
	Garrett Creek	0.1523
	Glover River	0.1224
	Gar1	0.1172
	Morris Creek	0.1033
	Hickory Creek	0.0674
	Cow Creek	0.0623
	Coal Creek	0.0529
	Ballard Creek	0.0494
	Kiamichi8	0.0429
	Mustang1	0.0412
	Crutcho1	0.0412
	Little Elm Creek	0.0394
	Illinois River	0.0380
	Blue River	0.0348
	Roanoke2	0.0228
	Chigley Sandy Creek	0.0205
	Piney M2	0.0198
	Baron Fork	0.0155
	South Fork Alum4	0.0145
	Salt Fork, Red River	0.0113
	Tyner Creek	0.0111
	Blaylock3	0.0024
	South Canadian River	0.0002
	Crooked5	0.0000

Note: For each community type, sites are arranged in descending order of predator:prey ratio such that communities with relatively more predators are at the top of the list.

most centrarchids have a deep-bodied morphology suited for maneuvering in slow-flowing water or around complex structure like columns of attached algae (which they use as shelter; WJM, pers. obs., snorkeling), whereas minnows are more streamlined and better able to swim continuously or hold position in substantial currents.

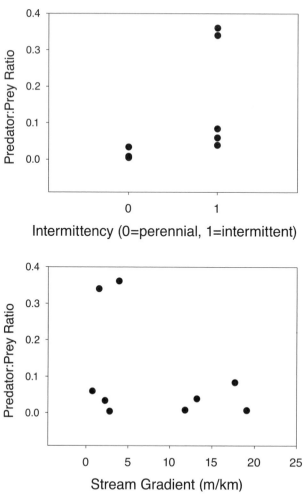

FIG. 5.5 Relationship of predator:prey ratio to intermittency (*top*) and stream gradient (*bottom*).

In the 31 local sites, we also found a wide range of predator:prey ratios (table 5.2), ranging from 3.366 in Lukfata Slough to 0 in Crooked Creek site 5. Lukfata Slough is a backwater in old borrow pits that are filled from Lukfata Creek when water is high, but it is usually cut off and nonflowing, with dense aquatic vegetation. In Lukfata Slough there were substantial numbers of Grass Pickerel (*Esox americanus*), Green Sunfish, and Largemouth Bass (many as juveniles but potential predators nonetheless) but no minnows. The small-bodied species in this slough were mostly Pygmy Sunfish, Blackstripe topminnow (*Fundulus notatus*), Western Mosquitofish, and Pirate Perch, with a few darters. This was the only one of our local sites where predators outnumbered prey.

In marked contrast to Lukfata Slough, Crooked Creek site 5 is on a steep hillside in the Ouachita Mountains just below a waterfall approximately 5 m high (see plate 5) and includes a plunge pool with substantial numbers of Creek Chub and a swift boulder-cobble rapids with large numbers of Orangebelly Darter (*Etheostoma radiosum*). We never found any sunfish, black bass, or other piscivores there (although a few of the Creek Chub, which we classified as prey, could have been large enough to eat smaller minnows). Those two extremes stand out substantially from the other 29 local sites in predator:prey ratios and thus are excluded from the graphs or treatments below.

At the other 29 sites there was a relatively smooth gradient from sites with predators more than half as abundant numerically as their potential prey (Choctaw2, Borrows along US 70, Little Glasses Creek) to sites such as Blaylock3 and the South Canadian River, where the predator:prey ratios were below 0.01. It is instructive to divide the local communities in table 5.2 into three arbitrary groups. Group 1 sites had predator:prey ratios >0.10; Group 2 had ratios between 0.02 and 0.07; and Group 3 had ratios below 0.02. Of the 11 sites in Group 1, with the highest predator:prey ratios, at least 8 are in relatively low-gradient terrain, and all but one (Glover River) have deep pools with low current speeds that favor centrarchids, large catfish, or gars.

The 12 sites in predator-prey Group 2 represent a mix of environmental conditions from low-gradient, slow-moving sites like Hickory, Cow, and Coal Creeks, to strongly flowing sites like the Illinois River below Lake Frances, Blue River, Kiamichi8, and the Roanoke River at Lafayette, Virginia. Six of the eight sites in predator-prey Group 3, with lowest ratios, were in upland streams with steep local gradients or strong flow, not favoring centrarchids. Two other sites in this latter group were quite different but had low numbers of predators for various apparent reasons. These sites on the Salt Fork of the Red River (predator:prey ratio = 0.0113) and on the South Canadian River (predator:prey ratio = 0.0002) are both on wide, shallow, sand-bed rivers, with harsh environmental variation (Hefley 1937; Matthews and Hill 1980) and extremely low (Matthews et al. 2005) to no flow during drought (see plates 11 and 12). In these habitats, temperature extremes or other physicochemical stressors make survival challenging for all fishes (Matthews and Zimmerman 1990). In addition, both are dominated by shifting sand substrates, without stable substrate where centrarchids could build nests for spawning.

Differing Predator Pressure over Time

In many stream systems there are cyclic and rather predictable seasonal changes in potential predator pressure. Young-of-year sunfish, for example, too small to be piscivorous, grow by autumn to be effective predators on minnows, darters, or other small fishes (Marsh-Matthews et al. 2011). There may be much more potential predation in late summer or autumn than in spring or early summer, when the year's new crop of sunfish are mostly too small to be piscivorous.

The presence of predators in a system may be seasonal. In Morris Creek (Poteau River drainage), for example, gars (Lepisosteidae) comprised only 0.03% of all individuals across all of our collections. But in one collection in Morris Creek in April 2002, our class captured (and released) seven large adult Longnose Gar (*Lepisosteus osseus*) in one pool that was about 1.5 m deep, and numerous others escaped attempts to net them. We had stumbled upon a breeding site with many adult gars in this one pool, above a swift riffle where we subsequently found large numbers of gar eggs attached to stones. But we found gars nowhere else in the reach of creek we sampled. Once our class disturbed the pool, gar began leaping about vigorously, and it looked much like a rodeo for the next 10 minutes as eager groups of students tried to outdo each other in attempts to catch these big fish! These gars were not likely residents of this pool and probably had migrated upstream to that spawning site, resulting in a concentration of gars atypical of small streams at most times. In April 2012, in a relatively large, shallow gravel riffle a bit downstream from our usual sampling site on Blue River, we also observed a spawning aggregation of about five large adult Longnose Gar, whereas we had not recorded gar in any of our regular Blue River seining samples. We do not know if gars feed while in spawning aggregations, but such a concentration of large piscivores could have the potential for more impact, at least transiently, on a local fish community than would be suspected from mere examination of raw numbers of individuals in our datasets.

Changes in temperature, water levels, or other environmental factors can influence the importance of predation. Temperature can have a strong effect on the level of feeding by piscivores or on their nonconsumptive effects on their prey. Harvey et al. (1988) showed that in summer Largemouth Bass had strong effects on the distribution of Central Stonerollers (*Campostoma anomalum*) in a manipulative experiment in Brier Creek pools. In trials in October and November, however, the bass became inactive as temperatures cooled. In October (water = 19°C) the bass remained active, and the stonerollers responded as they had in August, actively avoiding the bass. But by the last trial, in November with water at 11°C, Largemouth Bass were inactive and stonerollers showed little response to them.

Predation intensity greatly increases during dry periods (chap. 6) if streams stop flowing and piscivores and prey are crowded together in isolated pools. In Brier Creek, WJM followed changes in the fish in three remaining headwater pools during a severe dry period (Matthews and Marsh-Matthews 2007). In each of these pools during the first survey there were 32 to 40 Red Shiners and 152 to 382 centrarchids, many as adults. By a week later all but two Red Shiners were gone, but concentrations of centrarchids remained high (Matthews and Marsh-Matthews 2007). In another severe drought in 2006 we, with Nate Franssen, surveyed fish in isolated pools of Brier, Buncombe, and House Creeks in south Oklahoma. In all we found large numbers of piscivores, including adult gars, catfish, black bass, and sunfish, along with many juvenile sunfish, but few minnows. It was our impression that such piscivore concentrations could be directly responsible for loss of some minnows in creeks in south Oklahoma (Matthews and Marsh-Matthews 2007; and as described below).

Species-Specific Predator Potential

Piscivore identity, even within a genus, can be extremely important in their effect on prey species. On the basis of observations of minnow behavior in Brier Creek and in the environmentally contrasting Baron Fork of northeast Oklahoma, Bret Harvey, Bob Cashner, and WJM speculated that Largemouth Bass might have much stronger effects than Smallmouth Bass (*Micropterus dolomieu*) on behavior or distribution of Central Stonerollers. All observations in Brier Creek showed that Largemouth Bass strongly affected stoneroller distribution within or among pools (Power et al. 1985; Harvey 1991), but in Baron Fork we often observed Smallmouth Bass swimming near or through large shoals of stonerollers, in which individuals moved in a "fountain" fashion to open a pathway for the bass but apparently remained largely undisturbed by the presence of this potential predator in the same pool.

A comparison of Smallmouth Bass from the Baron Fork and Largemouth Bass from nearby habitats in south Oklahoma was used to examine their effects on the distribution of stonerollers in pools in Brier Creek (Harvey et al. 1988). To determine these effects, Harvey et al. transported Smallmouth Bass from the Baron Fork and placed them in a pool in Brier Creek for observations. (They carefully blocked the pool, which was already partly isolated by a small waterfall, with nets and recaptured all Smallmouth Bass at the conclusion of the observations, so no nonnative bass were introduced into Brier Creek.) Relative to piscivore identity, Largemouth Bass strongly affected stoneroller distribution in pools, as they actively avoided those bass by moving into shallows at pool edges, but the stonerollers responded much less to Smallmouth Bass.

Prey as Predators

Red Shiners, usually thought of as prey for larger sunfish or bass, can also be predators on small life stages of other fishes (Ruppert et al. 1993). They were shown by Gido et al. (1999) to be effective predators on small Red River Pupfish (*Cyprinodon rubrofluviatilis*). In large, outdoor experimental streams, the pupfish produced substantial numbers of larvae (as evidenced by light trap captures), but in units also containing Red Shiners, essentially none of the pupfish grew to recruit and presumably had been eaten by the Red Shiners. Following up on Gido et al. (1999), EMM's master's student Jacob Thompson (Thompson 2012) did an experiment in our outdoor mesocosms to determine whether Red River Pupfish could recruit more young in units with complex structure (artificial vegetation, wood debris, stones, and a cinderblock) to provide their juveniles with shelter from predation by Red Shiners. But in spite of the habitat complexity, Red Shiners depressed survival of juvenile pupfish in all treatments (Thompson 2012).

Effects of Mesopredators

"Apex predators" like black bass are well known to have strong effects in fish communities of small streams, influencing use of individual pools by their prey (Power and Matthews 1983; Power et al. 1985; Matthews et al. 1994) or controlling complicated, multitrophic-level cascades (Power and Matthews 1983; Harvey 1991). But there has been growing theoretical and empirical interest in the effects of "mesopredators" (rather than the larger-bodied apex predators) or combinations of various sizes of predators (Prugh et al. 2009). A fortuitous combination of historical data for streams in south Oklahoma, the serendipitous outcome of one experiment, monitoring of the fish communities, and several subsequent experiments enabled us to draw strong inference about the roles of mesopredators in declines of minnows we have observed.

Our long-term surveys and historical collections by Carl Riggs, George A. Moore, Tony and Alice Echelle, C. Lavett Smith, and Jimmie Pigg suggested that increased numbers of sunfish mesopredators (see plate 18) in creek tributaries to Lake Texoma caused depletion of several minnow species (Matthews and Marsh-Matthews 2007; Marsh-Matthews et al. 2011, 2013). We first noticed a sharp decline in numbers of Red Shiners about the same time that abundances of sunfish increased in Brier Creek in the late 1980s (Matthews and Marsh-Matthews 2007). We also detected declines in Bluntnose Minnow (*Pimephales vigilax*) and Blacktail Shiner (*Cyprinella venusta*), although these decreases were not as severe as for Red Shiners (WJM and EMM, unpub. data). In light of minnow declines in Brier Creek and in nearby Buncombe Creek (Lienesch et al. 2000), we sampled additional streams across south Oklahoma in 2005. Our samples in a total of 30 creeks showed sharp reduction in Red Shiners or other minnows in direct tributaries to Lake Texoma (Matthews and Marsh-Matthews 2007). In contrast, minnows remained abundant in creeks that were direct tributaries to the free-flowing mainstems of the Red or Washita Rivers. And across the 30 streams, sunfish and black bass were much more abundant in creeks that were direct tributaries to Lake Texoma, circumstantial evidence that greater predation pressure in Texoma-direct creeks could have contributed to the losses of minnows.

All of the field evidence, combined with information on extreme flood events in Lake Texoma (that affected geomorphology of the direct tributaries to the reservoir as described below), led us to hypothesize a complex scenario for the decline of minnows. In short, extreme high-water events in the reservoir resulted in standing water upstream for several to many kilometers in the tributary creeks (Matthews and Marsh-Matthews 2007; WJM, pers. obs.), elevating water in the creeks as much as 4 to 5 m above normal. This standing water, sometimes lasting weeks, softened the earthen banks of the creeks, causing extreme erosion or collapse of banks, with influx of large amounts of fine material. During an extreme reservoir flood in July 2007, we paddled about a kilometer up and down in the elevated, standing-water Buncombe Creek, continuously saw or heard slumping of the banks, and actually saw trees falling into the creek (plate 24), supporting the ideas we published in the 2007 paper.

Once floodwaters recede, flow resumes, but now over soft, erodible sand or silt substrates (where there were previously rock or gravel substrates), and deep scour pools develop. These pools, larger and deeper than historically, now become excellent habitat for apex predators (bass, gar, catfish) and mesopredators (sunfish, particularly

Green Sunfish and Longear Sunfish, which are capable of eating adult minnows, Marsh-Matthews et al. 2013; see plate 18). The region also has had several extraordinary droughts since the 1980s, in which long reaches of the normally flowing creeks are reduced to extremely small pools, with crowding of piscivores and their prey (Matthews and Marsh-Matthews 2007; Marsh-Matthews and Matthews 2010; Matthews et al. 2013). Such drought can definitely reduce minnow populations, and even after the creeks rewater we think that "rescue" of Red Shiners from populations in the reservoir can be inhibited by the densities of piscivores that now occupy pools in the lower parts of these creeks.

But the extreme flooding in summer 2007 seemed to offer an opportunity for minnows to potentially move upstream from the reservoir, essentially reinvading their native habitat. Samples we made in the Texoma-direct creeks in autumn 2007 showed no substantial influx of Red Shiners, however. In May–June 2008, in Brier Creek (Matthews et al. 2013) and Little Glasses Creek nearby, we found substantial numbers of Red Shiners, including 72 individuals with males in nuptial color and one young-of-year at our BR5 site, approximately 12 km upstream from the reservoir. The Red Shiners were back! Red Shiner is a hardy, invasive species with high reproductive potential (Marsh-Matthews et al. 2002), and we assumed that they would quickly reestablish populations. But we were wrong. Collections over the next several years showed two peaks in arrival of Red Shiners in Brier Creek that were followed by their disappearance. Propagules arrived but failed to become established. In surveys of Brier Creek through 2016 and Little Glasses Creek through 2014, we have seen one or two, but often none, and Red Shiner populations have never become reestablished (fig. 5.6).

In 2009 we began experiments to ask why, in spite of propagules, this hardy and invasive species regained no substantial presence in the creeks where they had historically been abundant. In our first experiment we established a native community of species common in Brier Creek, including natural densities of Bigeye Shiners, Blackstripe Topminnows, Orangethroat Darters, and a mixture of small young-of-year sunfish (not identified to species because of their small size) in large, outdoor mesocosms (Matthews et al. 2006) at the University of Oklahoma Aquatic Research Facility (ARF) in Norman. In May, we introduced adult Red Shiners, including females and males in nuptial color, at two natural densities to simulate different propagule pressures returning to Brier Creek, in this "invasion of the Red Shiner" (Marsh-Matthews et al. 2011).

The goal was to test the hypothesis from invasion biology that propagule pressure has an important effect on the ability of an invasive species to spread and become established. But to our surprise (here is the serendipity), the number of propagules had no effect in recruitment of Red Shiners, because over the summer the baby sunfish that we introduced in March grew to be small adults, fully capable of eating minnows! The result, when we ended the experiment in November, was a strong negative relationship between percentage survival of Red Shiners and numbers of adult sunfish per experimental unit (Marsh-Matthews et al. 2011).

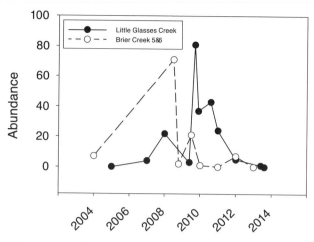

FIG. 5.6 Red Shiners captured in Brier Creek and Little Glasses Creek, 2004–2013.

The other interesting finding was that while Red Shiners had low survival in units with the most adult sunfish, Bigeye Shiners in the experiment showed no effects of predation. And our data (Matthews et al. 1994; Matthews and Marsh-Matthews, unpub. data) showed that, unlike Red Shiners, Bigeye Shiners had not declined in the creeks with the increases in centrarchids. We then turned attention to the role of sunfish in preventing Red Shiners from reestablishing populations in the Texoma-direct creeks and any differing vulnerability of the two minnow species to sunfish predation. In a mesocosm experiment in autumn 2010, we found that Bigeye Shiners survived better than Red Shiners in units with adult Green Sunfish and Longear Sunfish (Marsh-Matthews et al. 2013).

Finally, in a mesocosm experiment in 2012, we tested survival and recruitment by the two minnow species across different levels of sunfish predator pressure. In a predator-free control and at low predation pressure, Red Shiners survived well and produced young, but at medium and high predation pressure, they survived poorly and produced no young, whereas Bigeye Shiners again were apparently unaffected by the sunfish (Marsh-Matthews and Matthews, unpub. data). The upshot of all this is that not all minnows are equally affected when challenged by mesopredators. The Bigeye Shiners may be less vulnerable to predation by sunfish because of a greater swimming ability, better escape behaviors, or because in actual streams they tend to occupy upper parts of pools in shallow water just below riffles, where there are few sunfish or bass (Matthews et al. 1994). In contrast, snorkeling observations suggest that Red Shiners are more general in habitat use, occupying the water column throughout pools (or at least they did, before their decline), which may make them more accessible to sunfish or bass (Matthews et al. 2013).

The preponderance of our information from Brier Creek and other similar "centrarchid" creeks in south Oklahoma suggests that predation pressure as a driving force in community dynamics varies a great deal across time or water-level conditions and across life stages of potential prey and is not exerted equally across different prey species. Predator pressure, while clearly important in our streams, is difficult to summarize in one easy statement like "mesopredators control minnow density." The situation deserves more consideration depending on the circumstances (i.e., context-specific variation in predator effects).

Predator Threat

To this point we have mostly considered the direct effects of killing prey and removing them from the community. The threat of predation, as studied in detail for stream fish years ago by Doug Fraser and colleagues (Fraser and Cerri 1982; Fraser and Emmons 1984; Fraser et al. 1987), can have strong effects on behavior of potential prey. We have counts of fish, by species and by size for centrarchids, in 14 pools in Brier Creek from 1982 to 2012 (Matthews et al. 1994; Matthews and Marsh-Matthews 2006, unpub. data). Our first clue that predator threat was important in this system was in November 1982, when Mary Power and WJM surveyed fish and attached algae in these 14 pools (Power and Matthews 1983) and realized there was a strong dichotomy between pools with black bass (Largemouth Bass or Spotted Bass, not always distinguishable in snorkel observations) >100 mm TL and those with Central Stonerollers: those with bass (which contained few stonerollers) had more attached algae. Because bass affected the distribution of grazing stonerollers, they also indirectly affected the standing crop of algae via a classic trophic cascade (see more in chap. 9). WJM continued snorkel surveys over the next year, showing that the bass-stoneroller dichotomy existed, except shortly after floods disturbed the system. Power, Art Stewart, and WJM also carried out experiments showing that when bass were added to a pool, the stonerollers rapidly moved either into very shallow refuge space at pool edge or into the next pool (Power et al. 1985; Power 1987). In this case it was clearly the threat of predation, and not outright killing, that caused the changes in distribution of stonerollers, and Matthews et al. (1994) showed that several other minnow species also avoided pools that had juvenile or adult bass.

WJM, with EMM, continued snorkel surveys of the 14 pools in Brier Creek from 1995 to 2012. Matthews and Marsh-Matthews (2006) confirmed that some of the patterns seen in the earlier snorkel surveys (1982–1983) persisted into later decades, with negative associations between adult centrarchids and stonerollers in typical surveys. Now, with a total of 25 snorkeling surveys of the 14 pools from 1982 to 2012 (a total of 338 pool surveys), we find additional evidence that stonerollers avoid pools with bass >100 mm TL (product moment correlation of bass and stoneroller abundance per pool = −0.1827, $P = 0.0007$). And Red Shiners, in 122 pool surveys before they disappeared from the system, were also negatively correlated with bass >100 mm TL ($r = -0.238$; $P < 0.001$) and with Longear Sunfish >75 mm TL ($r = -0.125$; $P < 0.001$). However, as indicated in Matthews et al. (1994), Blackstripe Topminnows and Bigeye Shiners were positively correlated with both large Largemouth Bass and large

Longear Sunfish. Matthews et al. (1994) attributed the ability of the topminnows to live in pools with bass or other large centrarchids to the obliterative countershading of the topminnows, which are dark on the dorsum but almost completely white on the venter. Because topminnows typically swim at the surface of pools, the white underside is purported to make detection by predators difficult against the bright sky as viewed from below.

In addition to all the work by Power et al. on bass effects in Brier Creek, we have experimental evidence of the ability of Largemouth Bass to change behavior of prey. In large mesocosms at the University of Oklahoma Biological Station (Matthews et al. 2006), approximately 400 Red Shiners were allowed to freely move through an array of 12 pools connected by flowing riffles. They were consistently spaced rather evenly, with 25 to 35 individuals in each pool and only occasionally as many as 40 individuals (Matthews, unpub. data). Interestingly, this density approximated the density at which Red Shiners had fared well in a previous experiment (Matthews et al. 2001), whereas at higher densities in pools of this mesocosm system, we had detected substantial loss of condition or growth.

The observed dispersion of Red Shiners persisted from the first mapping of fish in September 2000 until March 2001, when three Largemouth Bass and one Spotted Bass ranging from about 18 to 24 cm TL were added to the system. Less than an hour after addition of the bass, Red Shiner distributions changed, with the minnows strongly clumped by the next day in a few pools. By a week later, all Red Shiners were in large aggregations in two of the twelve pools. The bass did not appear to patrol very actively and tended to hide near algae clumps. It appeared that the presence of the bass, even if only detected by their odor or some other cue (the streams were recirculating), resulted in immediate, rapid behavioral responses of the Red Shiners to form large groups. The arrival of a bass in any pool within a stream reach can have strong, potentially immediate effects on the dynamics of the fish community, if minnows change distributions or behaviors and thus their effects in the community or on ecosystem properties (Power et al. 1985).

Facilitation

Facilitation is one of the frequently overlooked factors in community structure, and research on positive interactions in communities has historically lagged behind that of other effects (Stachowicz 2001). Bruno et al. (2003) argued for the need for more research on mutualisms, facilitation, interspecific information transfer (Goodale et al. 2010), or other positive interspecific interactions. Matthews (1998, 512–525) outlined potential benefits to fishes of interspecific groups, particularly in enhanced detection of food or avoidance of predators. On numerous occasions we have observed minnows like Cardinal Shiners, and occasionally a small juvenile Smallmouth Bass, following large Northern Hogsuckers (*Hypentelium nigricans*), feeding actively on items suspended into the water column by the sucker (Matthews 1998), but this particular facilitation has been known for a long time (Reighard 1920). Matthews (1998) also documented the high proportion of minnows occurring in mixed-species groups, based on snorkeling observations or on finding multiple species together in small seine hauls. Matthews (1998) showed that 71% of a total of 787 observations that contained minnows included more than one species. As previously pointed out, finding fish in mixed-species groups does not really demonstrate that facilitation is going on. But such observations suggest that communication between species would be possible, especially in clear water.

Facilitation may also be difficult to detect because it is the result of complex interactions. Graduate students and postdocs at the University of Oklahoma designed and carried out an experiment in our outdoor mesocosms that showed indirect facilitation of Central Stonerollers (enhanced growth and body condition) by the presence of Orangethroat Darters (Hargrave et al. 2006; see also chap. 9), apparently due to the darters' reducing the standing crop of algivorous benthic macroinvertebrates, which in turn left more food for the algivorous stonerollers. A complex and unique suite of facilitative interactions were shown by WJM's doctoral student, Bret Harvey, to have strong effects on distribution or survival of different life stages of fish in the pools in Brier Creek. Harvey (1991) began by showing in pilot studies how, when adult Largemouth Bass were added to Brier Creek pools, juvenile sunfish moved into shallows, avoiding the bass. He also showed in a manipulative experiment that small sunfish readily ate minnow larvae in these pools and that larval fish survival was low when juvenile sunfish or bass were present. Then, across three creeks in south Oklahoma, Harvey showed that larval sunfish and larval minnows co-occurred with adult bass. Finally, another field experiment revealed a previously unknown, multilevel interaction among these fishes: larval fish that he released into experimental pools were found in the deep parts of pools, along with adult bass. The presence of adult bass forced juvenile sunfish and juvenile bass into shallower habitats, thereby creating a safe zone for larval fishes. Overall, Harvey's (1991) work was a real breakthrough in identifying a new multilevel interaction, confirming that Largemouth Bass could play a positive facilitative role in the structure of fish assemblages at the scale of individual pools.

Lack of Expected Biotic Interactions

For species that have been widely studied and for which interactions have been documented, the lack of an expected interaction can be instructive. Two fish species with which we have worked the most (Red Shiner and Western Mosquitofish) have reputations as being really bad actors in fish communities where they have been introduced outside their ranges in the United States or worldwide. The Red Shiner has been widely introduced in the American West, intentionally or through bait-bucket release, and has become the numerical dominant in many streams since the 1950s (Hubbs 1954; Minckley 1973). Where introduced in the West, they have been reported or suspected to be widely harmful to native species, through predation on native fish larvae (Ruppert et al. 1993), displacement of natives from preferred habitats (Douglas et al. 1994), or aggressive behavior (Karp and Tyus 1990), and are widely thought to contribute to declines of native fishes throughout the region (Minckley and Marsh 2009). Red Shiners are also considered an "aggressive colonizer" in the eastern United States, with the potential to displace other *Cyprinella* through hybridization (Herrington and DeVries 2008).

In spite of the bad reputation of Red Shiners outside their native range, Marsh-Matthews and Matthews (2000) found no evidence that within their native range this species had any detectable negative impacts on other native species with which it co-occurred in local communities. At 65 sites where we made community fish collections from north to south across most of the southern Great Plains to the central Texas Hill Country, Red Shiners were present at 50 sites. Our survey included 11 different major river basins and streams from small pasture creeks to broad, shallow river mainstems. Red Shiners were high in relative abundance throughout most of the region, ranking first in abundance in 25 of the 50 sites where they were present. But in spite of being widespread and often abundant, an analysis of the effects of Red Shiner on all of the other native fishes in these samples showed essentially no effect. Red Shiner relative abundance was not correlated with species richness (raw or by rarefaction), diversity, evenness, or complexity of the residual communities (i.e., including all species other than Red Shiner; Marsh-Matthews and Matthews 2000). The only component of the residual communities in which Red Shiner had apparent effect was that it was correlated with an increase (not a decrease) in relative abundance of benthic minnows. Overall, the results from Marsh-Matthews and Matthews (2000) suggest that while some invasive species may be bad actors in many new habitats, in their native ranges with coevolved species, they may be simply an ordinary and expected part of the fish community.

We found a similar situation with Western Mosquitofish, which, along with the similar Eastern Mosquitofish (*Gambusia holbrooki*), has been introduced worldwide since the 1930s, usually as an effort at mosquito control. In many parts of the world these mosquitofish have proliferated, however, and are widely accused of having deleterious effects on native fish populations. In fact, in Australia there is a website devoted to eradication of the "Damnbusia"! They have been called in print "fish destroyer" or "plague minnow" (Ling 2004; Pyke 2008) and are considered among the world's 100 worst invasive alien species (Lowe et al. 2000). Like Red Shiners, these gambusia are considered harmful to a large number of native fish species through aggression, nipping fins, competition for food, or eating eggs and juveniles of the natives (Matthews and Marsh-Matthews 2011).

To evaluate effects of Western Mosquitofish within their native range, we took an approach similar to that of Marsh-Matthews and Matthews (2000) for Red Shiners. We compiled data on whole fish communities at 154 different collection sites in Kansas, Arkansas, and Oklahoma and tested the effects of mosquitofish presence on species richness or Shannon diversity of the residual community. Western Mosquitofish were present at 102 of the 154 sites but accounted for <10% of total fish abundance at most. Residual species richness was actually higher in the presence of mosquitofish, and diversity did not differ between sites with or without mosquitofish. The relative abundance of mosquitofish was weakly negatively correlated with the number of species in the residual communities but accounted for only about 7% of the variance. The relative abundance of mosquitofish was not related to diversity of the residual community, and there were no negative correlations between its abundance and that of any of the most common other species (Matthews and Marsh-Matthews 2011). In Western Mosquitofish we had another example of a species considered really "bad" outside of, but innocuous within, its native range. We do not doubt most reports of harm to native species where this mosquitofish is introduced. But we think that this case and that of the Red Shiner both illustrate that interactions among fish species may vary greatly from place to place and that species in their native ranges may be normal, innocuous, and important components of coevolved fish communities.

Finally, what about interactions of invasive species with native species in the central United States? In spite of widely reported introductions of exotic species worldwide and in the United States, the systems that we work with in the lower Great Plains, Interior Highlands, or Texas Hill Country are relatively free of exotics. There have certainly been major introductions of sport or forage species in reservoirs of the southern United States, such as Striped Bass, Threadfin Shad (*Dorosoma petenense*), or the Florida strain of Largemouth Bass. But most of these species, introduced intentionally into artificial impoundments, have not spread far into free-flowing streams (e.g., Gelwick et al. 1995). We have compiled one data set of 509 whole-community fish collections that one or the other or both of us have made from 1975 to 2014 in streams of all sizes in the Arkansas and Red River basins. These include samples from a wide variety of sites from small spring-fed creeks to large rivers and swampy backwaters, with single samples at some sites and samples repeated over time in others. There are 142 species in the Arkansas-Red basin data set, but only 5 (numbers of sites in parentheses) were exotic or extralimital species, including Goldfish, *Carassius auritus* (1); Common Carp, *Cyprinus carpio* (30); Striped Bass (5); Tilapia (Probably Blue Tilapia), *Oreochromis aurea* (6); and Grass Carp, *Ctenopharyngiodon idella* (1). Of these, only Tilapia was ever abundant, with 338, 52, and 30 individuals, mostly juveniles, at three sites on the North Fork of the Canadian River in central Oklahoma, downstream from artificially heated waters of a power plant. There was no evidence that the Tilapia had spread successfully elsewhere in the system, and the presumption was that the Tilapia could not survive Oklahoma winters outside the thermal refuge provided by the power plant.

Common Carp, which we found at the most sites, was imported from Europe in the 1800s and spread eagerly over much of North America by ambitious fisheries management agencies. These carp are reported to have negative effects in aquatic ecosystems in general (Weber and Brown 2009). Specifically, they have high levels of excretion (Kulhanek et al. 2011) and behaviors that cause them to change physical characteristics of lakes by increasing turbidity, changing nutrient availability and macroinvertebrates, and destroying aquatic plants. They also have negative effects on native species (Weber and Brown 2011). Common Carp are considered "one of the world's worst 100 invasive alien species" (Lowe et al. 2000). There is no question that these carp can seriously disturb substrates. WJM once observed a large carp in Imhof Creek in Norman, Oklahoma, plowing through a mud bottom (presumably seeking invertebrates) with its snout and head buried in the mud up to its eyes, with substantial streams of mud passing out of its opercular openings as it worked the bottom. Years of electrofishing in Lake Texoma, a large reservoir on the Oklahoma-Texas border, showed Common Carp to be in substantial numbers, particularly obvious in shallows during their spawning season, as they vigorously splash water and probably disturb the substrates. But the good news is that, at least in the streams of our region, they are relatively scarce, occur in low numbers, and probably have few if any consequences for native fishes in streams.

We found Common Carp in only 30 of our 509 collections, with a maximum of 51 small individuals in one sample but usually with only 1 to a few individuals. And in 65 collections we made in streams of all sizes from Nebraska and Iowa to southern Texas in June 1995, we found Common Carp at only 14 sites, with a maximum of 12 individuals. Thus, although the Common Carp is widespread in North America, especially in lakes and reservoirs, the fact is that in the south-central United States it does not appear to be a major component of stream fish communities or a major factor in most stream ecosystems.

Overall, the central United States still seems to be a place where ecologists, biologists, and managers can find natural native stream fish communities to study or manage. This is not without caveats. While some centrarchids like Largemouth Bass and Bluegill are native to the region, for example, they have been widely stocked in ponds and lakes and also may be naturally invasive, and they may now occur in some streams in the region where they were formerly unknown (e.g., Matthews and Marsh-Matthews 2015). Lacking any early data, at any particular site in our region, there is no assurance that the suite of fishes now present is the same as it was a century ago. At present it is unknown whether or to what extent some potentially dangerous exotics, such as Northern Snakehead (*Channa argus*) or other Asian carps (Bighead Carp, *Hypophthalmichthys nobilis*, and Silver Carp, *Hypophthalmichthys molitrix*), that are so troublesome in the central Mississippi River drainages might eventually invade more broadly in streams of the Great Plains. Both of those Asian carps have been found in eastern Oklahoma (Patton and Tackett 2012) and eastern Kansas (US Geological Survey fact sheets). Although gravid female and sexually mature male Asian carp have been found in east Oklahoma, juvenile fish have not been documented (Barry Bolton, Oklahoma Department of Wildlife Conservation, pers. comm.). In more than 200 seining collections we have made in eastern Oklahoma in the last decade, we have found no *Hypophthalmichthys*. But the current lack of nonnative exotics in most streams in the Great Plains does suggest that continued vigilance is appropriate to ensure that additional exotics are not imported to the region and that action should be taken to control any that are detected.

Summary

This chapter has examined biotic interactions that can affect community structure and dynamics, including competition, predation, and facilitation. Competition has been studied, argued about, and fought over intensively for a century, across a huge number of animal and plant systems, so perhaps it is a bit presumptive to think that our limited studies would shed any new light on its importance in fish communities. But here we have tried to reach a consensus on what the evidence says about the potential role of competition in the systems we studied. Complementary distributions or abundances both at geographic scales and within individual streams offered the possibility that competition could affect minnow distributions, owing to the repulsed distributions of morphologically similar species in the Red River basin (Winston 1995) or to complementarity of fish distributions among Ozark streams (Taylor 1996), although Matthews (1982) argued that a random model could account for the observed mutual exclusions of minnows in one case. And within the Piney Creek watershed, minnows that were morphologically most similar actually occurred together more, not less, than expected, suggesting more influence of convergence to similar habitats (habitat filtering) than any influence of competitive exclusion. From complementary distributions at a geographic scale or tests for repulsion of morphologically similar species, we were left with two studies supporting competition as potentially important and two in which the importance of competition was not supported.

Within the Kiamichi River and Piney Creek there were complementary abundances of several pairs of minnow or darter species that were consistent with a simple longitudinal zonation explanation, thus shedding little light on the importance of competition. But other pairs of sculpins or minnows in Piney Creek had distributions not explained by simple upstream-downstream differences, as their relative abundances reversed once or more from site to site from downstream to upstream. These cases provide more support for the possibility that competition for some needed resource could be driving the observed complementarity.

Our studies of resource partitioning and coexistence of similar species showed for minnow and for darter species in the Roanoke River that niche separation was low but that these species occurred together nonetheless. We suspect that in the Roanoke River, invertebrates are sufficiently available as foods that resources for these insectivorous species are not typically limiting, but while we have year-round data on their resource use, we have no data on their use in "crunch" periods like severe drought. There was one similar situation in Blaylock Creek, in southwest Arkansas, with two abundant minnow species strongly overlapped in distribution within the creek and in food use. In Blaylock Creek a different minnow pair used extremely similar foods but were partly separated in habitat use up- and downstream, suggesting that competition could be important—but habitat preferences offered an equally plausible explanation for the pattern. Finally, for two darters in the Texas Hill Country, there was strong segregation in microhabitat and food use within the same stream reach, but the underlying reason for their separation could be extant competition that keeps the two species separated, or it could be their deeper, mutual evolutionary past.

It is our overall impression from all of the examples above that in some cases there is evidence that competition could be a factor in some fish distributions, but there is no compelling reason to consider contemporary competition as a primary factor in the stream fish communities we study. In all cases above, there are potential explanations other than ongoing competition that could account for the observed patterns. To actually tease apart competitive effects or assess mechanisms suggested by any of the complementary distributions or resource segregation patterns we have described, manipulative field experiments or well-designed experiments in the laboratory or in realistically large mesocosms are needed. Experiments could focus on specific examples from species we have studied, or future students might combine field observations in their own systems with experimental tests like the approach taken by Taylor (1996).

Carefully planned experiments with the species and systems we have described above might take knowledge about competition in these or other systems to the next level. Because distributions, microhabitat use, and food use by fishes can change with seasons or environmental conditions or between years, any short-term experiment under one set of conditions should ideally be repeated under alternative circumstances. Unfortunately, some of the most important funding agencies, such as the National Science Foundation, do not seem to encourage "incremental" research, so if an investigator has done an experiment with interesting results in one situation, it is likely to be hard to find funding to repeat the experiment in a different system or under different circumstances. Any comparative series of experiments on competition, repeated across a range of flow conditions, seasons, systems, or years, is going to be a challenge!

Our studies of predator effects on community structure have addressed the potential for predator impacts as well as direct and indirect predator effects. Across a wide range of stream fish communities, the potential predator threat (measured as the predator:prey ratio) varied by two or more orders of magnitude. Among 9 global communities, the predator:prey ratio was highest in intermittent streams and in some low-gradient systems. For 31 local communities, those with the highest predator:prey ratios typically had low gradients with deep pools and low current speeds, and those with the lowest ratios were upland systems with steep gradients and strong flow or systems with shifting sand substrates. The potential for predation to have strong impacts on community structure and function thus varies widely among stream fish communities and is at least potentially predictable from environmental factors.

Predator pressure also varies across time within communities: sunfish and other species that are insectivorous as juveniles grow large enough to be piscivorous by the end of a growing season; predators (such as gar) may move into spawning areas tem-

porarily, and predation pressure may vary with temperature or water level as predator activity or density changes.

Predator identity can also affect predation risk for prey: Central Stonerollers actively
avoided Largemouth Bass by moving into shallower habitat or out of pools in Brier
Creek but showed little reaction to the presence of Smallmouth Bass. Species other than
well-known "apex" predators may exert predation pressure on particular prey. Red Shiners, typically considered a prey species, are highly effective predators on larval pupfish
and on larvae of other minnows. Sunfish species, including Longear Sunfish, readily
prey on Red Shiners and thus act as "mesopredators" in stream fish communities.

A long-term study of the stream fish communities in Brier Creek and other direct
tributaries of the reservoir Lake Texoma illustrates the significant role that mesopredators may play in community dynamics. Despite being widespread, abundant, and
known for invasive success outside their native range, Red Shiners have disappeared
from several direct tributaries of Lake Texoma where they once were the single most
abundant species. Creeks throughout the region typically become intermittent in the
summer, and large reaches may dry completely. Although Red Shiners are known to
be able to rapidly move upstream in creeks as they rewater, changes in geomorphology at the reservoir tributaries at the creek-reservoir interface have altered habitat in
a way that supports large populations of mesopredators. These mesopredators have
probably prevented the reestablishment of Red Shiner populations in direct tributary
creeks. In a series of mesocosm experiments we have confirmed that mesopredators
affect Red Shiner survival and reproduction. But another minnow, the Bigeye Shiner,
is far less subject to predation by the same mesopredators—another example of the
importance of predator (or in this case, prey) identity.

Predators need not be lethal in order to affect prey. The presence of a predator may
alter prey distribution, which in turn may affect community or even ecosystem dynamics. Addition of Largemouth Bass to pools in Brier Creek causes Central Stonerollers
to vacate those pools. The absence of the grazing stonerollers allows extensive growth
of algae. The addition of bass to a mesocosm stream system with multiple pools resulted in aggregation of Red Shiners in a few of the pools at densities higher than those
observed in the absence of bass.

Not all species interactions involve a negative impact of one species on another.
The presence of multiple minnow species in shoals may enhance the benefits of shoaling for all species involved (although those specific benefits have not been explored
in detail in our systems). Presence of adult bass affected survival of juvenile minnows
by altering the distribution of juvenile sunfish that would otherwise prey on the minnow larvae.

Finally, the lack of expected interspecific interactions may reveal important insights
into stream fish community dynamics. Both Red Shiner and Western Mosquitofish
have reputations as highly invasive and destructive species where they have been introduced outside their native ranges. Within their native ranges, however, they have
no apparent effect on the composition or diversity of the "residual" (i.e., the other

species) in the communities where they occur. Similarly, other known invasive species that have been successful invaders in reservoirs have not had widespread impacts in the streams that we have studied.

References

Abrams, P. A. 1994. The fallacies of "ratio-dependent" predation. Ecology 75:1842–1850.

Abrams, P. A., and L. R. Ginzburg. 2000. The nature of predation: prey dependent, ratio dependent or neither? Trends in Ecology and Evolution 15:337–341.

Andrewartha, H. G., and L. C. Birch. 1954. The distribution and abundance of animals. University of Chicago Press, Chicago, IL.

Arditi, R., and L. R. Ginzburg. 1989. Coupling in predator-prey dynamics: ratio-dependence. Journal of Theoretical Biology 139:311–326.

Baltz, D. M., P. B. Moyle, and N. J. Knight. 1982. Competitive interactions between benthic stream fishes, Riffle Sculpin, *Cottus gulosus*, and Speckled Dace, *Rhinichthys osculus*. Canadian Journal of Fisheries and Aquatic Sciences 39:1502–1511.

Barrio, I. C., D. S. Hik, C. G. Bueno, and J. F. Cahill. 2013. Extending the stress-gradient hypothesis—is competition among animals less common in harsh environments? Oikos 122:516–523.

Berendzen, P. B., A. M. Simons, R. M. Wood, T. E. Dowling, and C. L. Secor. 2008. Recovering cryptic diversity and ancient drainage patterns in eastern North America: historical biogeography of the *Notropis rubellus* species group (Teleostei: Cypriniformes). Molecular Phylogenetics and Evolution 46:721–737.

Biehl, C. C., and W. J. Matthews. 1984. Small fish community structure in Ozark streams: improvements in the statistical analysis of presence-absence data. American Midland Naturalist 111:371–382.

Black, R. W., II, and N. G. Hairston Jr. 1988. Predator driven changes in community structure. Oecologia 77:468–479.

Britten, G. L., M. Dowd, C. Minto, F. Ferretti, F. Boero, and H. K. Lotze. 2014. Predator decline leads to decreased stability in a coastal fish community. Ecology Letters 17:1518–1525.

Brown, J. H. 1975. Geographical ecology of desert rodents. Pages 315–341 in M. L. Cody and J. M. Diamond, eds. Ecology and evolution of communities. Belknap Press, Cambridge, MA.

Brown, J. H., T. G. Whitham, S. K. Morgan Ernest, and C. A. Gehring. 2001. Complex species interactions and the dynamics of ecological systems: Long-term experiments. Science 293: 643–650.

Bruno, J. F., J. J. Stachowicz, and M. D. Bertness. 2003. Inclusion of facilitation into ecological theory. Trends in Ecology and Evolution 18:119–125.

Calsbeek, R., and R. M. Cox. 2010. Experimentally assessing the relative importance of predation and competition as agents of selection. Nature 465:613–616.

Chase, J. M. 2007. Drought mediates the importance of stochastic community assembly. Proceedings of the National Academy of Sciences USA 104:17,430–17,434.

Chase, J. M., P. A. Abrams, J. P. Grover, S. Diehl, P. Chesson, R. D. Holt, S. A. Richards, R. M. Nisbet, and T. J. Case. 2002. The interaction between predation and competition: a review and synthesis. Ecology Letters 5:302–315.

Cody, M. L., and J. M. Diamond, eds. 1975. Ecology and evolution of communities. Belknap Press, Cambridge, MA.

Connell, J. H. 1975. Some mechanisms producing structure in natural communities: a model and evidence from field experiments. Pages 460–490 in M. L. Cody and J. M. Diamond, eds. Ecology and evolution of communities. Belknap Press, Cambridge, MA.

Connell, J. H. 1980. Diversity and the coevolution of competitors, or the ghost of competition past. Oikos 35:131–138.

Cooper, H. R. 1975. Food and feeding selectivity of two cottid species in an Ozark stream. MS thesis, Arkansas State University, Jonesboro, AR.

Death, R. G. 2010. Disturbance and riverine benthic communities: what has it contributed to general ecological theory? River Research and Applications 26:15–25.

Dewey, S. L. 1988. Feeding relationships among four riffle-inhabiting stream fishes. Transactions of the Illinois Academy of Science 81:171–184.

Diamond, J. M. 1975. Assembly of species communities. Pages 342–444 in M. L. Cody and J. M. Diamond, eds. Ecology and evolution of communities. Belknap Press, Cambridge, MA.

Diamond, J. M., and T. J. Case, eds. 1976. Community ecology. Harper and Row, New York, NY.

Douglas, M. E., and W. J. Matthews. 1992. Does morphology predict ecology? hypothesis testing within a freshwater stream fish assemblage. Oikos 65:213–224.

Douglas, M. E., P. C. Marsh, and W. L. Minckley. 1994. Indigenous fishes of western North America and the hypothesis of competitive displacement: *Meda fulgida* (Cyprinidae) as a case study. Copeia 1994:9–19.

Duvernell, D. D., and J. F. Schaefer. 2014. Variation in contact zone dynamics between two species of topminnows, *Fundulus notatus* and *F. olivaceus*, across isolated drainage systems. Evolutionary Ecology 28:37–53.

Etnier, D. A., and W. C. Starnes. 1993. The fishes of Tennessee. University of Tennessee Press, Knoxville, TN.

Fausch, K. D. 1988. Tests of competition between native and introduced salmonids in streams: what have we learned? Canadian Journal of Fisheries and Aquatic Sciences 45: 2238–2246.

Fraser, D. F., and R. D. Cerri. 1982. Experimental evaluation of predator-prey relationships in a patchy environment: consequence for habitat use patterns in minnows. Ecology 63: 307–313.

Fraser, D. F., and E. E. Emmons. 1984. Behavioral response of Blacknose Dace (*Rhinichthys atratulus*) to varying densities of predatory Creek Chub (*Semotilus atromaculatus*). Canadian Journal of Fisheries and Aquatic Sciences 41:364–370.

Fraser, D. F., D. A. DiMattia, and J. D. Duncan. 1987. Living among predators: the response of a stream minnow to the hazard of predation. Pages 121–127 in W. J. Matthews and D. C. Heins, eds. Community and evolutionary ecology of North American stream fishes. University of Oklahoma Press, Norman, OK.

Gause, G. F. 1932. Experimental studies on the struggle for existence: I. mixed population of two species of yeast. Journal of Experimental Biology 9:389–402.

Gause, G. F. 1934. Experimental analysis of Vito Volterra's mathematical theory of the struggle for existence. Science 79:16–17.

Gelwick, F. P., E. R. Gilliland, and W. J. Matthews. 1995. Introgression of the Florida Largemouth Bass genome into stream populations of Northern Largemouth Bass in Oklahoma. Transactions of the American Fisheries Society 124:550–562.

Gido, K. B., J. F. Schaefer, K. Work, P. W. Lienesch, E. Marsh-Matthews, and W. J. Matthews. 1999. Effects of Red Shiner (*Cyprinella lutrensis*) on Red River Pupfish (*Cyprinodon rubrofluviatilis*). Southwestern Naturalist 44:287–295.

Glasser, J. W. 1979. The role of predation in shaping and maintaining the structure of communities. American Naturalist 113:631–641.

Goodale, E., G. Beauchamp, R. D. Magrath, J. C. Nieh, and G. D. Ruxton. 2010. Interspecific information transfer influences animal community structure. Trends in Ecology and Evolution 25:354–361.

Greenberg, L. A. 1988. Interactive segregation between the stream fishes *Etheostoma simoterum* and *E. rufilineatum*. Oikos 51:193–202.

Grossman, G. G., and J. L. Sabo. 2010. Incorporating environmental variation into models of community stability: examples from stream fish. Pages 407–426 in K. B. Gido and D. A. Jackson, eds. Community ecology of stream fishes: concepts, approaches, and techniques. American Fisheries Society Symposium 73. American Fisheries Society, Bethesda, MD.

Grossman, G. G., P. B. Moyle, and J. O. Whittaker Jr. 1982. Stochasticity in structural and functional characteristics of an Indiana stream fish assemblage: a test of community theory. American Naturalist 120:423–454.

Gurevitch, J., J. A. Morrison, and L. V. Hedges. 2000. The interaction between competition and predation: a meta-analysis of field experiments. American Naturalist 155:435–453.

Hardin, G. 1960. The competitive exclusion principle. Science 131:1292–1297.

Hargrave, C. W., R. Ramirez, M. Brooks, M. A. Eggleton, K. Sutherland, R. Deaton, and H. Galbraith. 2006. Indirect food web interactions increase growth of an algivorous stream fish. Freshwater Biology 51:1901–1910.

Harvey, B. C. 1991. Interactions among stream fishes: predator-induced habitat shifts and larval survival. Oecologia 87:29–36.

Harvey, B. C., R. C. Cashner, and W. J. Matthews. 1988. Differential effects of largemouth and smallmouth bass on habitat use by stoneroller minnows in stream pools. Journal of Fish Biology 33:481–487.

Hefley, H. M. 1937. Ecological studies on the Canadian River floodplain in Cleveland County, Oklahoma. Ecological Monographs 7:345–402.

Herrington, S. J., and D. R. DeVries. 2008. Reproduction and early life history of nonindigenous Red Shiner in the Chattahoochee River drainage, Georgia. Southeastern Naturalist 7: 413–428.

Hofbaeur, J., and K. Sigmund. 1989. On the stabilizing effect of predators and competitors on ecological communities. Journal of Mathematical Biology 27:537–548.

Holling, C. S. 1959. The components of predation a revealed by a study of small-mammal predation on the European pine sawfly. Canadian Entomologist 91:293–320.

Hubbs, C. L. 1954. Establishment of a forage fish, the Red Shiner (*Notropis lutrensis*), in the lower Colorado River system. California Fish and Game 40:287–294.

Hubbs, C. L., and K. F. Lagler. 1949. Fishes of the Great Lakes region. Bulletin No. 26. Cranbrook Institute of Science, Bloomfield Hills, MI.

Humphries, J. M., and R. C. Cashner. 1994. *Notropis suttkusi*, a new cyprinid from the Ouachita uplands of Oklahoma and Arkansas, with comments on the status of Ozarkian populations of *N. rubellus*. Copeia 1994:82–90.

Jackson, D. A., P. R. Peres-Neto, and J. D. Olden. 2001. What controls who is where in freshwater fish communities—the roles of biotic, abiotic, and spatial factors. Canadian Journal of Fisheries and Aquatic Sciences 58:157–170.

Karp, C. A., and H. M. Tyus. 1990. Behavioral interactions between young Colorado squawfish and six fish species. Copeia 1990:25–34.

Karr, J. R., K. D. Fausch, P. L. Angermeier, P. R. Yant, and I. J. Schlosser. 1986. Assessing biological integrity in running waters—a method and its rationale. Special Publication 5. Illinois Natural History Survey, Champaign, IL.

Kulhanek, S. A., B. Leung, and A. Ricciardi. 2011. Using ecological niche models to predict the abundance and impact of invasive species: application to the Common Carp. Ecological Applications 21:203–213.

Lang, N. J., and R. L. Mayden. 2007. Systematics of the subgenus *Oligocephalus* (Teleostei: Percidae: *Etheostoma*) with complete subgeneric sampling of the genus *Etheostoma*. Molecular Phylogenetics and Evolution 43:605–615.

Lepori, F., and B. Malmqvist. 2009. Deterministic control on community assembly peaks at intermediate levels of disturbance. Oikos 118:471–479.

Lienesch, P. W., W. I. Lutterschmidt, and J. F. Schaefer. 2000. Seasonal and long-term changes in the fish assemblage of a small stream isolated by a reservoir. Southwestern Naturalist 45:274–288.

Lima, S. L. 1998. Nonlethal effects in the ecology of predator-prey interactions. BioScience 48:25–34.

Ling, N. 2004. *Gambusia* in New Zealand: really bad or just misunderstood? New Zealand Journal of Marine and Freshwater Research 38:473–480.

Lowe, S., M. Browne, S. Boudjelas, and M. DePoorter. 2000. 100 of the world's worst invasive alien species: a selection from the Global Invasive Species Database. Invasive Species Specialist Group (ISSF) of the World Conservation Union (IUCN), Auckland, New Zealand.

MacArthur, R. H. 1958. Population ecology of some warblers of northeastern coniferous forests. Ecology 39:599–619.

Marsh-Matthews, E., and W. J. Matthews. 2000. Spatial variation in relative abundance of a widespread, numerically dominant fish species and its effect on fish assemblage structure. Oecologia 125:283–292.

Marsh-Matthews, E., and W. J. Matthews. 2010. Proximate and residual effects of exposure to simulated drought on prairie stream fishes. Pages 461–486 in K. B. Gido and D. A. Jackson, eds. Community ecology of stream fishes: concepts, approaches, and techniques. American Fisheries Society Symposium 73. American Fisheries Society, Bethesda, MD.

Marsh-Matthews, E., W. J. Matthews, and N. R. Franssen. 2011. Can a highly invasive species re-invade its native community? the paradox of the Red Shiner. Biological Invasions 13: 2911–2924.

Marsh-Matthews, E., W. J. Matthews, K. B. Gido, and R. L. Marsh. 2002. Reproduction by young-of-year Red Shiner (*Cyprinella lutrensis*) and its implications for invasion success. Southwestern Naturalist 47:605–610.

Marsh-Matthews, E., J. Thompson, W. J. Matthews, A. Geheber, N. R. Franssen, and J. Barkstedt. 2013. Differential survival of two minnow species under experimental sunfish predation: implications for re-invasion of a species into its native range. Freshwater Biology 58:1745–1754.

Matthews, W. J. 1982. Small fish community structure in Ozark streams: structured assembly patterns or random abundance of species? American Midland Naturalist 107:42–54.

Matthews, W. J. 1985. Critical current speeds and microhabitats of the benthic fishes *Percina roanoka* and *Etheostoma flabellare*. Environmental Biology of Fishes 12:303–308.

Matthews, W. J. 1998. Patterns in freshwater fish ecology. Chapman and Hall, New York, NY.

Matthews, W. J., and L. G. Hill. 1980. Habitat partitioning in the fish community of a southwestern river. Southwestern Naturalist 25:51–66.

Matthews, W. J., and E. Marsh-Matthews. 2006. Persistence of fish species associations in pools of a small stream of the southern Great Plains. Copeia 2006:696–710.

Matthews, W. J., and E. Marsh-Matthews. 2007. Extirpation of Red Shiner in direct tributaries of Lake Texoma (Oklahoma-Texas): a cautionary case history from a fragmented river-reservoir system. Transactions of the American Fisheries Society 136:1041–1062.

Matthews, W. J., and E. Marsh-Matthews. 2011. An invasive fish species within its native range: community effects and population dynamics of *Gambusia affinis* in the central United States. Freshwater Biology 56:2609–2619.

Matthews, W. J., and E. Marsh-Matthews. 2015. Comparison of historical and recent fish distribution patterns in Oklahoma and western Arkansas. Copeia 103:170–180.

Matthews, W. J., and E. G. Zimmerman. 1990. Potential effects of global warming on native fishes of the southern Great Plains and the Southwest. Fisheries 15:26–32.

Matthews, W. J., J. R. Bek, and E. Surat. 1982. Comparative ecology of the darters *Etheostoma podostemone*, *E. flabellare* and *Percina roanoka* in the upper Roanoke River drainage, Virginia. Copeia 1982:805–814.

Matthews, W. J., F. P. Gelwick, and J. J. Hoover. 1992. Food of and habitat use by juveniles of species of *Micropterus* and *Morone* in a southwestern reservoir. Transactions of the American Fisheries Society 121:54–66.

Matthews, W. J., K. B. Gido, G. P. Garrett, F. P. Gelwick, J. G. Stewart, and J. Schaefer. 2006. Modular experimental riffle-pool stream system. Transactions of the American Fisheries Society 135:1559–1566.

Matthews, W. J., K. B. Gido, and E. Marsh-Matthews. 2001. Density-dependent overwinter survival and growth of Red Shiners from a southwestern river. Transactions of the American Fisheries Society 130:478–488.

Matthews, W. J., B. C. Harvey, and M. E. Power. 1994. Spatial and temporal patterns in the fish assemblages of individual pools in a midwestern stream (U.S.A.). Environmental Biology of Fishes 39:381–397.

Matthews, W. J., E. Marsh-Matthews, R. C. Cashner, and F. Gelwick. 2013. Disturbance and trajectory of change in a stream fish community over four decades. Oecologia 173: 955–969.

Matthews, W. J., A. M. Miller-Lemke, M. L. Warren, D. Cobb, J. G. Stewart, B. Crump, and F. P. Gelwick. 2004. Context-specific trophic and functional ecology of fishes of small stream ecosystems in the Ouachita National Forest. Pages 221–230 in J. M. Guildin, ed. Ouachita and Ozark Mountains Symposium: ecosystem management research. General Technical Report SRS-74. US Department of Agriculture, US Forest Service, Southern Research Station, Washington, DC.

Matthews, W. J., W. D. Shepard, and L. G. Hill. 1978. Aspects of the ecology of the Dusky-stripe Shiner, *Notropis pilsbryi* (Cypriniformes, Cyprinidae) in an Ozark stream. American Midland Naturalist 100:247–252.

Matthews, W. J., C. C. Vaughn, K. B. Gido, and E. Marsh-Matthews. 2005. Southern Plains rivers. Pages 283–325 in A. C. Benke and C. E. Cushing, eds. Rivers of North America. Elsevier, Burlington, MA.

Metcalf, C., F. Pezold, and B. G. Crump. 1997. Food habits of introduced Rainbow Trout (*Oncorhynchus mykiss*) in the upper Little Missouri River drainage of Arkansas. Southwestern Naturalist 42:148–154.

Minckley, W. L. 1973. Fishes of Arizona. Arizona Game and Fish Department, Phoenix, AZ.

Minckley, W. L., and P. C. Marsh. 2009. Inland fishes of the greater Southwest: chronicle of a vanishing biota. University of Arizona Press, Tucson, AZ.

Morin, P. J. 2011. Community ecology, 2nd ed. Wiley-Blackwell, West Sussex, UK.

Moring, J. B. 1986. Resource partitioning between two darters, *Etheostoma spectabile* and *Etheostoma lepidum* (Pisces: Percidae) in the South Concho River. MS thesis, Angelo State University, San Angelo, TX.

Moulton, M. P., and S. L. Pimm. 1976. The extent of competition in shaping an introduced avivauna. Pages 80–97 in J. Diamond and T. J. Case, eds. Community ecology. Harper and Row, New York, NY.

Murdoch, W. W. 1969. Switching in general predators: experiments on predator specificity and stability of prey populations. Ecological Monographs 39:335–354.

Outten, L. M. 1958. Studies of the life history of the cyprinid fishes *Notropis galacturus* and *rubricroceus*. Journal of the Elisha Mitchell Society 74:122–134.

Paine, R. T. 1966. Food web complexity and community stability. American Naturalist 100:65–75.

Paine, R. T. 1969. A note on trophic complexity and community stability. American Naturalist 103:91–93.

Patton, T., and C. Tackett. 2012. Status of Silver Carp (*Hypophthalmichthys molitrix*) and Bighead Carp (*Hypophthalmichthys nobilis*) in southeastern Oklahoma. Proceedings of the Oklahoma Academy of Science 92:53–58.

Peckarsky, B. L. 1983. Biotic interactions or abiotic limitations? a model of lotic community structure. Pages 303–323 in T. D. Fontaine III and S. M. Bartell, eds. Dynamics of lotic ecosystems. Ann Arbor Science, Ann Arbor, MI.

Phillips, E. C., and R. J. Kilambi. 1996. Food habits of four benthic fish species (*Etheostoma spectabile, Percina caprodes, Noturus exilis, Cottus carolinae*) from northwest Arkansas streams. Southwestern Naturalist 41:69–73.

Power, M. E. 1987. Predator avoidance by grazing fishes in temperate and tropical streams: importance of stream depth and prey size. Pages 333–351 in W. C. Kerfoot and A. Sih, eds. Predation—direct and indirect impacts on aquatic communities. University Press of New England, Hanover, NH.

Power, M. E., and W. J. Matthews. 1983. Algae-grazing minnows (*Campostoma anomalum*), piscivorous bass (*Micropterus* spp.), and the distribution of attached algae in a small prairie-margin stream. Oecologia 60:328–332.

Power, M. E., W. J. Matthews, and A. J. Stewart. 1985. Grazing minnows, piscivorous bass and stream algae: dynamics of a strong interaction. Ecology 66:1448–1456.

Power, M. E., D. Tilman, J. A. Estes, B. A. Menge, W. J. Bond, L. S. Mills, G. Daily, J. C. Castilla, J. Lubchenco, and R. T. Paine. 1996. Challenges in the quest for keystones. BioScience 46:609–620.

Preisser, E. L., D. J. Bolnick, and M. F. Benard. 2005. Scared to death? The effects of intimidation and consumption in predator-prey interactions. Ecology 86:501–509.

Prugh, L. R., C. J. Stoner, C. W. Epps, W. T. Bean, W. J. Ripple, A. S. Laliberte, and J. S. Brashars. 2009. The rise of the mesopredators. BioScience 59:779–791.

Pyke, G. H. 2008. Plague minnow or mosquito fish? a review of the biology and impacts of introduced Gambusia species. Annual Review of Ecology, Evolution and Systematics 39: 171–191.

Reighard, J. 1920. The breeding behavior of the suckers and minnows. Biological Bulletin 38:1–32.

Robison, H. W., and T. M. Buchanan. 1988. Fishes of Arkansas. University of Arkansas Press, Fayetteville, AR.

Ross, S. T. 1986. Resource partitioning in fish assemblages: a review of field studies. Copeia 1986:352–388.

Ross, S. T. 2013. Ecology of North American freshwater fishes. University of California Press, Berkeley, CA.

Roughgarden, J., and J. Diamond. 1976. Overview: the role of species interactions in community ecology. Pages 333–343 in J. M. Diamond and T. J. Case, eds. 1976. Community ecology. Harper and Row, New York, NY.

Ruppert, J. B., R. T. Muth, and T. P. Nesler. 1993. Predation on fish larvae by adult Red Shiner, Yampa and Green Rivers, Colorado. Southwestern Naturalist 38:397–399.

Schoener, T. W. 1974. Resource partitioning in ecological communities. Science 185:27–39.

Sen, M., M. Banerjee, and A. Morozov. 2012. Bifurcation analysis of a ratio-dependent prey-predator model with the Allee effect. Ecological Complexity 11:12–27.

Sih, A., P. Crowley, M. McPeek, J. Petranka, and K. Strohmeier. 1985. Predation, competition, and prey communities: a review of field experiments. Annual Review of Ecology and Systematics 16:269–311.

Stachowicz, J. J. 2001. Mutualism, facilitation, and the structure of ecological communities. BioScience 51:235–246.

Surat, E. M., W. J. Matthews, and J. R. Bek. 1982. Comparative ecology of *Notropis albeolus, N. ardens* and *N. cerasinus* (Cyprinidae) in the upper Roanoke River Drainage, Virginia. American Midland Naturalist 107:13–24.

Taylor, C. M. 1996. Abundance and distribution within a guild of benthic stream fishes: local processes and regional patterns. Freshwater Biology 36:385–396.

Taylor, C. M., and P. W. Lienesch. 1996. Regional parapatry of the congeneric cyprinids *Lythrurus snelsoni* and *L. umbratilis*: species replacement along a complex environmental gradient. Copeia 1996:493–497.

Thompson, J. 2012. Influence of habitat complexity on reproduction and survival of Red River Pupfish in the presence of Red Shiner. MS thesis, University of Oklahoma, Norman, OK.

Trautman, M. B. 1957. The fishes of Ohio. Ohio State University Press, Columbus, OH.

Troia, M. J., and K. Gido. 2014. Towards a mechanistic understanding of fish species niche divergence along a river continuum. Ecosphere 5:1–18.

Volterra, V. 1928. Variations and fluctuations of the number of individuals in animal species living together. Journal du Conseil International pour l'Exploration de la Mer 3:3–51. [Translated by M. E. Wells.]

Weber, M. J., and M. L. Brown. 2009. Effects of Common Carp on aquatic ecosystems 80 years after "Carp as a dominant": ecological insights for fisheries management. Reviews in Fisheries Science 17:524–537.

Weber, M. J., and M. L. Brown. 2011. Relationships among invasive Common Carp, native fishes and physicochemical characteristics in upper Midwest (USA) lakes. Ecology of Freshwater Fish 20:270–278.

Whitney, J. E., K. B. Gido, and D. L. Propst. 2014. Factors associated with the success of native and nonnative species in an unfragmented arid-land landscape. Canadian Journal of Fisheries and Aquatic Sciences 71:1134–1145.

Wiens, J. A. 1977. On competition and variable environments. American Scientist 65:590–597.

Winston, M. R. 1995. Co-occurrence of morphologically similar species of stream fishes. American Naturalist 145:527–545.

Wood, R. M., R. L. Mayden, B. R. Kuhajda, and S. R. Layman. 2002. Systematics and biogeography of the *Notropis rubellus* species group (Teleostei: Cyprinidae). Bulletin of the Alabama Museum of Natural History 22:37–80.

CHAPTER 6

Disturbance
Weather Extremes, Flood and Drought,
and Fish Community Dynamics

An Extreme Example

On the afternoon of 1 July 1980, during an extraordinary heat wave in south Oklahoma, Eric Surat and two helpers found seven freshly dead Orangethroat Darters (*Etheostoma spectabile*) in one hot, shallow, isolated pool in Brier Creek, in a reach that was normally a flowing riffle with large numbers of these darters. No live darters were found in this or similar isolated pools, although live Orangethroat Darters were in substantial numbers in nearby shaded and deeper pools. Circumstantial evidence suggested outright heat death of these darters, and our follow-up field and laboratory investigations 2 days later confirmed that similar pools had temperatures as hot as 39°C and that this temperature exceeded the critical thermal maximum of darters from that population (Matthews et al. 1982). Ross et al. (1985) confirmed that maximum air temperatures in July 1980 were the hottest in 16 years, far exceeding the average, and WJM noted an unofficial measurement of 117°F (47°C) on his shaded porch in Madill, Oklahoma, about 12 km from the Brier Creek site. So, the summer of 1980 was extreme in temperatures for southern Oklahoma by any standard. Outright heat death of fish in natural environments is unusual, but such extreme events can clearly function as one of the "crunches" envisioned by John Wiens (1977) for variable environments.

Having considered the importance of the traits of individual species in chapter 4 and of interspecific interactions in chapter 5, we turn attention in this chapter to the ways that extreme disturbance events, like unusual heat and cold, drought, and flood, affect individual species and whole fish communities. In chapter 7 we focus on overall change in our 9 global and 31 local communities, which will include the intervals examined in the present chapter. But for now let's see how specific events have or have not affected the fish communities, especially comparing the effects of extreme floods and droughts. Matthews (1998, 326–379) reviewed the substantial literature on flood and drought, which will not be repeated here, and we focus instead on findings from our own research.

Across four decades, we have had opportunities for before-and-after assessment of fish community structure for two different hundred-year floods, decades apart, in Piney Creek (see plates 19–21) in the southern Ozark Mountains of northern Arkansas (Matthews 1986; Matthews et al. 2014); numerous extreme flood events in Brier Creek (see plate 22), in southern Oklahoma (Ross et al. 1985; Matthews et al. 1988, 2013), and two large floods in the Roanoke River in the uplands of western Virginia (Matthews 1998). Matthews et al. (1996) also studied the effects of engineered experimental floods in Sister Grove Creek in north Texas.

At the other hydrologic extreme, we have data on fish community responses to numerous droughts or extreme dry periods in Brier Creek (Matthews and Marsh-Matthews 2006; Matthews et al. 2013; see plates 23 and 25), some of which occurred in rapid succession in recent years. We also have data on drought effects on fishes in four small streams (see plates 7 and 8) in the Grand River basin in northeast Oklahoma (Franssen et al. 2006; Matthews, unpub. data) and in several southern Oklahoma creeks (in addition to Brier Creek) that are direct tributaries to a large impoundment (Lake Texoma) (Matthews and Marsh-Matthews 2007) and other streams in southwest Oklahoma (Matthews, unpub. data).

Disturbance

There is a long history of interest in the effects of disturbance in animal communities (Jewell 1927; Allee 1929; Wootton et al. 1996; Turner 2010). As twentieth-century confidence in equilibrium, niche-based, or competition-driven communities (Gause 1934; Hardin 1960; Schoener 1974) began to erode (Wiens 1977; Grossman et al. 1982; den Boer 1986), ecologists focused more and more on nonequilibrium communities driven by unpredictable "disturbances" of all kinds (Pickett et al. 1989), including hurricanes that devastate forest structure or coral reefs (Boose et al. 1994; Lugo et al. 2000), tree-falls that open light gaps on the forest floor (Hubbell et al. 1999), wave shock or battering by drift logs that remove intertidal communities (Dayton 1971), floods that restructure stream channels and their biotic communities (Resh et al. 1988), and droughts (Lake 2011) that severely affect many aquatic and terrestrial ecosystems. The shift from confidence in an overarching, equilibrium- and competition-based paradigm to embracing the possibility (or probability) that many communities are strongly influenced by disturbances that disrupt equilibrium was one of the major developments in ecology in the twentieth century. But disturbance is hard to study. Most disturbances are rather unpredictable (in spite of El Niño years or hurricane seasons or likelihood of late summer drought in many regions), so disturbance can be studied in one of three ways. The first is to get lucky, be in the right place, and have an event occur amidst an ongoing study, as illustrated by the response of small mammals to a catastrophic rainfall event during Jim Brown's 30-year study of a small-mammal community (Thibault and Brown 2008). The second is to keep a "SWAT team" of scientists ready to go, capturing data before (or between) events and shortly afterward (e.g., Grimm and Fisher 1989). (This is a nice idea, but it can be really expensive or logistically difficult. WJM tried the SWAT approach in 1990, keeping a team of colleagues ready to go during flood season in Brier Creek, but after starting to track the recovery of invertebrates and fishes from one major flood in the system, a series of additional stage rises over three weeks made it impossible to know which flood was being followed!). The third is to engineer a flood or drought in a natural stream (Matthews et al. 1996; Walters and Post 2008) or perform controlled experiments in large mesocosms (Bertrand 2007; Bertrand et al. 2009; Marsh-Matthews and Matthews 2010).

The first approach, emphasizing major disturbance events that happened serendipitously during our multidecade studies of Piney Creek, Arkansas, and Brier Creek, Oklahoma, is one focus of this chapter. We also review other opportunities we have had over shorter periods of time (weeks or months) to assess impacts of flood, drought, or unusual weather, to provide a collective synthesis about droughts and floods.

There is a huge, well-substantiated literature on the effects of disturbance in aquatic systems (Resh et al. 1988) and on the effects of floods and droughts on fishes. Matthews (1998) reviewed more than 80 references on the effects of floods on fishes and stream ecosystems with a wide range of both costs and potential benefits to organisms. Matthews (1998) also summarized the effects of droughts, and Matthews and Marsh-Matthews (2003) analyzed the effects of drought on fishes from more than 50 publications from across the world, beginning with historical summaries of effects of great droughts in the 1930s on fish in the midwestern United States. Lake (2011) provided a major overview of drought effects in aquatic ecosystems, including a nice review of recent papers on the effects of drought on stream fishes and the importance of the tolerances of different species, as noted in chapter 4 of this volume.

The Nature of Temperature Extremes, Floods, and Droughts

Extremes of heat and cold, droughts, and floods occur on quite different schedules, which may relate to the differences in their effects on stream fishes. Extremes of heat and cold in our region are transitory, from days to weeks. Floods rise and fall rapidly at the scale of hours or days. Droughts develop slowly and can last weeks, months, or years. Thus fish have different opportunities to resist or respond to the vicissitudes of these quite different kinds of disturbances.

Temperature Extremes

In our study region in the central United States, from about 30°N to 40°N latitude (central Texas to Nebraska), summers are hot and winters are cold. Maximum recorded water temperatures (US Geological Survey records) for rivers in south or central Oklahoma are about 40°C. Winters are cold enough that creeks and smaller rivers can ice over, and we lack the all-winter freeze-up of more northern streams, but survival of fish can be low in frozen stream pools under some situations (e.g., Labbe and Fausch

2000). Extreme cold temperatures (below 0°C) are relatively rare in the south-central United States, occurring in some winters and not others. Creeks and smaller rivers can ice over transiently, usually with only a few inches of ice at the most. During WJM's doctoral dissertation research on the South Canadian River and Pond Creek, in central Oklahoma, an unusually cold period caused thick ice cover on Pond Creek for 31 days. This reach of Pond Creek consisted of long pools of standing water approximately 0.5 m deep, and the reach was heavily shaded, preventing much warming from the sun. But such long periods of icing are unusual for our area. We have occasionally noted mortality of fish under ice, however (see below).

Floods

The late W. L. Minckley, of Arizona State University, once described a desert flash flood (pers. comm. to WJM) as "being almost alive—when the wall of water comes crashing down a canyon, with a front rising up loaded with debris—you don't want to be there." Desert flash floods clearly cause extensive change in physical and biotic structure of small streams (Fisher et al. 1982; Meffe and Minckley 1987). We have never personally witnessed the destructive effects of a desert flash flood. Floods in our region may not have the sharp erosive front of summer flash floods in canyons but nevertheless are impressive for their physical damage in the riparian zone and modification of stream habitats. We have been present during or immediately after a substantial number of powerful, erosive flood events, including several "great floods" or hundred-year floods of unusual magnitude (as well as severe to extraordinary droughts; Matthews 1986; Matthews et al. 2013, 2014). Most floods in natural streams in our area have fairly steep hydrographs, rising rapidly with powerful erosive force but then dropping almost as quickly (e.g., Wesner 2011). In Brier Creek (see plate 22) we have seen floods on numerous occasions that raised the water level 3–4 m in minutes to hours, with current speeds as much as 2 m/s (Harvey 1987; Matthews et al. 2013, supplementary material), but dropped almost as fast (Power and Stewart 1987), so that by the next day water levels were only 0.5 m or less above base flow.

Droughts

Droughts are a different matter. Droughts (or dry spells) develop gradually and last from weeks to months or sometimes years. Matthews (1998, 349–352) summarized generalized stages of drought. Figure 6.1, reproduced below from Matthews (1998), expressed a typical sequence, based largely on observations in Brier Creek, from a reach flowing freely (A); then dropping (B) so that current speeds decrease, riffles become shallow, and passage of fish from pool to pool would be limited; then drying further (C) so that pools become completely isolated from fish passage, although there can be flow of water from pool to pool through gravel. Development to this stage typically takes days or weeks. But if rains are lacking and summer heat exacerbates evaporation, the water table drops so low that shallow pools go dry (D), leaving only a few remaining pools as refugia for any fish that remain. Physical and chemical water-quality

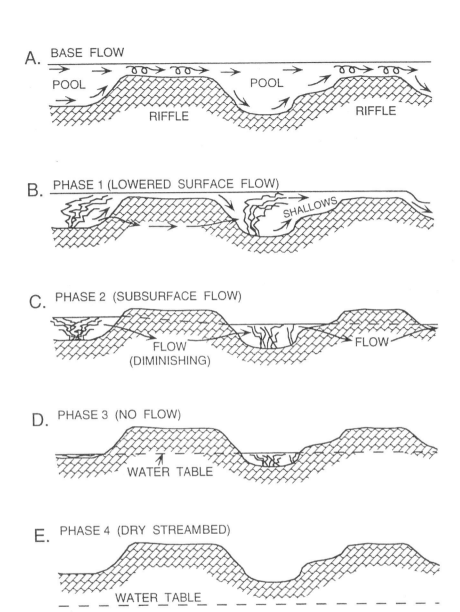

FIG. 6.1 Hypothetical stages of drought in a small, gravel-bottomed midwestern stream. From Matthews (1998). With permission of Springer Science+Business Media.

parameters become extreme in such isolated pools (Magoulick and Kobza 2003). Finally, long reaches of the creek may dry completely (*E*), leaving no local refugia for fish. Fish in the reach all die (e.g., Labbe and Fausch 2000).

In Brier Creek, extensive drought or drying-up of much of the creek occurred in summer or autumn of 1980, 1983, 1988, 1998, 2000, 2006, and 2012. During these periods, flow ceased in much or all of the creek for weeks (see plates 23 and 25) and in worst cases reduced typically large pools to scant remaining water (Marsh-Matthews and Matthews 2010; Matthews et al. 2013). During the dry period of 2000, for example, flow ceased throughout the creek, and water levels remained low for weeks (see fig. 6.3). In the normally flowing reach in our downstream site BR-5 (see fig. 2.1), the water was reduced on 27 August 2000 to one small pool that remained in a crevice in sandstone bedrock which was (with no meter stick handy) one "tennis shoe wide" by 13 "tennis shoes long," estimated to be about 13 cm deep (Marsh-Matthews and Matthews 2010).

Consequences for Fish
Extreme Temperature Effects on Fish
In 40-plus years of studying fishes in our region, the find described in Matthews et al. (1982) suggesting outright heat death of darters is the only such case we have observed directly. There have been other summertime fish kills in shallow rivers, like one reported by Ken Hobson (pers. comm. to WJM) in the South Canadian River (see plate 12) near Norman, Oklahoma, in August 2012, during hot, low-water conditions. In photographs or specimens provided by Dr. Hobson there were numerous Red Shiners (*Cyprinella lutrensis*) and Western Mosquitofish (*Gambusia affinis*) and at least one Plains Minnow (*Hybognathus placitus*) found dead in shallow water along the shore or in algae-encrusted side channels of the river. We suspect that this mortality was caused by a combination of heat and oxygen depletion, from extensive algae encrustations whose respiration could have lowered oxygen overnight in the extremely reduced river at that time, rather than the absolute water temperature. Regardless, the three species that can be confirmed in this fish kill are among the most tolerant of high temperatures and low oxygen (Matthews and Maness 1979; Matthews 1987), so even the hardiest of fishes sometimes reach their limits in the harsh conditions that are possible during drought in shallow prairie rivers.

During sampling for WJM's doctoral dissertation in summer 1976, the few minnows that remained in the main channel of the Canadian River (see plate 12), where water temperature reached 37°C because of high air temperatures and solar insolation over the shallow sand bed, often appeared stressed to near their limits of thermal tolerance and often died when trapped in the seine. At the same time, most minnows and other fishes were found in deeper, shaded pools near the banks of the river (Matthews and Hill 1980) and appeared to be healthy in these refugia. The ability to seek out and remain in the best of available conditions is probably crucial to survival of these fish during the worst of times.

Extreme winter cold can kill fish in our region if they become trapped in small pools with extensive ice cover, but evidence of outright death from cold or associated oxygen depletion in these pools is relatively rare. In the headwaters of Little River, central Oklahoma, WJM once found numerous minnows and bullhead catfish (*Ameiurus* spp.) dead under ice in one small pool. At the Tallgrass Prairie Preserve in north Oklahoma (Stewart et al. 1999), WJM, Chris Taylor, and Fran Gelwick sampled during an

PLATE 1 Piney Creek, Izard County, Arkansas, at site P-1, showing limestone bluffs, long pool, and large rapids in background. See chapters 2, 5, 7, and 8.

PLATE 2 Brier Creek, Marshall County, Oklahoma, at site BR-6, with Nathan Franssen and Edie Marsh-Matthews seining in one long pool. See chapters 2, 7, and 8.

PLATE 3 Kiamichi River near Muse, LeFlore County, Oklahoma, at site K-7, showing long pools interspersed by gravel bars with dense stands of water willow. See chapters 2, 5, 7, and 8.

PLATE 4 Roanoke River site 1, southwest of Salem, in Roanoke County, Virginia, showing wide pools and large riffle area. See chapters 2, 4, 5, and 7.

PLATE 5 Crooked Creek falls, below the uppermost sampling site on this stream and immediately above site CK-5. See chapters 2, 5, 7, and 8.

PLATE 6 Crutcho Creek, site CrOn5, on Tinker Air Force Base, Oklahoma County, Oklahoma, near a main thoroughfare on the base and crossed by a large utility pipe, bordered by a golf course. See chapters 2, 7, and 8.

PLATE 7 Little Elm Creek, Ottawa County, Oklahoma, at a site subject to frequent drying, where Western Mosquitofish dominate the fish community. See chapters 2, 3, and 6.

PLATE 8 Coal Creek, Ottawa County, Oklahoma, at a site with occasional drying but a diverse fish community, including numerous minnows, topminnows, Western Mosquitofish, Brook Silverside, sunfish, and black bass. See chapters 2 and 6.

PLATE 9 Illinois River, Adair County, Oklahoma, just downstream of a large rock dam forming Lake Frances, Arkansas. The fish community at this large, complex river site is highly diverse, with a total of 46 species detected in our collections. See chapter 2.

PLATE 10 Tyner Creek, Adair County, Oklahoma, with extremely clear, spring-fed water over gravel substrates. See chapter 2.

PLATE 11 Salt Fork of the Red River, Greer County, Oklahoma, with an ichthyology class seining fish in this shallow, sand-bottomed prairie river. See chapters 2, 5, and 6.

PLATE 12 Canadian River (also known as the South Canadian River), McClain County, Oklahoma, where the river is wide, shallow, and sand bottomed, but with one deep erosional pool near the base of the tree-lined bank, on the outside of a bend in the river. See chapters 2, 4, 5, 6, and 7.

PLATE 13 Blue River, on a preserve owned by The Nature Conservancy in Johnston County, Oklahoma. Within this reach, the spring-fed, marl-depositing Blue River is highly complex structurally, with clear water over substrates of gravel, cobble, or boulders. We have taken a total of 28 species at this site. See chapter 2.

PLATE 14 Red Shiner, *Cyprinella lutrensis*, a species extremely tolerant of harsh conditions and one of the most common minnows throughout the central United States. See chapter 4. Photo courtesy of Brandon Brown.

PLATE 15 Male Western Mosquitofish, *Gambusia affinis*, attempting copulation with a female. The Western Mosquitofish is a livebearer species highly tolerant of harsh environmental conditions, and it is able to rapidly repopulate rewatered streams after drought. Widespread in the central United States, often extremely abundant in small, variable, or vegetated habitats. See chapters 3, 4, and 7.

PLATE 16 Orangethroat Darter, *Ethesostoma spectabile*, a common darter species throughout the central United States, with large populations from highly stable, spring-fed upland streams, to some harsh, prairie-margin streams. See chapter 4. Photo courtesy of Brandon Brown.

PLATE 17 Typical members of a local fish community taken in one seine haul, including Western Mosquitofish, Blackstripe Topminnow, Highland Stoneroller, and a juvenile Largemouth Bass (with its head hidden). See chapter 1. Photo courtesy of Ana Hiott.

PLATE 18 Sunfish "mesopredators" in Brier Creek, including adult Green Sunfish (*left*) and adult Longear Sunfish (*right*). See chapters 4 and 5.

PLATE 19 Piney Creek at Boswell Road bridge, looking upstream toward site P-1. During the extreme floods of December 1982 and March 2008, water rose vertically more than 12 m and left debris on the undercarriage of the bridge. See chapters 4, 5, and 6.

PLATE 20 Site M-1 in Piney Creek after the great flood of December 1982. Large mature trees swept away by the flood and huge amounts of sand were deposited in the riparian zone as floodwaters receded. See chapters 4 and 6.

PLATE 21 Site M-3 in the headwaters of the Piney Creek watershed after the great flood of December 1982. This was the worst scoured of all our permanent study reaches. Before the flood, this riparian zone had been completely covered by grasses, shrubs, and small trees. See chapters 4, 6, and 8.

PLATE 22 Brier Creek headwater site BR-2, flooding after heavy rains. See chapters 4, 6, and 8.

PLATE 23 Brier Creek headwater site BR-2, nearly dry during extreme drought. See chapters 4, 6, and 8.

PLATE 24 Buncombe Creek, Marshall County, Oklahoma, with backwater flooding from Lake Texoma that raised the creek to more than 6 m above normal water level, with trees in the process of falling into the creek because the high waters eroded the earthen banks. See chapter 5.

PLATE 25 An extensive dried bedrock reach within site BR-2 of Brier Creek during drought, reducing what is normally a flowing reach to a few isolated pools. See chapters 4, 6, and 8.

PLATE 26 Grazing scars left by Central Stonerollers in the attached algae "felt" coating a cobble (approximately 18 cm wide) in Flint Creek, northeast Oklahoma. See chapters 4 and 9.

PLATE 27 Central Stoneroller, *Campostoma anomalum*, a male in reproductive condition with breeding tubercles and nuptial colors on the body and fins. This species has strong effects in stream ecosystems throughout the central United States. See chapters 4 and 9.

PLATE 28 A pool in Brier Creek that had been divided lengthwise by plastic sheeting, and the Central Stonerollers removed from the near side in the picture. Note proliferation of long strands of attached algae in the side (toward the rear) that was not grazed by stonerollers. See chapters 4 and 9.

PLATE 29 Arrowhead-shaped pens (pointing upstream) in Baron Fork, allowing Central Stonerollers and other algivorous fishes (such as Ozark Minnows) access to one side of the pen but protecting the other side from grazing fishes. See chapters 4 and 9.

PLATE 30 Close-up of one of the pens, with algae and silt accumulating on the nongrazed left side and cleanly grazed gravel and experimental clay tiles on the right side, which were exposed to grazing by Central Stonerollers and Ozark Minnows. See chapters 4 and 9.

PLATE 31 The authors, seining a small creek in the Fourche Maline drainage in Robbers Cave State Park, Latimer County, Oklahoma, in 2009. See chapter 1. Photo courtesy of Ana Hiott.

unusual November cold spell, breaking an inch of ice to seine. Field notes (WJM 2355b) indicated that under the ice in a large pool (10 m × 5 m) at least 7 species of fish were alive, including "many *Lepomis*." But in a smaller (5 m × 3 m) pool in the same stream reach, there were "lots of dead fish under ice, including 1 large *Percina caprodes* [Logperch] . . . mostly minnows and sunfish." Extreme winter cold can kill fish in our region, but typically only in small, isolated pools. More often when we have to sample in cold weather we find many fish alive under the ice. In the same November at the Tallgrass Prairie Preserve, for example, with water temperature 2°C under ice cover in a pool 50 cm deep and 5 m wide, field notes (WJM 2353b) indicated, "break and move ice to seine . . . (took live) many small *E. spectabile* and a few larger, *Labidesthes sicculus* [Brook Silversides], *F. notatus*, *N. boops* (20 observed but not caught), Campos [*Campostoma anomalum*], *P. notatus* [Bluntnose Minnow], *L. cyanellus*, *L. microlophus* [Redear Sunfish], *L. megalotis*." In another reach on the same day in pools 75 cm deep by 8 m wide, the field note (WJM 2356b) indicated, "all pools covered with ice," and many fish of at least 11 species were alive under the ice. We have found instances of fish apparently directly dead from summer heat or dead under ice in winter, but much more often, even in such temperature extremes, fish in our region seem to survive, partly through tolerance for extreme conditions and partly through habitat selection, seeking out the most favorable (or least unfavorable) microhabitats (Matthews and Hill 1979, 1980; Matthews 1987).

Flood Effects on Fish

Floods affect fish communities in many ways (reviewed in Matthews 1998, 326–340). But the effect of each flood differs, depending on factors like the timing relative to fish reproduction (Starrett 1951; Strange et al. 1992), the hydrograph ("flashy" and erosive, versus laterally spreading; Ross and Baker 1983), complexity of stream habitats (Pearsons et al. 1992), or the position of the study site in a watershed (Matthews 1986). And numerous studies have showed only limited impact of extreme erosive (Matthews 1986) or flash floods (Pires et al. 2008) on overall fish community dynamics. One study (Plath et al. 2010) showed that a flood of "catastrophic" proportions (50-year maximum) did little to break down differences among locally isolated gene pools in a Mexican cave and surface fish system.

Why do floods have somewhat minimal effects on fish in the streams we have studied (Wesner 2011) and in other systems (Meffe and Minckley 1987; Matheney and Rabeni 1995)? Floods can decimate fish larvae (Harvey 1987), but there is no evidence that adults generally are washed away or killed outright by erosive floodwaters. Typically, adult behaviors help them avoid the worst effects of floods. None of the midwater fishes in the Midwest can withstand currents in midchannel during erosive floods (which can reach speeds of 2 m/s; Matthews et al. 2013), and benthic fishes remaining in riffles or rapids could be crushed when stream substrates move. But adult fish in floods can move to stream edges or other low-current microhabitats like flooded forest (Matheney and Rabeni 1995), then move back into their usual habitats after water recedes, as documented extensively in the Des Moines River, Iowa, by Starrett (1951). (We note parenthetically that the work by William C. Starrett [1950, 1951] foresaw much about stream fishes that we have since "discovered" in our own studies.) Matthews (1998, 334–336) found that fish species typical of the main channel of the Roanoke River, Virginia, including darters and suckers, moved into slow-current habitats at the edge of the channel or were at the edge of floodwaters on urban lawns during two large, erosive floods (Matthews 1998). During a strong erosive flood in Brier Creek on 14 March 1990, WJM 2437 noted, "seined along edge in water slightly protected from fast current and in small backwater. On edge of main channel in slow water [were] a *N. boops* [Bigeye Shiner], several *L. cyanellus* [Green Sunfish] and *L. macrochirus* [Bluegill] and numerous *E. spectabile* [Orangethroat Darter] males with milt . . . far more darters than would usually be at very edge of stream." Franssen et al. (2006) and Wesner (2011) suggested that erosive floods allowed rapid colonization of fish into formerly dry areas, isolated pools, or pools that were experimentally defaunated. And flooding can sweep substrates clean of silt (Matthews et al. 2014) or deepen pools and expose rock substrates, potentially favoring fish that are egg attachers or gravel nesters, so flooding may have benefits as well as costs for various fish species.

As an example of different flood effects within one system, we studied effects of two hundred-year floods in Piney Creek, in December 1982 and March–April 2008 (Matthews et al. 2014). Each included a rapid stage rise of about 12 m at the lowermost site (see plate 19). In-stream scouring (see plate 21) and movement of the streambed were extensive, and there was much destruction of riparian forests, with substantial input of large woody debris (whole trees) into the creek or deposited in the riparian zone (see plate 20).

But the two floods differed in that one was in early winter, outside the reproductive season for most fishes, while the other occurred when many fishes had begun reproductive activity. They also differed in that the 2008 event consisted of two flood peaks in Piney Creek. The first, an erosive flood in March, was from extreme rainfall, but in April there was a second flooding as waters from the nearby White River surged into Piney Creek as river levels were raised by water releases from upstream reservoirs. While the 1982 flood peaked and subsided fairly rapidly, the 2008 events resulted in prolonged high water. These floods have differed in the time course of recovery of the fish communities. After the December 1982 flood, our survey in January 1983 showed decreases in numbers of species present at several sites (Matthews 1986), but by August 1983, eight months after the flood, the overall fish community was similar to its structure in August of the year before (Morisita-Horn, or MH, index = 0.980; table 6.1), and the numbers of species had recovered to be equal to or even greater than before the flood (Matthews 1986). In contrast, the spring flood in 2008 was followed by more substantial change in the Piney Creek fishes (fig. 6.2). Surveys of five of the Piney Creek study sites in August 2008, August 2010, and July 2012 indicated some changes in the community immediately after the flooding and that even four

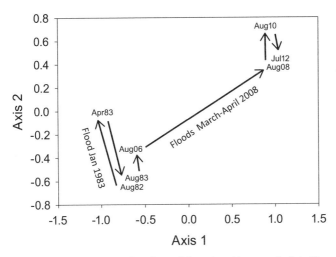

FIG. 6.2 Nonmetric multidimensional scaling of five sites (data pooled) in Piney Creek before and after two hundred-year floods. From Matthews et al. (2014). Copyright by the American Society of Ichthyologists and Herpetologists.

years later (fig. 6.2) community structure was only partly recovered (Matthews et al. 2014).

Experimental Floods

A large experiment based on engineered floods in a natural creek was made possible by a major construction project by the North Texas Municipal Water District to transport large volumes of water from Lake Texoma, Oklahoma-Texas, to Lake Lavon in north-central Texas, to supply water for the Dallas metro area (Golladay and Hax 1995). A pumping station was built near the downstream end of Lake Texoma to deliver large volumes of water through a pipeline to the headwaters of Sister Grove Creek, which then became an open-channel conveyance system for the water down to Lake Lavon. We sampled fish throughout Sister Grove Creek before and after 3 experimental flows of 10–14 days each. Each trial period included at least 8 days of high-velocity pumping to maintain discharge in the headwaters of the creek slightly above 3 m³/s, much greater than the discharge of <0.1 m³/s in the headwaters before initiation of trial flows, and current speeds in some parts of the creek were as much as 100 cm/s (Matthews et al. 1996). Fish were abruptly exposed to more than a 30-fold increase in discharge, from nonflowing pools to these very high flows (Matthews et al. 1996).

Samples taken a week before and a week after the trial flows showed little overall change in the abundance of individual fish species, but at some sites there were substantial changes in the local fish communities. Across the three periods of experimental high flows, some sites had relatively high similarity before and after (percentage similarity 80%–90%), but others had percentage similarity values <50% before and after, suggesting substantial rearrangement of some fishes by the trial flows.

Many sites in the samples after the trial floods were similar to samples before the event. In 1990, only site 2 (most downstream) was substantially displaced from "before." In 1991 and 1992, this site was again the one most displaced by the flood. Matthews et al. (1996) concluded that "the overall impression was that the experimental discharges resulted in little marked or directionally consistent change in the fish assemblages." And there was no significant effect (measured by ANOVAs) of discharge on numbers of species collected, numbers of individuals, or diversity for the creek as a whole. They concluded that Sister Grove Creek, as a naturally dynamic system, with extended periods of natural intermittency but sharp natural flood peaks (Matthews et al. 1996), has a fish fauna that is probably well adapted to rapid changes in hydrology. The fish fauna of Sister Grove Creek consisted of 39 species in 12 families, dominated by rather hardy species such as Western Mosquitofish, Red Shiner, Blackstripe Topminnow (*Fundulus notatus*), Bullhead Minnow (*Pimephales vigilax*), Longear Sunfish (*Lepomis megalotis*), and Bluegill (*Lepomis macrochirus*). The results of these experimental floods reinforce our idea that floods probably have less effect than drought on our native fish communities.

Drought Effects on Fish

Droughts have a wide range of effects on stream fishes (Lohr and Fausch 1997; Matthews and Marsh-Matthews 2003; Magoulick and Kobza 2003; Lake 2011), including outright death from desiccation; depletion of small species by piscivores in crowded refuge pools; or differential mortality of species from high temperature, low oxygen, or generally poor water quality (Magoulick and Kobza 2003). Species differ in patterns of recovery of individuals that do survive drought (Marsh-Matthews and Matthews 2010) or in their ability to recolonize rewetted reaches after drought (Matthews 1987). Prolonged drought can strongly affect fish or aquatic communities (Grossman et al. 1998; Bêche et al. 2009; Galbraith et al. 2010). And there can be short-term (Rutledge et al. 1990) or extremely long-term (Douglas et al. 2003) changes in the genetic composition of populations as results of drought-caused bottlenecks and genetic drift. We have several case histories of extreme crowding of fish during droughts in south Oklahoma streams.

Case 1

We found extreme crowding of fish in three isolated, shrinking pools in the headwaters of Brier Creek in late summer 1982 (Matthews and Marsh-Matthews 2007). On 10 September 1982 (WJM 1297) WJM, Scott Shellhaass, and Art Stewart seined and fin-clipped all fish sufficiently large for that treatment in 1 remaining pool 50 m×5 m×30 cm deep, about 0.5 km downstream from our usual BR-2 study reach, which was completely dry. On 16 September 1982, WJM and Shellhaass fin-clipped (different marks) and released fish in 2 additional pools (11 m×5 m×30 m; and 10 m×1 m×20 cm) upstream from the BR-2 reach. The plan was to await rains, then determine whether recolonization in BR-2 was greater from upstream or down-

Table 6.1.   Fish found in three isolated pools in Brier Creek on 10 or 16 September 1982

Species	Downstream Pool	Upstream Pool 1	Upstream Pool 2	Total
Red Shiner	32	32	40	104
Blacktail Shiner	2	0	0	2
Bigeye Shiner	164	0	1	165
Bullhead Minnow	19	6	35	60
Fathead Minnow	4	0	1	5
Golden Shiner	0	0	1	1
Common Carp	0	0	4	4
River Carpsucker	2	0	2	4
Spotted Sucker	1	0	2	3
Black Bullhead	0	7	29	36
Yellow Bullhead	4	1	7	12
Blackstripe Topminnow	1	1	11	13
Largemouth Bass	13	2	20	35
Green Sunfish	51	175	151	400
Orangespotted Sunfish	15	10	25	50
Longear Sunfish	212	46	66	324
Redear Sunfish	16	8	11	35
White Crappie	4	0	14	18
Slough Darter	0	0	1	1
Freshwater Drum	1	1	0	2
Total Fish	643	343	474	1485

stream. A total of 1482 fish of 21 species was found in the 3 pools, collectively (table 6.1). Many of the centrarchids, including large Largemouth Bass (*Micropterus salmoides*), Green Sunfish (*Lepomis cyanellus*), and White Crappie (*Pomoxis annularis*), were big enough (100 to 380 mm TL) to eat minnows or smaller sunfish, and the small sunfish could have been competitors with minnows for food in these isolated pools.

Densities of fish were extremely high in these pools, with predators, prey, and potential competitors all confined to the little water that remained in about 1 km of Brier Creek. A week later, many of the centrarchids remained but most minnows had disappeared (Matthews and Marsh-Matthews 2007). As for the results of the "recolonization" trial, drought was prolonged, all three pools dried completely, and all the fish died. So it goes in field experiments—sometimes you get lucky, sometimes you don't. The reach of Brier Creek was recolonized the following spring, after rains finally filled the creek (Matthews 1987), but not by any of the fish we fin-clipped!

Case 2

By 27 August 2000, a normally flowing reach of Brier Creek (described in the introduction to drought, above) was reduced to 1 tiny pool, with an estimated volume of 0.176 m³ (Marsh-Matthews and Matthews 2010). We seined this small pool until no more fish were caught, yielding 35 Central Stonerollers (*Campostoma anomalum*),

30 small Yellow Bullheads (*Ameiurus natalis*), 17 Orangethroat Darters, 2 Green Sunfish, and 1 Largemouth Bass, for an overall density of 85 fish in this small volume. The degree of crowding can be indicated by (obviously inappropriate) extrapolation of this density to 482 fish per square meter. Regardless, this example also shows that there can be extremely high densities of fish in shrinking pools.

Case 3

Another example of concentration of fish in the same extreme drought in 2000 was from sampling by EMM, Becky Marsh, and Andrew Marsh, when Brier Creek had been in severe drought for about six weeks. The driest period in recorded state history for the months of August and September for the South Central Climate District (including Brier Creek) was 1 August to 30 September 2000, with only 2.3 cm of rain, almost 16 cm below normal, and heat in that period was intense (Johnson et al. n.d., Oklahoma Climatological Survey fact sheet J3.4). On 10 September 2000, they made 2 standard small seine hauls (each about 4–5 m long) with short (1.8 m) seines in each of 3 "small" pools (Pools 3, 7, 9) and 3 "large" pools (Pools 4, 5, 12), all severely reduced by drought (see fig. 6.3, showing Pool 7 a few weeks later), in the snorkeling reach of Brier Creek north of State Highway 32 (Matthews et al. 1994). A third or fourth haul was made in each pool with slightly longer coverage to obtain additional specimens, but in no case was the entire pool seined or were fish removed to depletion.

FIG. 6.3 Keith Gido at Brier Creek Pool 7 during drought in October 2000. Pool 7 normally filled all of the gravel-bedded area in the image, bank to bank.

A snorkeling survey in July 1999 by WJM was the most recent before the 2000 drought, so we used those data to approximate the kinds of fish in these pools the year before and during the extended 2000 drought. These comparisons should not be taken too literally, because snorkeling is a nearly complete census of fish in a pool, and the seining in 2000 was limited to a small part of the pools that remained; that is, seining did not result in nearly as complete a census as snorkeling provides. And fish often move between pools, so there is no implication that an individual in a pool in 2000 had been in that pool in 1999. In the 3 "small" pools (28–78 m long; maximum width 7–10 m; maximum depth 75–97 cm), 790 individuals of 11 species were counted by snorkeling in 1999. In the same pools in 2000, much reduced by drought (e.g., Pool 3 was by then not more than about 10 m long, and Pool 7 appeared similarly reduced in size), seining produced 661 individuals of 10 species. In the 3 "large" pools (75–92 m long; maximum width 8–13 m; maximum depth 80–143 cm), 317 individuals of 7 species were counted in 1999. Smaller numbers of species and total individuals in these larger pools related to more large bass and fewer Central Stonerollers: 426 in small pools compared to only 11 stonerollers in large pools. In these same 3 large pools in the 2000 late-drought seine hauls, we took 769 individuals of 10 species. The take-home message is that during the drought of 2000, as many or more individuals and species remained in the shrinking pools as had been present in the same pools at their normal base-flow level in the previous summer.

Case 4

We summarized in Matthews and Marsh-Matthews (2007) our sampling fish with Nate Franssen in isolated pools in Buncombe, Brier, and House Creeks, all tributaries to Lake Texoma, during extreme drought in summer 2006 (see fig. 6.8). In 5, 5, and 4 small isolated pools in these three creeks, respectively, embedded within long, dry reaches of the creeks, we found few minnows of any kind but large concentrations of potential piscivores (see fig. 1.1), including large sunfish of various species, an adult Spotted Gar (*Lepisosteus oculatus*), and a large Channel Catfish (*Ictalurus punctatus*), and many small sunfish, bass, or bullheads (totaling 557, 363, and 173 individuals in the three creeks, respectively) that could have competed with minnows for food (Matthews and Marsh-Matthews 2007). Small, isolated pools in creeks are tough places to be a fish during droughts or dry spells.

Whole-Community Changes in Structure after Drought

Fish communities can potentially be affected long after drought ends. Four examples highlight sustained negative impacts of drought on fish communities we have studied. Much of Oklahoma was in "severe" (Palmer drought index below −3.0) or "extreme" (Palmer drought index below −4.0) drought much of 2011 and 2012 (National Oceanic and Atmospheric Administration data). In Cow Creek in June and September 2010 before these droughts, we took 104 and 118 individuals, comprising 8 and 9 species, respectively. During the droughts of 2011–2012, the reach of creek we normally

FIG. 6.4 Edie Marsh-Matthews and Nick Shepherd in pool in Cow Creek, October 2012, after the creek had rewatered following drought of 2011–2012.

seined was completely dry (according to the landowner). By October 2012, when we and Nick Shepherd sampled Cow Creek, rains had restored the creek to normal water levels (fig. 6.4), but extensive seining in the reach produced absolutely no fish! Apparently the period of rewatering since the end of the drought had been insufficient for recolonization of fish from downstream (or upstream?) refugia. However, in Cow Creek samples in December 2014 and June 2015, with Nick Shepherd, Aaron Geheber, and Zach Zbinden, we found 179 individuals of 14 species, including (in June 2015) numerous Red Shiner males in bright nuptial color and numerous Largemouth Bass young-of-year. So the local fish community in Cow Creek did eventually recover from the 2011–2012 drought, but recovery had not been immediate.

For nearby Coal Creek (see plate 8), we had no explicit information that the reach had been dry in 2011 or 2012, but seining in our fixed reach in that stream in October 2012 also produced no fish, whereas in June and September 2010 we had captured 137 and 354 individuals, of 8 and 7 species, respectively. In October 2012 we finally did locate a few fish in a large pool downstream of our usual sampling reach, but for both of these creeks the drought of 2011–2012 apparently depleted all fish in our fixed study reaches. In December 2014 and June 2015 samples from the Coal Creek reach, we took a total of 220 individuals from 14 species, also with Red Shiner males in peak color, numerous young-of-year Largemouth Bass, and very large female Western Mosquitofish (livebearers) that were extremely swollen with embryos. Fishes in both

FIG. 6.5 Ichthyology students sampling in the Salt Fork of Red River south of Mangum, Oklahoma, spring 2013. Note that the stream had good flow at this time, after a long period of no flow (see also plate 11).

Coal and Cow Creek local reaches recovered from the drought, but it may have taken two to three years to match earlier community complexity.

Two other cases in which the 2011–2012 droughts caused serious depletion of fish were in southwest Oklahoma, which had had even worse drought conditions than the state as a whole. Salt Fork of Red River (fig. 6.5; see also plate 11), south of Mangum, Oklahoma, which was at zero flow for months during the drought, had been sampled by our ichthyology classes in 2000, 2002, 2004, 2008, and 2013, always with several groups collecting thoroughly for about an hour. In all samples before the 2011–2012 droughts, we caught from 418 to 1918 individuals of 10 to 14 species at the Salt Fork site. In spring 2013 there was again water flowing at the site, but extensive effort produced only 23 individuals of 4 species, including Red Shiner, Plains Minnow, Red River Pupfish (*Cyprinodon rubrofluviatilis*), and Western Mosquitofish. Most minnow species and all the centrarchids we had found previously were not found in 2013. We suspected that in the extreme droughts there might have been major fish losses in that river and that it could take years for rewatered stream segments to repopulate.

We were equally concerned about West Cache Creek, which usually has substantial clear-water flow from the granitic Wichita Mountains, west of Lawton, Oklahoma.

Historically, West Cache Creek has had a relatively abundant and diverse fauna for a small stream, with darters, minnows, sunfish, and black bass. In 3 samples by ichthyology classes from 1975 to 2002, from 39 to 650 individuals were taken, with 8 species detected in each survey. But in spring 2013, our class collected only 18 individuals of 3 species, in spite of our keeping them there longer than usual and working hard to find any fish at all. In October 2015, we revisited both sites with our ichthyology class and found that these communities had returned to their former richness and complexity in the intervening two years.

In addition to the anecdotal cases above, three droughts (two only two years apart) in Brier Creek each resulted in marked changes in this well-documented fish community (Matthews et al. 2013). The community eventually returned toward its average condition, however. Our data on Brier Creek suggest that successive disturbance events within a short period of time may have greater effects on a fish community than any single event. On balance, the timing or sequences of events may be the most critical aspect of disturbances—which may become more frequent in the predicted changed climate of the future.

In the future, too many stressors in close proximity, with harsher or more extended droughts, could cause permanent loss of native fish species. The examples above suggest that frequent monitoring of stream fish communities (e.g., Kiernan and Moyle 2012) is needed if we are to assess future effects of droughts (or, of course, climate change in general; Covich et al. 1997) or protect native fishes and devise innovative conservation strategies.

Responses to Manipulated Droughts in Artificial Streams

In the summers of 2000 and 2001, we created experimental droughts in large outdoor mesocosms (fig. 6.6) to examine proximate and residual effects of drought on a typical stream fish community. Fishes from Brier Creek, Oklahoma, were stocked in experimental streams consisting of three pool units connected by two riffle units at densities mimicking those that occur naturally.

In June 2000, five species comprised the experimental communities: Central Stoneroller, Bigeye Shiner (*Notropis boops*), Blackstripe Topminnow, juvenile Longear Sunfish, and Orangethroat Darter. In 4 of 8 large, outdoor experimental stream systems, each with 3 pools and 2 riffles (Matthews et al. 2006), flow was maintained by recirculating pumps throughout a 40-day experiment. In the other four, water was removed gradually over a number of days to simulate natural stages of drought (fig. 6.1) until pools were isolated and flow ceased. The "drought" pools further dried by desiccation over the 40 days, trapping fish in small, shallow pools. At the end of the experiment (fig. 6.6), fish were removed from all the units, enumerated, and preserved for dissection and fat analysis. Of the five species in this experiment, two (Blackstripe Topminnow and Central Stoneroller) had significantly lower survivorship in the drought relative to the nondrought units. Although the other three species did not display differential survivorship, there were differences in size of Central Stonerollers,

FIG. 6.6 Conditions in one pool of the modular stream system at the end of an experimental drought.

with smaller individuals retrieved from the drought units; in percentage body fat of both Bigeye Shiners and Orangethroat Darters, with leaner individuals found in drought units; and in reproductive investment of Bigeye Shiners, such that those from drought units had lower gonosomatic indices (Marsh-Matthews and Matthews 2010).

In June 2001 we set up a similar experiment to examine residual effects of drought. In this experiment we stocked units with only three species: Central Stoneroller,

Bigeye Shiner, and Orangethroat Darter. As in the previous experiment, flow was maintained in four of the eight stream units, and water was removed in stages, mimicking the onset of drought in the other four. At the end of 40 days, rather than ending the experiment, water was restored (by a fortuitous rainstorm) to the drought units, and all units were returned to flowing condition for an additional 221 days. When the experiment was ended in March 2002, all units were similar in flow and water-quality conditions. None of the three species displayed significant differential mortality between those units subjected to drought and those that were not, but there was a marginal reduction in survivorship of Orangethroat Darters in the drought units. As in the previous experiment, Central Stonerollers exposed to drought were smaller, but Orangethroat Darters exposed to drought were actually larger, had higher percentage body fat, and had higher gonosomatic indices than those in the units that had not been exposed to drought. It is likely that the reduced density of Orangethroat Darters (albeit a small effect) allowed the survivors access to more resources to support growth, fat storage, and reproduction (Marsh-Matthews and Matthews 2010). Thus different species varied not only in their proximal responses to simulated drought but also in the residual effects of drought exposure. The bottom line from both experiments is that (1) different species in a community may show differential survival or different immediate effects of drought, but (2) individuals that survive a severe drought may thrive afterward, make up for lost condition, and have substantial reproductive value. In other words, individuals that do survive a stressful period may be "good for something," after all.

Mechanisms of Resistance to Disturbance

Fish resist disturbances by species-specific physiology, behavior, and ecological traits. We evaluated thermal and oxygen tolerances of numerous individual species in chapter 4, which related directly to their differential survival during stress periods. And microhabitat selection—seeking out shaded, cooler pools during harsh periods of high temperature—may also help these species survive drought (Matthews and Hill 1979, 1980; Matthews et al. 1994; Matthews 1998). We also found that occupying deeper pools with cooler water, as thermal refugia, was important to survival of the federally threatened Leopard Darter (*Percina pantherina*) in the Glover River, Oklahoma, when solar insolation heated shallow parts of the river beyond limits of tolerance for the species (Schaefer et al. 2003). In one study, Matthews and Styron (1981) found in laboratory tests that fish from intermittent headwaters, which often had higher temperatures or lower oxygen concentrations than the main Roanoke River, were more tolerant of abrupt physicochemical changes than were fish from the river mainstem. Other field studies have showed differential recolonization rates of fish into the headwaters of Brier Creek after drought ended, as related to their tolerance for low-oxygen conditions (Matthews 1987). Finally, reproductive behaviors or increased output after extreme floods or severe droughts (Matthews 1986; Marsh-Matthews et al. 2002; Marsh-Matthews and Matthews 2010) may be important for some fish species.

This latter phenomenon is probably important, although the responsible factors are not understood and mechanisms remain speculative. However, we have found remarkably similar responses of some fish (suckers in particular) after both major droughts and floods. The three (or four) most extreme events for which we have samples soon afterward (within a year) include the Piney Creek flood of December 1982 (followed by survey in August 1983), the Piney Creek flood of spring 2008 (followed by survey in August 2008 at five sites), and the extreme droughts in Brier Creek in 1998 and 2000 (followed by samples throughout the creek in 1999 and 2001). After all three events, there were two- to threefold increases in total fishes collected, relative to samples before the events, mostly related to strong production of young-of-year or juveniles.

In August 1983, after the December 1982 flood, we caught a total of 10,524 individuals across all 12 permanent sites, compared to 5916 in August 1982. Minnows increasing from 1.5- to 5-fold included: Stoneroller spp., Bigeye Chub (*Hybopsis amblops*), Ozark Minnow (*Notropis nubilus*), Hornyhead Chub (*Nocomis c.f. biguttatus*), Bigeye Shiner, Duskystripe Shiner (*Luxilus pilsbryi*), Sabine Shiner (*Notropis sabinae*), and Bluntnose Minnow (*Pimephales notatus*), and Northern Studfish (*Fundulus catenatus*) increased by threefold (Matthews 1986, table 2). Golden Redhorse (*Moxostoma erythrurum*) and Black Redhorse (*Moxostoma duquesnei*) each increased threefold from 1982 to 1983. Most of these increases were the result of many more young-of-year and juveniles in 1983 than in 1982. Other species remained in approximately the same abundance in August 1983 and in 1982, and no common species showed any noteworthy decrease in raw abundance.

In August 2008, after the springtime Piney Creek flood, total individuals in our samples pooled across 5 sites increased from 1684 in 2006 to 4167 (Matthews et al. 2014). Increases in total abundance were shown by Ozark Minnows (twentyfold) and Telescope Shiner (*Notropis telescopus*) (fourfold), and Black Redhorse increased from 3 individuals in August 2006 to 189 individuals in August 2008. Much of the overall increase in summer 2008, months after the springtime flood, was in young-of-year and juveniles (Matthews et al. 2014, table 4). Figure 6.7 shows dramatic increase in numbers of juveniles for five minnow species and the Black Redhorse (sucker) in August 2008 compared to their numbers in August 2006.

No common species showed any noteworthy decrease after the 2008 flood, with the possible exception of Longear Sunfish, which decreased from 66 individuals in 2006 to 27 in 2008. We (Matthews et al. 2014) speculated that increases in juveniles after both floods occurred because these erosive events removed much sand and fine material from the streambed, depositing it in riparian woods or fields (see plates 20 and 21) after floodwaters overtopped the banks and currents slowed, decreasing power of the stream to retain particles. The result after both floods appeared to be much cleaner in-stream substrates, with interstitial spaces free of silts and fines, offering spaces for egg-depositing minnows or suckers to place eggs, for larvae to develop, or for postlarvae to shelter. Increased interstitial space also could have favored production of benthic aquatic invertebrates, on which most of these minnows and suckers

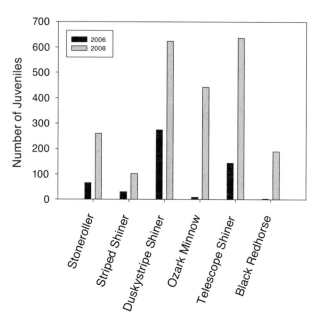

FIG. 6.7 Raw number of juveniles of five minnow species and one sucker in August 2006 and August 2008 (after the floods of April and May 2008), across five sites in Piney Creek. Data from Matthews et al. (2014, table 4).

depend for food. These ideas are speculative but suggest opportunities for future studies of erosive floods in stony-bed streams.

Our samples in Brier Creek in 1999 and 2001, the years following two of the most severe drought on record (1998 and 2000), present an enigma: after the 2000 drought we found huge increases in abundance of Golden Redhorse and Spotted Sucker (*Minytrema melanops*), much like the increases we saw in Black Redhorse suckers in Piney Creek after catastrophic floods! In 10 previous surveys of Brier Creek we had never caught more than 10 individuals of ether species, but both numbered well over 200 in 2001 (Matthews et al. 2013, table 1). Figure 6.8 shows increases in total numbers of these suckers, as well as Central Stoneroller, Largemouth Bass, and Orangethroat Darter, mostly attributable to large numbers of juveniles in the year(s) after drought. For the latter three species, samples in 2001 (after both droughts had occurred) represented big increases relative to previous years and the second-largest number of each of these species ever detected. Curiously, the largest numbers of these three species detected in any of our surveys were all taken in 1988 when the creek was also in drought condition! We have no idea about the mechanism(s) that resulted in huge numbers of young for these species in the summers following drought. Starrett (1951) speculated that harsh periods in Great Plains rivers might reduce standing crops of adults, allowing "space" for new fish, but in 1999 and 2001 we still found substantial numbers of adults in Brier Creek, so that seems an unlikely explanation for the mystery.

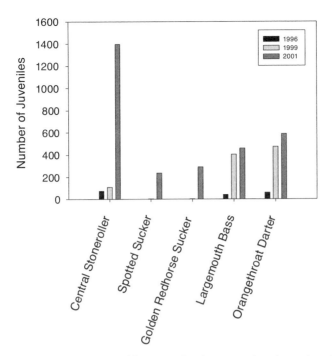

FIG. 6.8 Total number of juveniles of five Brier Creek species that showed marked increases in abundance in 1999 or 2001, after droughts, relative to their abundance in 1996. Data from Matthews et al. (2013, table 1).

Species Associations before and after Extreme Events

Here we are concerned specifically with the effects of major floods or droughts on the associations among species within a watershed, asking whether the associations among species within a watershed are altered by extreme events. Power et al. (1985) offers an example of disturbance changing associations between ecologically important species, showing in repeated snorkel surveys that the negative association between Central Stonerollers and predatory black bass (Largemouth Bass and Spotted Bass, *Micropterus punctulatus*) in 14 adjacent pools of Brier Creek was interrupted by 2 floods. In five of the seven snorkel surveys, stonerollers and bass were negatively associated, but in surveys on 5 May and 24 June 1983, each of which followed a major flood, this association was not significant (Power et al. 1985). Floods can transiently interrupt associations between species, but the associations may be rapidly reestablished after streams return to base flow.

Now we take a broader approach to test for changes in associations of common species in whole communities after drought or flood events. We calculated correlations (product moment correlation; NtSys Version 2.2) across fixed sampling sites in each watershed for all of the common species before and after an event. The triangular matrices of interspecific correlations before and after each event were then compared by

Mantel tests of congruence, based on 10,000 iterations. A Mantel test that was positive and significant was interpreted to mean that associations among the species had not changed in the system as a result of flood or drought.

Our first test was for associations of the 20 most abundant species (based on pre-flood data) across all 12 permanent collecting sites on Piney Creek in August 1982 and August 1983, before and after the great flood of December 1982 (Matthews 1986). The Mantel-normalized statistic z was 0.656, highly significant at $p = 0.0001$. In other words, in spite of the physically catastrophic flood of December 1982, which drastically rearranged the physical structure of many of our study sites in the short term (Matthews 1986), patterns of association among the 20 most abundant species across all sites in Piney Creek remained similar before and after the flood.

A similar analysis for species associations in Piney Creek across our 12 permanent study sites compared associations in summer 2006 with those in summer 2012, before and after the second "great flood" in that system in spring 2008. Comparing triangular correlation matrices for the distribution of 25 common species across the 12 permanent collecting sites, the Mantel z was 0.616, highly significant at $p = 0.001$. Even though there was more evidence of change in the Piney Creek fish community after the 2008 flood than the 1982 flood (Matthews et al. 2014), neither flood resulted in any persistent changes in the associations among species. It appeared that even though these hundred-year floods were dramatic in their physical impacts, the morphometry of most sites recovered rapidly as riffles and pools reformed and aquatic vegetation was reestablished (Matthews et al. 2014). Our analyses suggest that in spite of such major disturbances, the common fishes in the system settled back into their typical habitats, causing post-flood interspecific associations congruent with associations before the floods.

In the same way, we evaluated seine samples in Brier Creek for two periods (arbitrarily selected): 1981–1985, which included three floods and a dry period, and 1999–2001, with extreme drought in 2000. Associations of the 14 most common species from July 1981 to June 1985 were shown by Mantel tests to be congruent in spite of 3 major floods in Brier Creek (47 cm of rain in a 3-day period in autumn 1981, with scouring of the creek and destruction of some streamside forest) and other large floods in 1983 and 1985. However, the congruence of species associations in 1981 and 1985 was not as strong as we had found for the 2 floods in Piney Creek, as the normalized Mantel z was only 0.283 ($p = 0.017$)—still significant but not on par with the consistency of congruence in the more generally benign Piney Creek. We note anecdotally that whole, large trees that were washed into Brier Creek in the extraordinary flood of October 1981 (fig. 6.9) were deposited within our permanent BR-6 reach (the most downstream), and the trunks and large limbs of those trees remained in one of the pools in BR-6 for decades, forming persistent shelter for sunfishes and black bass where none existed previously. So it is no surprise that a flood of that magnitude, by actually rearranging structure in the watershed, might be followed by differences in spatial associations of some species.

We also compared species associations in Brier Creek seine samples before (1999) and after (2001) the drought in 2000 (Marsh-Matthews and Matthews 2010; Mat-

FIG. 6.9 Trees deposited within the stream channel at BR-6 by the October 1981 flood. From supplementary material of Matthews et al. (2013). With permission of Springer.

thews et al. 2013). For this comparison only nine species were sufficiently abundant that their associations could be evaluated. This was in part because some of the more abundant minnow species had been depleted in Brier Creek during the 1980s or early 1990s. Additionally, 1999 was the year after a drought in 1998, so the year we chose as "before" drought for this analysis was actually the year following a different drought. Regardless, for those 9 species, a Mantel comparison of correlation matrices was significant (Mantel $z = 0.614$; $p = 0.014$). While these values, again, were not as strongly significant as comparisons across flood episodes in Piney Creek, all of the cases we evaluated above, for both flood or drought, showed that associations among common species, even in these two sharply contrasting creeks (environmentally harsh Brier Creek and environmentally benign Piney Creek; Ross et al. 1985), were persistent.

Comparative Synthesis across Floods and Droughts

We close this chapter with a synthesis across all of the floods and droughts we have evaluated with before-and-after data. Matthews et al. (2013) found in Brier Creek that intervals with floods showed no more change in the community than intervals without events, but that intervals with drought(s) had more change than intervals lacking events. The floods and droughts we have studied were in a wide range of stream types, from larger (Roanoke River) to smaller systems (Piney Creek, Brier Creek), from environmentally harsh (Brier Creek, Tar Creek tributaries) to relatively stable or "benign" (Peckarsky 1983) systems (Piney Creek, Roanoke River), and had data intervals surrounding the events from weeks (Brier Creek, Roanoke River) to years (Piney and Brier Creeks floods and droughts). To compare events across all these systems, we used two metrics: (1) the number of species detected by similar collecting efforts before and after events, and (2) the Morisita-Horn similarity index, an abundance-based measure of similarity recommended by Jost et al. (2011) for its "density-invariant" property. As a first comparison of effects of floods and droughts on fish, consider the numbers of species detected before and after events (table 6.2).

Table 6.2. Numbers of species detected before and after flood or drought events

Streams and Surveys Compared	Event	Species Detected Before-After	MH Index	Time Lag
Piney Creek, Arkansas: five sites				
Aug. 1982 to Aug. 1983	hundred-year flood, Dec. 1998	32-37	0.980	8 months
July 2006 to Aug. 2008	hundred-year flood, Mar.–Apr. 2008	40-39	0.807	4 months
July 2006 to Aug. 2010	hundred-year flood, Mar.–Apr. 2008	40-38	0.857	2 years
July 2006 to July 2012	hundred-year flood, Mar.–Apr. 2008	40-41	0.787	4 years
Brier Creek, Oklahoma: seine surveys of six sites				
1976–1981	extreme drought, 1980	18-18	0.570	1 year
1981–1985	extreme floods, 1981, 1983, 1985	22-24	0.828	up to 4 years
1986–1988	very dry during survey, 1988	16-22	0.303	none
1988–1991	floods = 4-m stage rise, 1990	22-23	0.361	1 year
1996–1999	extreme drought, 1998	18-19	0.529	1 year
1999–2001	extreme drought, 2000	19-23	0.596	1 year
2008–2012	worst drought on record 2011	21-20	0.592	1 year

Brier Creek, Oklahoma: snorkel surveys in 14 pools in a 1 km reach

14 Mar. to 5 May 1983	severe flood	20-19	0.977	1 month
10–24 June 1983	severe flood	20-21	0.867	2 weeks
July 1997 to July 1999	extreme drought, 1998	17-15	0.742	1 year
July 1999 to June 2001	extreme drought, 2000	15-15	0.596	1 year
May 2006 to May 2008	drought in summer, 2006	15-17	0.819	1 year
May 2008 to May 2012	worst regional drought on record	17-15	0.929	1 year

Tar Creek watershed, Oklahoma: seine surveys at four sites before and after 2006 drought

Coal Creek, 2005–2007	often dry in 2006	20-13	0.834	1 year
Cow Creek, 2005–2007	often dry in 2006	22-16	0.936	1 year
Garrett Creek, 2005–2007	often dry in 2006	13-5	0.558	1 year
Little Elm Creek, 2005–2007	often dry in 2006	10-6	0.991	1 year

Roanoke River, Virginia: seine surveys at mainstem and tributary sites

23 Apr. to 18 May	largest flood in six years, 29 Apr. 1978			
	mainstem (Dixie Caverns)	15-15	0.928	3 weeks
	tributary (Brake Branch)	4-5	0.692	3 weeks

In table 6.2, six postflood surveys showed increases in number of species, compared to only three cases with fewer species detected postflood. Surveys after four droughts showed increases in number of species detected, compared to seven cases of apparent loss of species. These findings provide additional support for the idea that drought can have more serious consequences for a fish community than flood. But note in table 6.2 that the only substantial losses (of four to seven species) after drought were in the four relatively small streams in the Grand River drainage of Ottawa County, Oklahoma. As in chapter 3 and above, these four streams are different from most other sites we have studied because they are strongly dominated by Western Mosquitofish, which appeared to have survived and repopulated, whereas other species, always in small numbers, may have been eliminated.

As another estimate of comparative effects of floods and droughts on fish community structure, we compared MH similarity values for intervals before and after events (table 6.2). For the 10 flood and 13 drought intervals with before-and-after data, MH similarity values were lower on average for drought intervals (mean MH = 0.692) than for flood intervals (mean MH = 0.807), indicating greater change in fish communities following droughts (table 6.2). But there was a wide range of MH similarity values for both drought (0.303–0.991) and flood intervals (0.361–0.980), so caution must be used in making generalizations. We acknowledge that the events characterized in table 6.2 differed greatly in the kinds of streams or fish communities that were represented. Each flood or drought, even within the same system, is unique, and thus warrants evaluation for its similarity or differences to other events. Table 6.2 suggests a trend, however, at least toward greater effects from drought than flood.

In table 6.2, three of the four sites in the Grand River drainage had high MH similarity values between samples in 2005 and 2007, bracketing the severe drought year of 2006. But these three sites with high MH values were ones dominated by Western Mosquitofish (Cow, Coal, and Little Elm Creeks; table 6.2). These creeks illustrate what may be an interesting generalization, in that small, harsh habitats may be dominated by only one or a few species that are particularly resistant (via tolerance) or resilient to disturbance (e.g., by rapid postevent reproduction). Such species may dominate the local community before and after events, resulting in strong similarity between surveys. For example, the proportions of the fish communities consisting of Western Mosquitofish before and after the dry year of 2006 were: Cow Creek, 54.7% before, 95.3% after; Coal Creek, 66.8% before, 82.5% after; Little Elm Creek, 87.0% before, 97.5% after. We noted in chapter 4 the propensity of Western Mosquitofish to have rapid reproductive output because it is a livebearer, and a pregnant female might birth several dozen young (as many as 70 in a big female; EMM, pers. obs.) and produce a brood as often as every 21 days, depending on temperature. Additionally, this species is noteworthy for its tolerance of harsh conditions (Pyke 2008). In these three creeks this one hardy and highly fecund species appeared to recover rapidly from harsh conditions and thus strongly dominate these small, drought-prone streams.

There is a similar situation in the extreme headwaters of Brier Creek, Oklahoma, which we have sampled on many occasions before and after dry periods. At our BR-1 site there is one pool about 1 m deep and a few smaller and shallower pools, in 200 m of mostly treeless pasture. We have found this reach completely dry several times in the last 30 years. Since the late 1980s this site has been strongly dominated by Green Sunfish, which was often the only species present, with a few adults and large numbers of young-of-year or juveniles. Green Sunfish, like Western Mosquitofish, are tolerant of harsh conditions (chap. 4) and apparently good at reinvading these harsh headwaters any time rains rewater the reach. In three samples at that headwater site from 1981 to 1986, Green Sunfish made up 89.6% of all fishes, but 7 other species were also detected. In our 1988 survey, site BR-1 was completely dry, and no sample was made there. After 1988, in 11 surveys from 1991 to 2012, we found a total of 2010 Green Sunfish, and the only other species found in all that time was Central Stoneroller, with 4 individuals once. Both the Ottawa County creeks and this headwater site on Brier Creek support the idea that small, environmentally harsh sites can actually appear to have very "stable" communities, consisting of or strongly dominated by a few (or only one) hardy species.

Summary

Extreme events—including unusual heat or cold, erosive floods, and harsh or prolonged droughts—can be important in the short- or long-term dynamics of stream fish communities, with negative effects on some species but apparently positive effects on others. Episodes of heat and cold, floods, and droughts occur on different schedules, with heat or cold episodes usually brief in our region and erosive floods arising rapidly but lasting only hours to days, whereas droughts develop over weeks and may last for months or years. Each provides fish with different kinds of challenges. Anecdotes from our research provide evidence of death of fish from extreme heat or cold, but such cases are relatively rare in streams of the south-central United States. Most native fish in our region seem adapted, through physiological tolerance or behavior, to survive episodes of summer heating or the limited icing of streams that occurs in winter.

Floods can be powerfully erosive, altering morphometry within the streambed and destroying riparian vegetation or forests. Larval fish can be killed and washed away by erosive floods, but adults seem to survive flooding by moving to low-flow refugia. Our evidence suggests that adult fish can resume normal use of habitats and normal behaviors, including reproductive activity, soon after floodwaters recede. Floods can alter the structure of stream fish communities, but their effects seem to be transitory, with recovery of the communities to preflood structure within months; but after one erosive flood we studied, the fish community had not fully returned to preflood structure after four years. Three large, engineered experimental floods in a typically intermittent north Texas creek resulted in some changes in local fish communities, particularly low in the watershed, but overall the postflood communities were much like those before the floods. Each flood seems to be different in its effects on fish community dynamics, depending on factors like magnitude or length of the flood, timing relative to reproduction by fishes, or vulnerability of the stream to alteration of morphometry.

Droughts may have longer-lasting consequences than floods in the dynamics of fish communities. Droughts develop in stages, eventually trapping fish in high concentrations in isolated pools. Under these conditions, extremes of heating and lower oxygen can result in differential death or survival of fish, depending on species-specific tolerances or tolerances of local populations. Small fish are often trapped at high densities with large numbers of adult piscivores (bass, sunfish) in drought-shrunken pools, and our field observations suggest that mortality of prey is high under those conditions.

Two experimental droughts in large, outdoor mesocosms showed differential survival or condition of native species from a highly variable Oklahoma stream but that individuals that survived the drought could regain condition and have reproductive value in the future community.

Some floods and droughts were followed by major increases in young-of-year or juveniles for numerous fish species, including several minnows, suckers, and some centrarchids. Erosive floods in Piney Creek removed large quantities of sand and fines from the streambed, depositing them in riparian pastures and forests. The result appeared to be more clean gravel, shale, or cobble substrates, with more interstitial spaces, which might have favored deposition of eggs and protection of larval fish or enhanced production of benthic macroinvertebrates, which are food for many kinds of juvenile fishes. It is unclear what mechanisms are responsible for the great increase in juvenile suckers or bass in Brier Creek after two extreme droughts.

A critical synthesis across all of the floods and droughts we have studied across several different stream systems suggested that neither resulted in any clear pattern of reduction in numbers of species, with the exception of drought in four small streams that were dominated by Western Mosquitofish. In some cases, numbers of species were greater in the local or global fish communities in the months or years after extreme flood or drought. Similarity indices comparing fish communities before and after all events showed a trend for postflood communities to be more like the preflood communities than postdrought communities were like those before drought. Overall, all of our evidence—anecdotal, quantitative, and experimental—suggests that droughts may have stronger effects than floods on stream fish community dynamics.

References

Allee, W. C. 1929. Studies in animal aggregations: natural aggregations of the isopod, *Asellus communis*. Ecology 10:14–36.
Bêche, L. A., P. G. Connors, V. H. Resh, and A. A. Merelender. 2009. Resilience of fishes and invertebrates to prolonged drought in two California streams. Ecography 32:778–788.

Bertand, K. N. 2007. Fishes and floods: stream ecosystem drivers in the Great Plains. PhD dissertation, Kansas State University, Manhattan, KS.

Bertrand, K. N., K. B. Gido, W. K. Dodds, J. N. Murdock, and M. R. Whiles. 2009. Disturbance frequency and functional identity mediate ecosystem processes in prairie streams. Oikos 118:917–933.

Boose, E. R., D. R. Foster, and M. Fluet. 1994. Hurricane impacts to tropical and temperate forest landscapes. Ecological Monographs 64:369–400.

Covich, A. P., S. C. Fritz, P. J. Lamb, R. D. Marzolf, W. J. Matthews, K. A. Poiani, E. E. Prepas, M. B. Richman, and T. C. Winter. 1997. Potential effects of climate change on aquatic ecosystems of the Great Plains of North America. Hydrological Processes 11:993–1021.

Dayton, P. K. 1971. Competition, disturbance, and community organization: the provision and subsequent utilization of space in a rocky intertidal community. Ecological Monographs 41:351–389.

den Boer, P. J. 1986. The present status of the competitive exclusion principle. Trends in Ecology and Evolution 1:25–28.

Douglas, M. R., P. C. Brunner, and M. E. Douglas. 2003. Drought in an evolutionary context: molecular variability in Flannelmouth Sucker (*Catostomus latipinnis*) from the Colorado River basin of western North America. Freshwater Biology 48:1254–1273.

Fisher, S. G., L. J. Gray, N. B. Grimm, and D. E. Busch. 1982. Temporal succession in a desert stream ecosystem following flash flooding. Ecological Monographs 52:93–110.

Franssen, N. R., K. B. Gido, C. S. Guy, J. A. Tripe, S. J. Shrank, T. R. Strakosh, K. N. Bertrand, C. M. Franssen, K. L. Pitts, and C. P. Paukert. 2006. Effect of floods on fish assemblages in an intermittent prairie stream. Freshwater Biology 51:2072–2086.

Galbraith, H. S., D. E. Spooner, and C. C. Vaughn. 2010. Synergistic effects of regional climate patterns and local water management on freshwater mussel communities. Biological Conservation 143:1175–1183.

Gause, G. F. 1934. Experimental analysis of Vito Volterra's mathematical theory of the struggle for existence. Science 79:16–17.

Golladay, S. W., and C. L. Hax. 1995. Effects on an engineered flow disturbance on meiofauna in a north Texas prairie stream. Journal of the North American Benthological Society 14: 404–413.

Grimm, N. B., and S. G. Fisher. 1989. Stability of periphyton and macroinvertebrates to disturbance by flash floods in a desert stream. Journal of the North American Benthological Society 8:293–307.

Grossman, G. D., P. B. Moyle, and J. O. Whitaker Jr. 1982. Stochasticity in structural and functional characteristics of an Indiana stream fish assemblage: a test of community theory. American Naturalist 120:423–454.

Grossman, G. D., R. E. Ratajczak Jr., M. Crawford, and M. C. Freeman. 1998. Assemblage organization in stream fishes: effects of environmental variation and interspecific interactions. Ecological Monographs 68:395–420.

Hardin, G. 1960. The competitive exclusion principle. Science 131:1292–1297.

Harvey, B. C. 1987. Susceptibility of young-of-the-year fishes to downstream displacement by flooding. Transactions of the American Fisheries Society 116:851–855.

Hubbell, S. P., R. B. Foster, S. T. O'Brien, K. E. Harms, R. Condit, B. Wechsler, S. J. Wright, and S. Loo de Lao. 1999. Light-gap disturbances, recruitment limitations, and tree diversity in a tropical forest. Science 283:554–557.

Jewell, M. E. 1927. Aquatic biology of the prairie. Ecology 8:289–298.

Johnson, H. L., D. S. Arndt, G. D. McManus, and M. A. Shafer. n.d. The Oklahoma Mesonet: a mesoscale tool for drought recognition and monitoring. Fact Sheet J3.4. Oklahoma Climatological Survey, Norman, OK.

Jost, L., A. Chao, and R. L. Chazdon. 2011. Compositional similarity and beta diversity. Pages 66–84 in A. E. Magurran and B. J. McGill, eds. Biological diversity—frontiers in measurement and assessment. Oxford University Press, Oxford, UK.

Kiernan, J. D., and P. B. Moyle. 2012. Flows, drought, and aliens: factors affecting the fish assemblage in a Sierra Nevada, California, stream. Ecological Applications 22:1146–1161.

Labbe, T. R., and K. D. Fausch. 2000. Dynamics of intermittent stream habitat regulate persistence of a threatened fish at multiple scales. Ecological Applications 10:1774–1791.

Lake, P. S. 2011. Drought and aquatic ecosystems—effects and responses. Wiley-Blackwell, West Sussex, UK.

Lohr, S. C., and K. D. Fausch. 1997. Multiscale analysis of natural variability in stream fish assemblages of a western Great Plains watershed. Copeia 1997:706–724.

Lugo, A. E., C. S. Rogers, and S. W. Nixon. 2000. Hurricanes, coral reefs and rainforests: resistance, ruin and recovery in the Caribbean. Ambio 29:106–114.

Magoulick, D. D., and R. M. Kobza. 2003. The role for refugia for fishes during drought: a review and synthesis. Freshwater Biology 48:1186–1198.

Marsh-Matthews, E., and W. J. Matthews. 2010. Proximate and residual effects of exposure to simulated drought on prairie stream fishes. Pages 461–486 in K. B. Gido and D. A. Jackson, eds. Community ecology of stream fishes: concepts, approaches, and techniques. American Fisheries Society Symposium 73. American Fisheries Society, Bethesda, MD.

Marsh-Matthews, E., W. J. Matthews, K. B. Gido, and R. L. Marsh. 2002. Reproduction by young-of-year Red Shiner (*Cyprinella lutrensis*) and its implications for invasion success. Southwestern Naturalist 47:605–610.

Matheney, M. P., IV, and C. R. Rabeni. 1995. Patterns of movement and habitat use by Northern Hog Suckers in an Ozark stream. Transactions of the American Fisheries Society 124:886–897.

Matthews, W. J. 1986. Fish faunal structure in an Ozark stream: stability, persistence, and a catastrophic flood. Copeia 1986:388–397.

Matthews, W. J. 1987. Physicochemical tolerance and selectivity of stream fishes as related to their geographic ranges and local distributions. Pages 111–120 in W. J. Matthews and D. C. Heins, eds. Community and evolutionary ecology of North American stream fishes. University of Oklahoma Press, Norman, OK.

Matthews, W. J. 1998. Patterns in freshwater fish ecology. Chapman and Hall, New York, NY.

Matthews, W. J., and L. G. Hill. 1979. Influence of physico-chemical factors on habitat selection by Red Shiners, *Notropis lutrensis* (Pisces: Cyprinidae). Copeia 1979:70–81.

Matthews, W. J., and L. G. Hill. 1980. Habitat partitioning in the fish community of a southwestern river. Southwestern Naturalist 25:51–66.

Matthews, W. J., and J. D. Maness. 1979. Critical thermal maxima, oxygen tolerances and success of cyprinid fishes in a southwestern river. American Midland Naturalist 102: 374–377.

Matthews, W. J., and E. Marsh-Matthews. 2003. Effects of drought on fish across axes of space, time and ecological complexity. Freshwater Biology 48:1232–1253.

Matthews, W. J., and E. Marsh-Matthews. 2006. Persistence of fish species associations in pools of a small stream of the southern Great Plains. Copeia 2006:696–710.

Matthews, W. J., and E. Marsh-Matthews. 2007. Extirpation of Red Shiner in direct tributaries of Lake Texoma (Oklahoma-Texas): a cautionary case history from a fragmented river-reservoir system. Transactions of the American Fisheries Society 136:1041–1062.

Matthews, W. J., and J. T. Styron Jr. 1981. Tolerance of headwater vs. mainstream fishes for abrupt physicochemical changes. American Midland Naturalist 105:149–158.

Matthews, W. J., R. C. Cashner, and F. P. Gelwick. 1988. Stability and persistence of fish faunas and assemblages in three midwestern streams. Copeia 1988:945–955.

Matthews, W. J., K. B. Gido, G. P. Garrett, F. P. Gelwick, J. G. Stewart, and J. Schaefer. 2006. Modular experimental riffle-pool stream system. Transactions of the American Fisheries Society 135:1559–1566.

Matthews, W. J., B. C. Harvey, and M. E. Power. 1994. Spatial and temporal patterns in the fish assemblages of individual pools in a midwestern stream (U.S.A.). Environmental Biology of Fishes 39:381–397.

Matthews, W. J., E. Marsh-Matthews, G. L. Adams, and S. Reid Adams. 2014. Two catastrophic floods: similarities and differences in effects on an Ozark stream fish community. Copeia 2014:682–693.

Matthews, W. J., E. Marsh-Matthews, R. C. Cashner, and F. Gelwick. 2013. Disturbance and trajectory of change in a stream fish community over four decades. Oecologia 173: 955–969.

Matthews, W. J., M. S. Schorr, and M. R. Meador. 1996. Effects of experimentally enhanced flows on fishes of a small Texas (U.S.A.) stream: assessing the impact of interbasin transfer. Freshwater Biology 35:349–362.

Matthews, W. J., E. Surat, and L. G. Hill. 1982. Heat death of the Orangethroat Darter *Etheostoma spectabile* (Percidae) in a natural environment. Southwestern Naturalist 27: 216–217.

Meffe, G. K., and W. L. Minckley. 1987. Persistence and stability of fish and invertebrate assemblages in a repeatedly disturbed Sonoran desert stream. American Midland Naturalist 117:177–191.

Pearsons, T. N., H. W. Li, and G. A. Lamberti. 1992. Influence of habitat complexity on resistance to flooding and resilience of stream fish assemblages. Transactions of the American Fisheries Society 121:427–436.

Peckarsky, B. L. 1983. Biotic interactions or abiotic limitations? a model of lotic community structure. Pages 303–323 in T. D. Fontaine III and S. M. Bartell, eds. Dynamics of lotic ecosystems. Ann Arbor Science, Ann Arbor, MI.

Pickett, S. T. A., J. Kolasa, J. J. Armesto, and S. L. Collins. 1989. The ecological concept of disturbance and its expression at various hierarchical levels. Oikos 54:129–136.

Pires, A. M., M. F. Magalhaes, L. Moreira da Costa, M. J. Alves, and M. M. Coelho. 2008. Effects of an extreme flash flood on the native fish assemblages across a Mediterranean catchment. Fisheries Management and Ecology 15:49–58.

Plath, M., B. Hermann, C. Schroder, R. Riesch, M. Tobler, F. J. Garcia de Leon, I. Schlupp, and R. Tiedemann. 2010. Locally adapted fish populations maintain small-scale genetic differentiation despite perturbation by a catastrophic flood event. BMC Evolutionary Biology 10:256.

Power, M. E., and A. J. Stewart. 1987. Disturbance and recovery of an algal assemblage following flooding in an Oklahoma stream. American Midland Naturalist 117:333–345.

Power, M. E., W. J. Matthews, and A. J. Stewart. 1985. Grazing minnows, piscivorous bass and stream algae: dynamics of a strong interaction. Ecology 66:1448–1456.

Pyke, G. H. 2008. Plague minnow or mosquito fish? a review of the biology and impacts of introduced Gambusia species. Annual Review of Ecology, Evolution, and Systematics 39: 171–191.

Resh, V. H., A. V. Brown, A. P. Covich, M. E. Gurtz, H. W. Li, G. W. Minshall, S. R. Reice, A. L. Sheldon, J. B. Wallace, and R. C. Wissmar. 1988. The role of disturbance in stream ecology. Journal of the North American Benthological Society 7:433–455.

Ross, S. T., and J. A. Baker. 1983. The response of fishes to periodic spring floods in a southeastern stream. American Midland Naturalist 109:1–14.

Ross, S. T., W. J. Matthews, and A. A. Echelle. 1985. Persistence of stream fish assemblages: effects of environmental change. American Naturalist 126:24–40.

Rutledge, C. J., E. G. Zimmerman, and T. L. Beitinger. 1990. Population genetic responses of two minnow species (Cyprinidae) to seasonal stream intermittency. Genetica 80:209–219.

Schaefer, J. F., E. Marsh-Matthews, D. E. Spooner, K. B. Gido, and W. J. Matthews. 2003. Effects of barriers and thermal refugia on local movement of the threatened leopard darter, *Percina pantherina*. Environmental Biology of Fishes 66:391–400.

Schoener, T. W. 1974. Resource partitioning in ecological communities. Science 185:27–39.

Starrett, W. C. 1950. Distribution of the fishes of Boone County, Iowa, with special reference to the minnows and darters. American Midland Naturalist 43:112–127.

Starrett, W. C. 1951. Some factors affecting the abundance of minnows in the Des Moines River, Iowa. Ecology 32:13–27.

Stewart, J. G., F. P. Gelwick, W. J. Matthews, and C. M. Taylor. 1999. An annotated checklist of the fishes of the Tallgrass Prairie Preserve, Osage County, Oklahoma. Proceedings of the Oklahoma Academy of Science 79:13–17.

Strange, E. M., P. B. Moyle, and T. C. Foin. 1992. Interactions between stochastic and deterministic processes in stream fish community assembly. Environmental Biology of Fishes 36:1–15.

Thibault, K. M., and J. H. Brown. 2008. Impact of an extreme climatic event on community assembly. Proceedings of the National Academy of Science USA 105:3410–3415.

Turner, M. G. 2010. Disturbance and landscape dynamics in a changing world. Ecology 91: 2833–2849.

Walters, A. W., and D. M. Post. 2008. An experimental disturbance alters fish size structure but not food chain length in streams. Ecology 89:3261–3267.

Wesner, J. S. 2011. Shoaling species drive fish assemblage response to sequential large floods in a small midwestern U.S.A. stream. Environmental Biology of Fishes 91:231–242.

Wiens, J. A. 1977. On competition and variable environments. American Scientist 65:590–597.

Wootton, J. T., M. S. Parker, and M. E. Power. 1996. Effects of disturbance on river food webs. Science 273:1558–1561.

CHAPTER 7

Temporal Dynamics of Fish Communities and the Loose Equilibrium Concept

Overview of Studies on Stream Fish Community Dynamics

Community Concepts

For the first two-thirds of the twentieth century, theory in community ecology focused on equilibrium communities, presumably regulated by interspecific competition. Theoretical works such as the Lotka-Volterra equation predicted that populations should reach a carrying capacity (K) and thenceforth remain stable, albeit perhaps oscillating around K, with N for the population varying slightly but always changing back toward K. Later, the addition of competition coefficients to the Lotka-Volterra equation (e.g., Strobeck 1973) incorporated competition among multiple species into equilibrium community dynamics. Experiments in the 1930s (Gause 1937) supported the idea that competition between species was the driving force in a community, leading to community stability (once the "losers" had lost). The influential Charles Elton (1946) concluded in a large review that communities contained fewer species within genera than expected mathematically and attributed low species:genus ratios to competition among similar species. The extension of this concept was the "competitive exclusion principle," summarized by Garrett Hardin (1960), suggesting that communities stabilized by "resource partitioning." The balanced equilibrium community, hypothetically regulated by resource partitioning, probably reached its pinnacle of acceptance in ecology as a result of the concepts of G. E. Hutchinson (1957), describing the fundamental niche as an "n-dimensional hypervolume" with many resource axes; the detailed observations of Robert MacArthur (1958), who concluded that differences in feeding location, behavior, habitat use, or time allowed coexistence of five warbler species; and the idea that resource partitioning existed on the three major axes of space, food, and time, by Thomas Schoener (1974).

Fish ecologists found support for resource partitioning in demonstrations by Keast and Webb (1966) that differences in mouth and body shape led to food and habitat specialization that reduced interspecific competition and in the study of a Panama fish community in which Zaret and Rand (1971) argued that species segregated by space use and food habits. A comprehensive review of resource partitioning in fish assemblages worldwide by Ross (1986) showed that, overall, "trophic separation is more important than habitat separation," but he pointed out that the patterns differed among different "global assemblages," ranging from marine to freshwater. The review by Ross (1986) effectively brought Schoener's resource axes into focus for fish ecologists.

Equilibrium communities were also predicted as the endpoint of temporal "succession." Allee et al. (1949) devoted pages 562–580 to "community succession and development," emphasizing that ecological succession is "an orderly, progressive sequence of replacement of communities over a given point, area, or locality." In the "green version" of *Fundamentals of Ecology*, by Odum and Odum (1959), pages 257–270 were devoted to "Ecological Succession" or "Concept of the Climax." They described succession as "the orderly process of community change . . . the sequence of communities which replace one another in a given area . . . [it] begins with pioneer stages which are replaced by a series of more mature communities until a relatively stable community is evolved which is in equilibrium with the local conditions." Pioneering work cited by Odum and Odum (1959) included Cowles's (1899) studies on the plants of the Indiana Dunes of Lake Michigan and Shelford's (1913) review of "Animal Communities in Temperate America," which was actually predated, for fishes, by Shelford's (1911) noteworthy paper on "ecological succession in stream fishes." Thus, even in the context of community change expected during succession, the assumption persisted that early, pioneering communities of plants and their companion animals moved regularly toward a mature climax (although Odum and Odum [1959] made clear that the same climax community would not always occur at different sites within a broad ecological region, given the influence of local conditions).

In the 1970s, authors began to question whether equilibrium communities structured by competition were important in the real world. Wiens (1977) promoted the idea that disturbance or resource shortages ("crunches"), at rare intervals, could be more important in community structuring than ongoing, essentially daily, competition. Following Wiens (1977), there was growing emphasis on community responses to disturbance, ranging from daily disturbance in the rocky intertidal (Lubchenco and Menge 1978) to hurricane devastation (Lugo et al. 2000) to catastrophic disturbances to forests (Bormann and Likens 1979) and to flood, drought, or periods of no flow in streams (e.g., Resh et al. 1988; Lake 2011). Grossman et al. (1982) clearly shook confidence in equilibrium communities for fish with their conclusion that stochastic rather than deterministic processes dominated the long-term structure of an Indiana stream fish community.

Two important reviews on competition both appeared in 1983. In an evaluation of more than 150 field experiments across many taxa, in the August issue of *The Ameri-

can Naturalist, Schoener (1983) concluded that there was good evidence for the importance of competition in 90% of the studies, indicating "its pervasive importance in ecological systems." In the November issue of the same journal, Connell (1983) found in a review of more than 500 field experiments in the literature that competition was evident in most.

But by the 1980s, as ecologists grappled with the fact that communities are constantly changing to some degree, Davis (1986) noted that "the view of community structure that I have presented contrasts with models that . . . explain species abundances . . . under equilibrium or climax conditions," and that "the climatic-instability view emphasizes the dynamic nature of biotic communities, with species frequencies (and sometimes species composition) changing continually, even during the lifetimes of individual organisms." In fact, since at least the analyses of Grossman et al. (1982), concepts about habitat harshness by Peckarsky (1983), and other general treatises on disturbance (Resh et al. 1988; Pickett et al. 1989), far more emphasis has been placed on understanding the responses of communities over short or long periods of time to disturbance, disaster, catastrophe (see chap. 6), or climate change than to seeking stable "climax" communities in streams. For fishes in general, the questions now focus more on whether there are long-term directional trajectories of change in communities (Collie et al. 2008; Magurran and Henderson 2010; Matthews et al. 2013, 2014; Matthews and Marsh-Matthews 2016) or whether a fish community, once disturbed by exogenous factors, returns over time to resemble its predisturbance structure (Matthews et al. 2013; chap. 8, this volume).

Since the 1970s, community theory had also moved away from a strictly equilibrium focus. May (1973) predicted that communities might exhibit nonequilibrium patterns, varying substantially at times but eventually returning toward a central, average structure. DeAngelis et al. (1985) developed a theoretical model of community dynamics intermediate between equilibrium and nonequilibrium, which they described as "loose equilibrium." In the DeAngelis et al. model, the community tended to return over time toward a central position in community structure, but not specifically to any one average. The hallmark of loose equilibrium is that a community may change substantially from one time to the next, but that given sufficient time (i.e., enough sampling in the real world), the community will return toward some central condition, rather than remaining in some alternate state. Collins (2000) found evidence across years that mammal, insect, and plant communities on the Konza Prairie, Kansas, changed from year to year but remained over longer spans of time in the kind of loose equilibrium predicted by the DeAngelis et al. hypothesis. Following Collins's (2000) model, we found evidence that fish communities in Brier Creek (Matthews et al. 2013) and Piney Creek (Matthews et al. 2014; Matthews and Marsh-Matthews 2016) have, over spans of 40 years or more, behaved in patterns consistent with loose equilibrium. In Matthews and Marsh-Matthews (2016, Appendix C) we summarized the development of the loose equilibrium concept (LEC) and suggested guidelines for identifying communities that are in a loose equilibrium.

Summaries of Temporal Changes in Fish Communities

Matthews (1998, 104–129) summarized temporal changes in stream and lake fish communities through the mid-1990s as related to disturbance (Matthews 1998, 320–326), so most of that material will not be repeated here. Matthews (1998, table 3.4) compared 9 studies (including 4 sites on Brier or Piney Creeks) of "long-term change" in stream fish communities that spanned (at the time) a range of 9 to 30 years. Most of these systems were dominated by fishes with relatively short life spans of 2–5 years, and results might be different in streams like some in the western United States that are dominated by species such as large minnows or suckers with life spans of 30–45 years. But these particular 9 studies had a range of variation in their mean time-to-time percentage similarity index (PSI) from 0.469 to 0.803, with a grand mean PSI across all 9 studies of 0.639 (Matthews 1998, 121). The PSI indicates the minimum similarity between any 2 communities based on relative abundance of all species, with a range from 0 to 1.00 if their relative abundances are a perfect match. Index values like the PSI have little meaning in and of themselves in isolation, but the outcome of these nine studies was sort of left as a "cup half full." Matthews (1998, 122–125) also emphasized that there can be substantial change in any community from one time to the next (i.e., a low similarity value), but during a subsequent time the community can change back toward its previous state, and it may be that the trajectory of a community over numerous samples is more important than a raw index of similarity from one time to the next. The index indicates change from one time to the next; the trajectory tells a more complete story about how a community has changed, or not, across time.

Numerous contributions in other recent books, symposia, and individual papers have summarized or provided empirical information on temporal dynamics of fish communities. An American Fisheries Society symposium in 2001 (Rinne et al. 2005) summarized changes in the fishes of large rivers (most substantially larger than the streams we address in this book), providing a wealth of detail on changes in individual species or whole-river faunas over time. An American Fisheries Society symposium in 2008 (Gido and Jackson 2010) addressed changes in stream fish communities, especially chapters by Gido et al. (2010), Grossman and Sabo (2010), Rahel (2010), Roberts and Hitt (2010), and Taylor (2010), among others. A 2013 symposium at the annual meeting of the American Society of Ichthyologists and Herpetologists focused on the theme of "Fish Out of Water," and numerous papers and the summary by S. T. Ross (2015) focused on temporal changes in fish communities or individual species from the perspective of "eco-evolutionary" dynamics between ecology and evolution. Ross (2013, 91–116) devoted an entire chapter to "persistence of fish assemblages in space and time," with his table 6.1 summarizing change in 25 stream fish assemblages in "long-term" studies (2 or more years), with half to about three-fourths of the studies interpreted as showing high persistence or stability in some community measure. Ross and Matthews (2014) show similar results for streams, ranging from benign to harsh environmentally.

The strength of all the overviews above is that they considered a wide range of communities across many types of systems. But the original authors of these diverse studies sampled fish in many different ways for various reasons and analyzed data in different ways, and there are potentially other unknown differences among the systems, fish communities, and investigator protocols that may have influenced outcomes. To the extent that patterns emerged nevertheless (e.g., Ross's conclusions), the detected patterns may be robust. In the present chapter, we take a different approach, using our 9 global community data sets and 31 local data sets, analyzing all of them in the same way. This approach should allow us to detect similarities of temporal dynamics of these warm-water stream fish communities. But first we review our published work on changes in fish communities.

Our Studies of Temporal Dynamics of Stream Fish Communities

Studies of temporal dynamics of stream fish communities by WJM began with documenting seasonal differences in fish communities in collections in Piney Creek (1972–1973) for his MS thesis (see plate 1) and subsequently studying changes in the abundances of different species year-round in the Canadian River (see plate 12) during his PhD sampling in 1976 and 1977 (Matthews and Maness 1979; Matthews and Hill 1980). All of WJM's sampling in the Roanoke River (see plate 4) with students at Roanoke College in 1977–1979 reinforced the impression of large differences in local fish communities between years, before and after floods, or from month to month.

Interest in "long-term" change began for WJM when he and Bob Matthews revisited all of the 1970s Piney Creek sites in 1981. Then WJM, with Tom Heger, Bruce Wagner, and Mike Lodes, sampled Piney Creek before and after the huge flood of December 1982 (Matthews 1986; see chap. 6), which took his research program down the "catastrophic events" route. EMM plunged deeply into long-term change when we began collaboration in December 1994 to sample all the Piney Creek sites and when we drove about 9000 miles throughout the Midwest in June 1995 to resample about 80 sites that WJM had sampled in 1978, from Nebraska and Iowa as far south as the Texas Hill Country and the Big Bend region. (We also got married in July 1995, shortly after completing the Midwest sampling, and days before returning to Piney Creek to spend our honeymoon at the Jenkins Motel and Trout Dock while seining all 12 sites in the watershed, again).

Many of our publications (summarized in Matthews et al. 2013 and Matthews and Marsh-Matthews 2016) are based on surveys at multiple sites in Piney Creek and Brier Creek (see plates 1 and 2), the two streams that we have studied the longest (since 1972 and 1976, respectively). Papers addressing long-term changes in Piney Creek include Matthews (1986), which assessed the effects of a catastrophic flood in December 1982 (as described in chap. 6). In spite of the physically devastating flood, the summer fish communities in the years before and after the flood were quite similar, and Marsh-Matthews and Matthews (2002) showed that minnow species associations in Piney Creek changed little across 10 surveys at our 12 fixed sample sites from the 1970s to the 1990s. Matthews et al. (2014) found that in spite of 2 catastrophic floods (in 1982 and 2008), fish in Piney Creek tended to recover to earlier community structure (albeit on different timescales, depending on the details of a flood), and Matthews and Marsh-Matthews (2016) provided a comprehensive summary of dynamics of global and local fish communities at 12 sites in the watershed from the 1970s to 2012.

Papers on temporal dynamics of the Brier Creek fish community include Matthews et al. (1994), showing consistency of fish species distribution among 14 adjacent pools for which we now have snorkel surveys from 1982 to 2012. But an ordination showed overall differences in fish species composition in the Brier Creek pools between the 1980s and the 1990s, during which time cattle were introduced to the property, with resultant trampling of some stream banks (Matthews and Marsh-Matthews 2003). Using an experimental mesocosm approach, Matthews and Marsh-Matthews (2006a) found that initially identical fish communities consisting of common Brier Creek species diverged significantly in composition or in their temporal trajectories over 388 days. In the 14 pools, interspecific associations of many but not all common species, as evaluated by snorkel surveys, were persistent from 1982–1983 to more recent surveys in 1995–2003, and one particularly important minnow species (Red Shiner, *Cyprinella lutrensis*) was essentially lost from Brier Creek (Matthews and Marsh-Matthews 2006b). There were two known reinvasions of this formerly important species into Brier Creek from 2007 to 2009, with the subsequent failure of these invasions to reestablish populations of Red Shiners (Marsh-Matthews et al. 2011, fig. 1). That pattern has persisted as recently as 2014 (Marsh-Matthews and Matthews, unpub. data).

Matthews et al. (2013) provided a comprehensive review of the dynamics of all fishes and of the global fish community of Brier Creek from the 1960s to 2008. Brier Creek is environmentally harsh and variable relative to the more environmentally benign Piney Creek (see chap. 2), and several publications have contrasted fish community dynamics in these different systems. Based on surveys of five sites each on Brier Creek from 1969 (Smith and Powell 1971) to 1981 and Piney Creek from 1972 to 1981, Ross et al. (1985) showed that the fish communities of the environmentally variable Brier Creek were less stable than those in Piney Creek. Common species remained common in both creeks; their rank abundances were statistically similar across time in Piney Creek but not Brier Creek. Also, upstream local fish communities in Brier Creek were less stable than communities in larger, deeper, downstream sites (Ross et al. 1985). A summer (1980) with extremes of temperature resulted in no persistent changes in the global community in Brier Creek, however. Ross et al. (1985) speculated that the harsh environment of Brier Creek has selected for a subset of species that are adapted to such conditions. Our later work supports this hypothesis, as we found in the artificial droughts in experimental streams that some Brier Creek fishes were quite resistant to conditions in isolated pools, and that once drought ended they were able to compensate for lost condition and have good reproductive potential (Marsh-Matthews and Matthews 2010).

Matthews et al. (1988) quantified long-term variation in the global and local fish communities of Piney Creek, Brier Creek, and Kiamichi River, based on surveys at multiple sites through 1986 in all three systems. In all three systems, abundant species remained so over 14, 17, and 5 years, respectively, and their global communities appeared stable based on similarity indices and rank abundance of common species. The environmentally benign Piney Creek had greater faunal similarity across time than did the harsh Brier Creek, however, and Kiamichi River (also considered to have a benign environment, but less so than Piney Creek) was intermediate between the other two in community stability.

Two other studies from our labs have addressed short- or long-term changes in fish communities in other systems. The late A. P. (Pat) Blair (1959) summarized his samples of darters (Percidae) from 1948 to 1955 in 546 collections at 272 sites throughout the Arkansas River drainage in northeastern Oklahoma. In October 2006 and November 2007, the FishLab at the University of Oklahoma (combined graduate students and faculty in labs of Richard Broughton, Ingo Schlupp, EMM, and WJM) undertook resampling 123 of Blair's sites where he had collected at least 20 individual darters, attempting to mimic Blair's approach. By the time we undertook the project, Dr. Blair had passed away, so we had to assume the manner in which he probably sampled and used standardized sampling of 10 kicksets per site. David Gillette, then a PhD student, took the lead in relocating all of Blair's sites, planned sampling, coordinated all logistics, and was the lead author on the resulting paper (Gillette et al. 2012).

On the same weekend in October 2006, five teams from the FishLab went to assigned areas of northeast Oklahoma to resample Blair's sites. All 123 sites were visited, but only 71 could be sampled because of limited access or habitat alteration (several sites were drowned under reservoirs since Blair's sampling). A similar, albeit less coordinated, effort was made in November 2007. Overall, darters were collected at 60 of Blair's sites. The results showed marked differences between the historical and contemporary distributions of darters or numbers of darter species, with overall loss of 5 of his species not detected by us and the average number of darter species per site decreasing from 3.8 in Blair's study to 2.6 in our surveys. All species occurred at a smaller proportion of sites in recent as compared to historical samples. There was no evidence of homogenization, as range expansions were not observed, and extirpation at local sites was much more common than increases in numbers of darters by immigration. Gillette et al. (2012) closed with a cautionary note that species with highly specialized habitats, like the benthic darters, may be particularly at risk from anthropogenic alteration of habitat.

But a different spatially broad comparison suggested that across much of Oklahoma the native fish fauna of free-flowing streams (outside the actual pools of reservoirs) has not changed greatly since widespread surveys of fishes were made in the 1920s

(Matthews and Marsh-Matthews 2015). A. I. Ortenburger of the Oklahoma Biological Survey at the University of Oklahoma led large parties of biologists across much of Oklahoma and western Arkansas for weeks in the summers of 1925–1927, setting up base camps from which fish, herps, mammals, birds, and plants were surveyed. Ortenburger's work, done in collaboration with Carl Hubbs, resulted in three important papers that are the foundation for knowledge about the fishes in Oklahoma in the early twentieth century, before most dams were built in the state (Ortenburger and Hubbs 1926; Hubbs and Ortenburger 1929a,b). Comparing our collections from 1975–1999 to Ortenburger's, there were only minor changes in minnow species associations (Marsh-Matthews and Mathews 2002). We also compiled the historical distributions of all fish species from the 3 publications by Ortenburger and Hubbs at 86 sites for which we also have more than 300 "recent" collections from 1975 to 2009 in the same general areas, but not the exact sites (Matthews and Marsh-Matthews 2015). In our collections, we found 81 of the 95 species that Ortenburger had collected in the 1920s, missing only species that have always been rare. Species associations were congruent between the two eras, and broad geographic patterns in fish distribution were similar between the two time periods (Matthews and Marsh-Matthews 2015). Within US Geological Survey hydrologic unit codes (HUCs), most had faunal similarities >75% between the eras. We did detect a weak signal of faunal homogenization, mostly from the contemporary presence of some sunfishes (*Lepomis* spp.) in western Oklahoma that Ortenburger did not detect. But these include species such as Bluegill (*Lepomis macrochirus*) that have been stocked widely across the state. On balance, even though there were noteworthy changes within some HUCs, the general distribution of fishes in Oklahoma seems to have changed little between the 1920s and now (Matthews and Marsh-Matthews 2015), and most of the native stream fishes in Oklahoma or western Arkansas remain available for future positive management of the streams.

Assessing Temporal Dynamics of Fish Communities
Methods for Assessing Temporal Dynamics and Community Change
There are many ways to assess temporal dynamics of a fish community. Presence-absence data across surveys can be used to ask if any species have been lost or if new species have become established. With abundance data it is also possible to track changes in the numerical dominance of species over time. Abundance data additionally can be used to generate indices of similarity (or difference) to compare communities across space or change in a given community across time. For our analyses, we compared community similarity of consecutive surveys using the Morisita-Horn (MH) similarity index based on abundances. The MH index is based on the proportion of each species in a sample, emphasizing the more abundant species. Species comprising a small proportion of samples have essentially no influence on the MH values, and readers interested in rare species or their effects on community structure may wish to download our data sets (which are publicly available through DataDryad) and run

other kinds of analyses. We do not use rarefaction, for reasons outlined earlier in this book, and we focused more on abundance than presence-absence because we think that the more abundant species typically have more influence on a fish community than do very rare species. That is not always true, because a single adult Largemouth Bass (*Micropterus salmoides*) patrolling a pool can modify behavior or abundance of small-bodied prey. On balance, however, we consider that our abundance data, at least relative abundance as used in the MH index, are more reliable than presence-absence data would be, given the unknown detectability of some of the rare species. And we cannot fathom that one or two Streamline Chubs (*Erimystax dissimilis*) or an occasional Wedgespot Shiner (*Notropis greenei*) would have nearly the biological influence in the Piney Creek fish community or the ecosystem that the hundreds to sometimes thousands of minnows such as Duskystripe Shiner (*Luxilus pilsbryi*) or Telescope Shiner (*Notropis telescopus*) would have, although in presence-absence data, each would be of equal importance.

We established previously why we use the Morisita-Horn similarity measure, for its "density-invariant" and other desirable mathematical properties (Jost et al. 2011). Based on comparisons of various index performances for our data sets and Jost et al (2011), we elected to use MH similarity to make standardized comparisons across all of our data sets for this book. We prefer MH to the often-used Bray-Curtis dissimilarity measure, which can be influenced substantially by raw abundances of species. The use of MH similarity indices to evaluate time-to-time similarity of samples allows comparison of the magnitude of change but does not address the direction of change over time. For examination of directional patterns of change, we used nonmetric multidimensional scaling (NMDS) to visualize and quantify trajectories of community change in multivariate space. NMDS begins with a triangular matrix of MH similarity comparisons among all possible pairs of samples, and it places all samples in multivariate space in a way that best preserves the patterns of similarity in the triangular matrix. The number of axes in multivariate space is not predetermined but depends on the final "stress" (a measure of fit to the original data), with lower stress indicating better fit. Stress levels between 0.10 and 0.20 are typical for ecological data, and values in this range are generally acceptable, particularly in the lower part of the range (McCune and Grace 2002). In the final solution, all axes are essentially weighted equally. The proximity of samples in multivariate space can be measured in Euclidian distance (because of the equal weighting of the axes) and approximates their similarity in the triangular matrix of MH values. We used this approach here and in recent papers (e.g., Matthews et al. 2013, 2014; Matthews and Marsh-Matthews 2016), as well as detrended correspondence analysis (the first axis of which remains valuable as a first estimator of overall variation in a community, following Gauch 1982). For data sets in this book, NMDS of MH values gave satisfactory two-dimensional solutions in most cases, with Stress 1 values typically lower than 0.15. One exception was on Piney Creek (Piney Local = site M-2), for which different runs of the NMDS gave rather different solutions, as described in detail below.

As to what we consider a best overview of dynamics of a community, points in multivariate space for consecutive samples can be connected by successional vectors to plot the trajectory of community structure over time. We then follow hypothetical trajectory patterns in Matthews et al. (2013) or use other visual estimates to determine the degree to which a community has changed directionally over time or, alternatively, returned over time toward earlier structures in patterns consistent with the loose equilibrium of Collins (2000) and Matthews et al. (2013, 2014).

Detecting Loose Equilibrium

DeAngelis et al. (1985) coined the term "loose equilibrium" but did not provide specific criteria for its detection. Collins (2000) and Matthews et al. (2013, 2014) provided evidence of communities that appeared to fit a loose equilibrium model. However, specific criteria for identifying loose equilibrium were not provided until Matthews and Marsh-Matthews (2016) suggested objective criteria for evaluation of trajectories based on (1) the number of steps (vectors between surveys) in which the location of a community on an NMDS biplot returned toward the centroid, and (2) the existence of any unusually long steps as shown by statistical tests for outliers ("saltatory" rather than gradual change) or as judged by eye in some cases.

Matthews et al. (2013) suggested six hypothetical "types" of trajectories of community change (fig. 7.1). We concur with a caveat of DeAngelis et al. (1985) that "no scheme [to categorize models] is ideal for all purposes, and few models fit perfectly into any such scheme." No single criterion or set of criteria can assign a community with certainty to one of the six hypothetical trajectory types, but combinations of criteria should help classify real-world communities and identify them as being in loose equilibrium or not. Four (A, C, D, and F) of the six hypothetical trajectory types de-

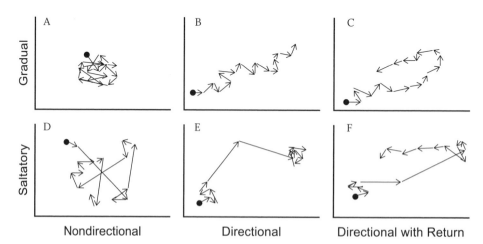

FIG. 7.1 Six hypothetical community trajectory types. From Matthews et al. (2013). With permission of Springer.

fined in Matthews et al. (2013) are consistent with loose equilibrium, and two (B and E) are not because they suggest consistent movement away from an earlier community structure.

We use an outlier analysis to determine whether any of the intervals between samples represents large or saltatory changes. We calculate Euclidean distances between the x-y coordinates for consecutive surveys on NMDS biplots. Euclidean distances can then be tested by outlier analyses like a Tukey test to identify vector(s) longer than 1.5 times the interquartile distance on a box plot of vector distances or a generalized ESD (extreme studentized deviate) test. If no vector is a markedly long outlier, changes in the community are considered gradual.

It is actually less important whether or not a community has saltatory steps (by our outlier analyses, below, or by eye) than whether or not it shows substantial numbers of steps with "return" toward an earlier state. The real issue is whether over time the community trajectory shows persistent directional movement away from its earlier state (not in loose equilibrium) or, alternatively, tends to return over time toward some central or average condition, represented by the centroid in NMDS biplots (in loose equilibrium). The returns are the more critical test of loose equilibrium, and the presence or absence of saltatory steps provides an indication of how "loose" the loose equilibrium is.

Sometimes subjective criteria help in evaluating patterns of community change. For example, the trajectories in NMDS biplots can be inspected for long periods without changes in direction or for extended sequences of directional displacement. Substantial numbers of returns from displacement, spaced rather evenly in time, are indicative of returns toward an average community. Sometimes it is the most recent surveys that are important, that is, if the recent steps tend to move the community back toward the centroid of points or to an earlier state. While the two quantitative criteria (number of return steps and presence of outliers) are important, subjective inspection of the actual trajectories on the NMDS biplot should be used to help identify community trajectory type and to determine whether a community exhibits loose equilibrium. The following, from Matthews and Marsh-Matthews (2016, Appendix C), are examples of using these combined criteria to determine whether a community is in loose equilibrium.

In trajectory 1A of Matthews et al. (2013), 5 of the 14 temporal steps moved back toward the centroid (arrows in fig. 7.2, *top*) after a displacement in the step immediately preceding, suggesting repeated movement back toward average, that is, nondirectional. We count a step as moving back toward the centroid only if it immediately follows a step that had moved farther away from the centroid (or if in the first step the point moves closer to the centroid than the position of the first survey). Because of the requirement that a return follow a displacement, the maximum number of return steps equals half of the total steps. So, for a potential maximum of eight returns, if the pattern were a perfect "away followed by return" in alternating steps, the pattern in 1A from our *Oecologia* paper showed five returns. Most of the time there is a return

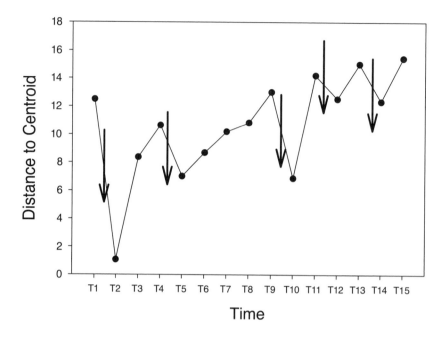

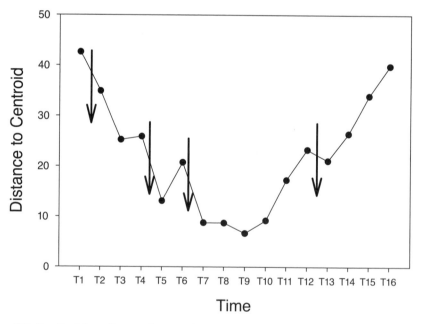

FIG. 7.2 *Top*, Euclidean distance from each point to the centroid of all points in a biplot representing successive survey times (T) in hypothetical community 1A in Matthews et al. (2013, fig. 1). *Bottom*, Same presentation for hypothetical community 1B. From Matthews and Marsh-Matthews (2016). Copyright by the Ecological Society of America. Reprinted with permission.

toward the centroid from a previous displacement. Additionally, no outliers were detected for hypothetical trajectory type 1A. Thus, based on the frequent movement of points back toward the centroid after having been displaced and the lack of outliers, we considered this type 1A trajectory (Matthews et al. 2013) to represent a community with temporal pattern of "gradual change, nondirectional." Such a pattern is consistent with expectations for a community that is in loose equilibrium.

As a contrasting example with a more complicated pattern, trajectory type 1B in Matthews et al. (2013) showed movement toward the centroid from a more distant position in 3 of 15 steps, but figure 7.2, *bottom*, shows that during the first 8 steps the community moved gradually and rather consistently closer to the centroid but away from the centroid thereafter. The long-term temporal pattern or sequence of movement toward or away from the centroid is important because there is no evidence of any substantial return toward the centroid in the later surveys. There were no outliers, so the changes were considered to be gradual. The hypothetical community type 1B was thus identified as "gradual, directional," not consistent with loose equilibrium.

We end this section with a caution about use of NMDS. Each run of NMDS on a data set will give slightly different solutions. Most of the time with our actual data sets, repeated runs of NMDS based on MH similarities gave consistent results and similar positions of surveys on biplots. But in one case, repeated NMDS runs gave inconsistent patterns, and no single solution was clearly best. For Piney Local (site M-2 in Piney Creek) a two-dimensional NMDS run in Matthews and Marsh-Matthews (2016, fig. 6) and a two-dimensional NMDS run for this chapter, along with four additional runs, all gave rather different outcomes. In one of these the positions of the first and last surveys were fairly close together in the biplot (Matthews and Marsh-Matthews, fig. 6), whereas in a different run of the same triangular MH matrix (the NMDS in this chapter) the first and last surveys were relatively distant from each other. For some data sets, different NMDS runs of the same data may give solutions that appear quite different, so it is advisable to make several runs of NMDS for any community to determine a consistent solution. We made multiple reruns for five other NMDS biplots (Roanoke, Little Glasses, and Borrows Local, and Brier and Piney Global) that are used in this chapter. In each case the multiple reruns produced two-dimensional NMDS solutions that were either identical or very similar. NMDS may need to be used with caution, ensuring that outcomes of multiple runs are consistent.

Changes in Presence of Individual Species in Our Study Systems

Our simplest analysis of community change used our Midwest data set to ask if species that were detected in the first survey, in 1978, were also present in 1995. We found that the percentage of species remaining (%REM), at a site in 1995, of those originally captured in 1978 varied widely, from <40% at 3 sites to >90% at 15 sites, with a mean of 77.2% (fig. 7.3). Overall, this suggests that most species that were present at sites in 1978 were again present in 1995, 17 years after the first survey. Most of the 1978 survey species that were not found again at the sites in 1995 were

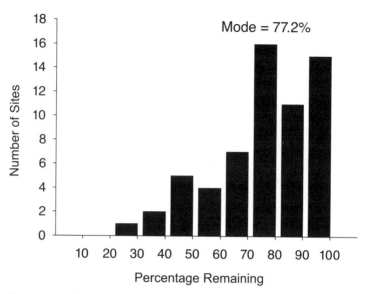

FIG. 7.3 Percentage of species found in 1978 that were present in 1995 samples at 61 sites in the midwestern United States, from Iowa and Nebraska to south Texas.

relatively scarce in 1978, often detected as only one or a few individuals. In general, species that were abundant at a site in 1978 remained at least present, and often abundant, in 1995.

For our nine global streams and in Brier Creek snorkel surveys, we examined patterns of species' occurrences across all surveys. We assigned each species to one of the following occurrence categories:

1. Frequent: occurred in 75% or more of all surveys.
2. Occasional: occurred in multiple surveys but not in 75% or more.
3. Infrequent: occurred once (for the 4 survey sites) or very few times over all surveys.

For species assigned to the Frequent and Occasional categories (in systems with more than four surveys), we examined the temporal pattern of occurrence for signals that suggested that any species was consecutively absent from later surveys and therefore may have been lost from the system over the survey period or any had been detected only in later surveys and possibly was "gained" in the system (table 7.1). For the species suggested by inspection as possibly lost, several were always found in low numbers when they did occur. These included Mimic Shiner (*Notropis volucellus*), with 1 individual in each of 2 collections (0.03%) in the Roanoke and White Crappie (*Pomoxis annularis*) (0.04% to 0.15%) in 5 snorkel surveys in Brier Creek. Yellow Bullhead (*Ameiurus natalis*), which ranged from 0.2% to 0.9% of total individuals in 3 collections in the Kiamichi, was present in the first 3 collections in that system and absent in the last 2, but this species, like other catfishes, is benthic and prone to hide

Table 7.1. Comparison of patterns of species presence-absence in surveys of the nine global streams and the snorkel surveys in Brier Creek pools

Study System (Total Number of Surveys)	Total Species	Frequent Species	Occasional Species	Infrequent Species	Species Potentially Lost	Species Potentially Gained
Alum (4)	19	13	1	4	NA	NA
Blaylock (4)	16	11	2	3	NA	NA
Bread (4)	14	13	1	0	NA	NA
Brier Seine (17)	30	17	6	7	*Pimephales vigilax*	none
Brier Snorkel (23)	27	8	12	7	*Cyprinella lutrensis* *Cyprinella venusta* *Pimephales vigilax* *Pomoxis annularis* *Aplodinotus grunniens*	none
Crooked (4)	9	7	1	1	NA	NA
Crutcho (5)	26	13	5	8	none	*Notropis atherinoides*
Kiamichi (5)	48	27	14	7	*Ameiurus natalis*	*Notemigonus crysoleucas* *Moxostoma erythrurum* *Esox americanus*
Piney (11)	48	34	7	7	*Notropis sabinae*	*Gambusia affinis*
Roanoke (5)	33	24	4	5	*Notropis volucellus*	none

Note: "Lost" or "gained" was not assessed for sites with only four surveys. NA means not applicable.

beneath structure, and the "loss" of this species should be interpreted with caution. In fact, recent surveys (2014–2015) of the Kiamichi drainage confirm presence of this species.

Other species identified as potentially lost may in fact have decreased in abundance below levels of detectability. In the Brier Creek seine surveys, Bullhead Minnow (*Pimephales vigilax*) was missing from the last 3 collections but comprised as much as 1.23% of individuals captured in previous collections. The snorkel surveys in Brier Creek covered a much smaller reach of stream than the seining surveys, so loss from the snorkeling reach does not necessarily mean that a species has disappeared from the entire system. Interestingly, however, loss from the snorkeling reach does correspond to declines within the system for three minnow species: Bullhead Minnow, Blacktail Shiner (*Cyprinella venusta*), and most notably Red Shiners (fig. 7.4). For these minnow species, abrupt declines or losses were first detected in samples in the late 1990s, which followed a period of severe drought in the area but also coincided with increases in mesopredators (Matthews and Marsh-Matthews 2007), as described in chapter 5. The other species, Freshwater Drum (*Aplodinotus grunniens*), that "disappeared" from the snorkeling reach was never captured in seining collections. All sightings of Freshwater Drum were in the early surveys (1982–1995), during which time Freshwater Drum accounted for between 0.04% and 1.13% of all individuals observed.

The other species potentially lost from our study systems was Sabine Shiner (*Notropis sabinae*), in Piney Creek. This species never comprised more than 0.04% to 1.7% of the total individuals captured but was reliably present, with dozens to a hundred or more individuals taken at one site. This species was not encountered in the last two collections in Piney Creek despite targeted efforts to collect it.

Of the species potentially gained in our study systems, most were in low numbers when they were detected, and their absence in early collections may reflect low detectability. Emerald Shiner (*Notropis atherinoides*) accounted for <0.04% of all individuals collected in every survey in which they were detected in Crutcho Creek. All species potentially gained in the Kiamichi were always a small fraction of the total number of individuals collected: Golden Shiner (*Notemigonus crysoleucas*) made up 0.05% and 0.08% of the total in 2 occurrences; Grass Pickerel (*Esox americanus*) accounted for between 0.02% and 0.05%, in 3 collections; and Golden Redhorse (*Moxostoma erythrurum*) ranged from 0.05% to 0.4% of the total in 3 collections.

The only species that might truly have been gained in Piney Creek is Western Mosquitofish (*Gambusia affinis*). This species was not detected in the first 6 of the 11 surveys but was present in the subsequent 5 surveys, and it increased progressively in relative abundance over time (ranging from 0.07% to 4.8% of the total catch from the time it was first detected to the time of the last survey). This increase in Western Mosquitofish (see plate 15) is coincident with increases in emergent vegetation (as noted in WJM field notes) at several sites in the drainage.

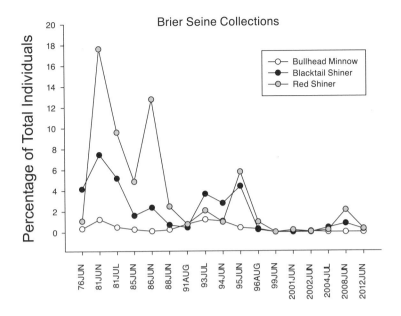

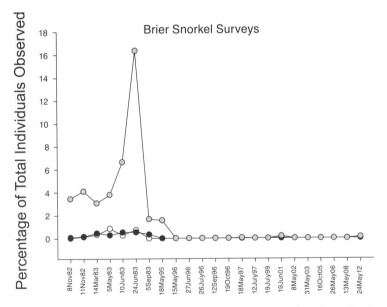

FIG. 7.4 Percentage of total number of individuals in a given sample in Brier Creek attributable to each of three minnow species. *Top*, Seine collections. *Bottom*, Snorkel surveys.

Changes in Species Dominance

Here we consider turnover of the numerically dominant individual species over time. We recognize that using raw abundance to determine the numerically dominant species has limitations and that numerical dominance is only one measure of the potential effects a species might have on a community. We limit our assessment to insectivorous species of the family Cyprinidae, because this trophic group of minnows always numerically dominated the creek. In Piney Creek (table 7.2), the numerically dominant minnow species was any one of five different species at least once. The dominant species remained dominant from one survey to the next only three times and changed from one species to another seven times. The Piney Creek minnow community is complex and clearly not dominated numerically by any one species. However, Duskystripe Shiners or Telescope Shiners were the numerical dominants in the global Piney Creek community in 8 of 11 surveys, whereas the others that dominated (Bigeye Shiner, *Notropis boops*; Ozark Minnow, *Notropis ozarcanus*; Bluntnose Minnow, *Pimephales notatus*) did so only once each. So, one could argue that in Piney Creek either Duskystripe Shiners or Telescope Shiners, or both, numerically dominate the minnow community most of the time and may well make the greatest sustained contribution to the biology of minnows in the creek.

All of the above for Piney Creek is in sharp contrast to the situation for the Brier Creek global community, in which Bigeye Shiner was the numerically dominant minnow in 15 of 16 surveys, often by a large margin. Now, comparing minnows in these 2 creeks could be comparing apples and oranges, in that Piney Creek had 18 water-column minnow species, whereas Brier Creek had only 7 native minnows in our surveys. The potential for 1 minnow species or another to take the lead (i.e., become dominant) in Piney Creek versus Brier Creek might be a bit like comparing lead changes in 2 horseraces with 18 versus 7 horses at the starting gate. Regardless, it appears that Bigeye Shiner has a persistent numerical dominance in Brier Creek, whereas no single species of minnow has a clearly established dominance in the overall fish community of Piney Creek.

In our other global systems (table 7.3), Red Shiner was the numerical dominant in each of five surveys in Crutcho Creek (see plate 6), and one species (Redfin Shiner,

Table 7.2. The numerically dominant water-column minnow species in Piney Creek in 11 surveys

Species	w72	a73	s82	w83	a83	s83	w94	a95	s95	s06	s12
LUXPIL	660	471	**1141**	536	**747**	**1920**	**1777**	952	**1095**	538	822
NOTTEL	594	**642**	854	411	400	1046	811	417	751	**541**	**1698**
NOTNUB	**692**	280	193	212	150	714	1299	1051	552	379	1297
NOTBOO	47	58	243	310	209	174	1659	**1716**	659	372	286
PIMNOT	131	91	234	**782**	227	566	931	767	306	300	270

Note: Boldface indicates the numerically dominant species within each survey.

Table 7.3. Number of different numerically dominant water-column minnow species, number of transitions from one dominant species to another, total number of surveys, and the species that was the numerical dominant most often

Site	Number of Dominant Species	Number of Transitions between Dominants	Number of Surveys	Dominant Most Often
Global				
Alum	1	0	4	Redfin Shiner
Crooked	1	0	4	Creek Chub
Crutcho	1	0	5	Red Shiner
Blaylock	2	2	4	Striped Shiner
Bread	2	1	4	Creek Chub
Brier	2	2	16	Bigeye Shiner
Roanoke	2	2	5	Crescent Shiner
Kiamichi	3	2	5	Bigeye Shiner
Piney	5	7	11	Duskystripe Shiner
Local				
Alum	1	0	4	Redfin Shiner
Blaylock	1	0	4	Striped Shiner
Choctaw	1	0	5	Red Shiner
Crooked	1	0	4	Creek Chub
Crutcho	1	0	5	Red Shiner
Glover	1	0	5	Bigeye Shiner
Baron Fork	2	3	4	Ozark Minnow and Cardinal Shiner
Blue River	2	1	5	Blacktail Shiner
Brier	2	2	16	Bigeye Shiner
Chigley Sandy	2	1	4	Red Shiner
Gar	2	1	5	Red Shiner
Hickory	2	3	5	Red Shiner
Kiamichi	2	1	5	Bigeye Shiner
Morris	2	1	5	Red Shiner
Mustang	2	2	5	Red Shiner
South Canadian	2	3	4	Red Shiner and Arkansas River Shiner
Tyner	2	2	5	Cardinal Shiner
Ballard	3	3	5	Cardinal Shiner
Illinois River	3	2	5	Cardinal Shiner
Piney	3	5	11	Whitetail Shiner
Roanoke	3	4	6	White Shiner and Crescent Shiner
Salt Fork Red River	3	3	5	Red Shiner and Red River Shiner

Note: Stonerollers were not included in this analysis because they are a benthic, algivorous species. Streams are sorted by increasing number of dominant species.

Lythrurus umbratilis) was the dominant in Alum Creek in all four of our surveys. One species (Creek Chub, *Semotilus atromaculatus*) was dominant in every survey of Crooked Creek (see plate 5) and in three of four surveys in Bread Creek. Bigeye Shiner was numerically dominant in the Kiamichi River (see plate 3) in three of five surveys, and Striped Shiner (*Luxilus chrysocephalus*) was the dominant in Blaylock Creek in three of four surveys. In the Roanoke River (see plate 4), 2 minnow species (out of 13 species in the community) alternated in their dominance, with Crescent Shiner (*Luxilus cerasinus*) dominant in 3 monthly surveys and White Shiner (*Luxilus albeolus*) dominant in the other 2.

In summary, six of our nine global streams, particularly those with less complex minnow communities (sometimes as few as three species), tended to be dominated consistently by a single species. One of our more complex systems (Kiamichi River, with 16 minnow species) was also dominated 60% of the time by 1 species. The Roanoke River minnow community was dominated by only two species. And only in the most complex minnow community of Piney Creek (total of 18 water-column minnow species) did we see frequent transitions in the identity of the numerical dominant. Maybe simpler systems tend to be dominated by a single species, especially in systems with potentially harsh environmental conditions (like Brier Creek, Alum Creek, or Crutcho Creek), whereas more benign upland streams such as Piney Creek and Roanoke River may provide more diverse habitats or environmental conditions, leading to more opportunities for different species to reach high population numbers at different times.

At 22 of the 31 local sites, Cyprinidae was the dominant family (table 7.3). In 6 local sites, only 1 minnow species was ever numerically dominant, with that species consistently the most abundant across 4 or 5 surveys. These 6 sites tended to be smaller streams with relatively simple fish communities, with Glover River being the exception, where the Bigeye Shiner was always the local dominant in spite of its having a relatively diverse minnow fauna. At 11 of the local sites, 2 minnow species were dominant at different times, and in 5 of these sites there was only a single transition from 1 dominant species to another. But in the 6 other sites there were either 2 or 3 transitions, as the numerically dominant species changed back and forth over time. The local sites with two dominants ranged widely, from very small sites like Mustang and Tyner Creeks and from stony-bottomed upland to mud-bottomed lowland streams, with no apparent pattern. Finally, at 5 local sites (Ballard Creek, Illinois River, Piney Creek, Roanoke River, and Salt Fork of the Red River) there were 3 different minnow species that dominated at various times, with frequent (3 to 5) transitions from one dominant to another. The first four of the sites are all upland streams with complex minnow communities (and fish communities in general), where, as we suspected for the global communities, it was not likely that any one species would remain most abundant across time. The Salt Fork of Red River is interesting in that in four of our surveys, either Red Shiner or Red River Shiner (*Notropis bairdi*) was dominant, but in the last survey, in 2013 after severe drought in 2012, we caught few fish of any kind and the numerically dominant species (if it can be called such) was the Plains Minnow (*Hybognathus placitus*), of which we caught a total of seven individuals. Overall it appeared that most of the local sites having frequent transitions in dominant species had complex minnow communities, with a larger number of species that potentially waxed or waned in abundance as environmental conditions varied.

Complex Ways to Examine Dynamics of Community Structure
Morisita-Horn Comparisons of Midwest Sites from 1978 to 1995

Morisita-Horn values comparing composition of the local fish communities at 61 sites we sampled throughout the lower Midwest in 1978 and 1995 (fig. 7.5) ranged from 0.992 at 1 site (CHK2) on the Chikaskia River in southern Kansas to 0.204 in the Pecos River near Sheffield, Texas (SHEF). Many of the 61 sites had intersurvey MH values of 0.750 or greater, suggesting that they had changed little in community structure in the 17 years between our surveys. Nine sites had MH similarity <0.400, usually for reasons that were obvious in comparing species identities or abundances between decades.

A Chikaskia River site that had strong similarity between the 2 Midwest surveys was a shallow, sand-bed river with an extremely simple fauna (8 species detected in 1978 and only 3 in 1995, but in both cases the local community was dominated strongly by Red Shiners (86% and 88% Red Shiners in the 2 surveys, respectively), resulting in the high MH index of 0.992. The samples in the Pecos River also had few species (six and nine, respectively), and Red Shiners were relatively abundant in both samples. In contrast to more stable sites, however, we found in 1995 that the study reach of Pecos River had been heavily invaded by Sheepshead Minnows (*Cyprinodon variegatus*) (Echelle and Connor 1989), which were absent in 1978 but comprised 49% of all fishes collected in 1995. In addition, in 1995 in the Pecos River we caught 120 Gulf Killifish (*Fundulus grandis*), comprising 30% of the sample, whereas we found none at that site

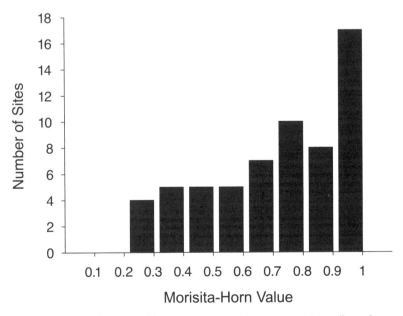

FIG. 7.5 Distribution of Morisita-Horn values comparing communities collected at individual sites in 1978 and 1995 for 61 sites from Nebraska and Iowa to south Texas.

in 1978. The Gulf Killifish has recently invaded and become established in numerous west Texas drainages (Cheek and Taylor 2015) and apparently became established at our site between 1978 and 1995. That almost 80% of the 1995 sample was "new" at the Pecos River site resulted in the very low MH similarity of 0.204. These extreme examples are the unusual ones among all 61 of our repeated sites across the Midwest, however, and at most sites the local community in 1995 was similar to that in 1978.

Morisita-Horn Values for Consecutive Surveys in Nine Global Communities

Because means of MH values for our 9 global (and 31 local) communities were sometimes bimodal (or otherwise not a good representation of central tendency), we used the median of MH values between consecutive surveys as a quantitative index of temporal variation with which to examine factors related to the magnitude of temporal change. The nine global communities showed wide temporal variation in similarity (fig. 7.6). The figure facilitates the assessment of the range of variation in time-to-time change but does not explain factors in differences among the systems. For example, Crooked Creek, the community with the highest median MH and lowest variation in MH (as indicated by the interquartile range, or IQR, the range of values between the 25th and 75th percentile) is in the same drainage as the community (Bread Creek) with the lowest median MH and the greatest IQR.

To examine factors potentially affecting temporal variation, we used linear regressions of median MH and IQR separately against factors related to the schedule of sampling (time span, total number of surveys), environmental traits ("intermittency"

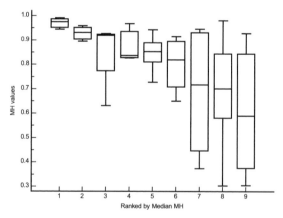

FIG. 7.6 Box-and-whisker plots summarizing values of consecutive Morisita-Horn (MH) values for each of nine global communities. Communities are ranked in descending order of the median MH value. Boxplots depict the following: lower and upper limits of whiskers show range of MH values; lower and upper boundaries of box represent the 25th and 75th percentiles; horizontal line within box is median value. Community ranks are: (1) Crooked Creek, (2) Alum Creek, (3) Roanoke River, (4) Blaylock Creek, (5) Piney Creek, (6) Kiamichi River, (7) Crutcho Creek, (8) Brier Creek, and (9) Bread Creek.

from none to occasional, mean temperature, elevation, stream width, stream gradient), and properties of the communities themselves (total species). None of the regressions were significant, although intermittency was related to median MH values at $P = 0.082$. Overall, however, none of the factors we tested—including sampling schedule, environment, or community structure—was strongly predictive of the magnitude of community similarity from time to time.

Morisita-Horn Values for Consecutive Surveys at 31 Local Sites

The 31 local communities showed a wide range of median MH values and substantial differences in IQRs (fig. 7.7). Median MH values were >0.90 for 3 local sites (Alum, Crooked, and Little Elm Creeks). This observation contrasts with historical views of the diversity-stability concepts because these sites, which were highly similar across time, also had quite simple community structure. In a previous chapter, however, we addressed the fact that very simple communities, like in the extreme headwaters of Brier Creek, can appear mathematically to be quite stable if only one or a few unusually hardy species dominate a community in a stressful habitat. At the other extreme, the median MH values were <0.30 at Salt Fork of the Red River (MH = 0.2770, where community dominants such as Red Shiner and Red River Shiner were lost or dramatically reduced during drought periods), Hickory Creek (MH = 0.1482, where at least 7 species varied dramatically in abundance among surveys), and Hauani Creek (MH = 0.1450, where Red Shiner, Bullhead Minnow, Longear Sunfish (*Lepomis megalotis*), and Orangethroat Darter (*Etheostoma spectabile*) were drastically reduced in recent surveys, and Western Mosquitofish and Bluegill became much more abundant). The median consecutive MH values varied widely across all other sites, however, between these examples of high and low values. Additionally, the amount of variation in consecutive MH values (depicted by the IQR, or the size of the "box") differed widely across all 31 sites.

To explore factors that might explain the differences in medians or in the interquartile ranges for the 31 local communities (fig. 7.7), we plotted both median and IQR against metrics like those for the global streams (above). Linear regressions showed that the IQR and the number of surveys were positively related ($P = 0.02$), indicating that conducting more surveys provides more opportunity to detect excursions from some average community structure, because doing more surveys means a better chance of detecting changes related to events such as droughts or floods. Time span of surveys was marginally related ($P = 0.056$) to a lower median of MH values, also suggesting that longer studies have a better potential to detect changes that could be related to events.

Of the environmental factors, the median MH ($P = 0.047$) and IQR ($P = 0.027$) were both significantly related to local stream gradient, with higher-gradient sites having greater MH similarities and lower IQR between surveys. Both of these results suggest that communities in higher-gradient streams vary less from time to time. For other traits—intermittency of the stream, stream depth, and stream width—there were no

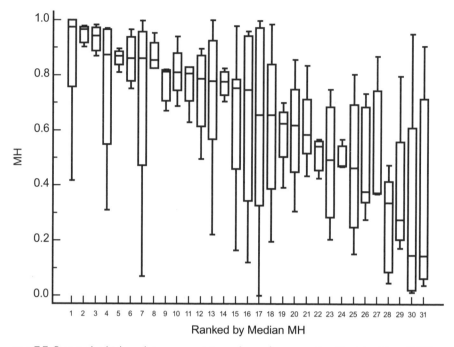

FIG. 7.7 Box-and-whisker plots summarizing values of consecutive Morisita-Horn (MH) values for each of 31 local communities. Communities are ranked in descending order of the median MH value. Boxplots depict the following: lower and upper limits of whiskers show range of MH values; lower and upper boundaries of box represent the 25th and 75th percentiles; horizontal line within box is median value. Community ranks are: (1) Little Elm Creek, (2) Crooked Creek, (3) Alum Creek, (4) Choctaw Creek, (5) Glover River, (6) Blaylock Creek, (7) Garrett Creek, (8) Kiamichi River, (9) South Canadian River, (10) Tyner Creek, (11) Crutcho Creek, (12) Gar Creek, (13) Coal Creek, (14) Blue River, (15) Piney Creek, (16) Morris Creek, (17) Cow Creek, (18) Brier Creek, (19) Lukfata Slough, (20) Illinois River, (21) Little Glasses Creek, (22) Bread Creek, (23) Ballard Creek, (24) Baron Fork, (25) Mustang Creek, (26) Borrows, (27) Chigley Sandy Creek, (28) Roanoke River, (29) Salt Fork Red River, (30) Hickory Creek, and (31) Hauani Creek.

significant relationships at $P = 0.05$ with either the median MH or the IQR of MH values for consecutive surveys.

The assessments above for both global and local communities show that it is not easy to predict which streams or kinds of environments may relate to variation in a fish community. But our global and local data do show, as noted by Ross et al. (1985), that fish communities within local stream reaches typically vary more than communities of whole watersheds.

Community Trajectories in Multivariate Space

For the two global and eight local communities for which we had six or more surveys and for Brier Creek snorkeling, we used NMDS of their MH indices to produce bi-

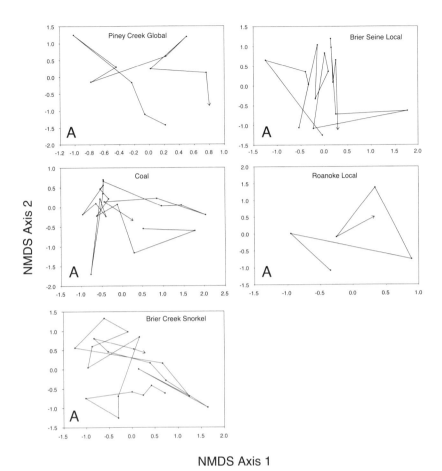

FIG. 7.8A Temporal trajectories for communities in one global stream, three local communities, and the community determined by snorkeling in Brier Creek, all of which were considered type A (gradual, nondirectional). Arrowheads indicate the last vector and the final position of each in NMDS biplots.

plots showing the location of surveys in multivariate space. Consecutive surveys were connected to produce successional vectors (fig. 7.8A,B). These temporal trajectories provide a visual overview of dynamics of these communities and can be used to assess the likelihood that they were in loose equilibrium, per Collins (2000), Matthews et al. (2013), and Matthews and Marsh-Matthews (2016).

For all these communities, a two-dimensional NMDS solution produced satisfactory Stress 1 values (table 7.4), and for all but one, multiple repeated runs of NMDS produced trajectories with similar patterns. For Piney Local (site M-2), repeated NMDS runs failed to produce solutions with similar trajectories, so we omitted that site from further consideration. For the other communities, we evaluated the trajectories by criteria outlined above and in Matthews and Marsh-Matthews (2016, Appendix C), emphasizing the number of returns and presence or lack of outliers.

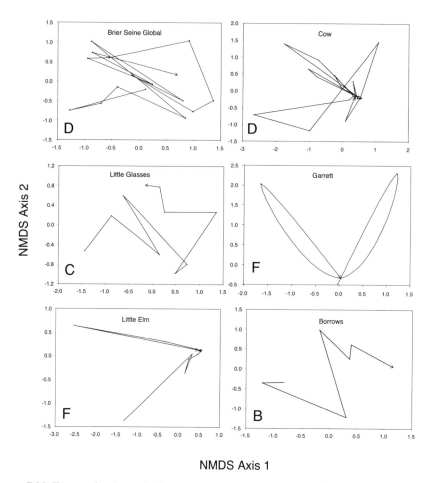

NMDS Axis 1

FIG. 7.8B Temporal trajectories for communities in one global stream, and five local communities, with trajectories of four different types. Arrowheads indicate the last vector and the final position of each in NMDS biplots. In Cow Creek the arrowhead is embedded within the cluster of points on the right of the figure, making it difficult to see.

The trajectories for all global and local communities for which we had six or more surveys, omitting one that failed to achieve a stabilized NMDS solution, are shown in fig. 7.8A and 7.8B, respectively. The Piney Creek global community, three local communities (Brier, Coal, and Roanoke), and the snorkel surveys of Brier Creek pools all were classified as trajectory type A (gradual, nondirectional). For each there were no outliers in vector lengths, but there was a relatively high proportion of return steps (fig. 7.8A) toward the biplot centroid (50% to 80% out of the number possible; remember that there can only be a return for a step that follows a previous departure from the centroid, so that the maximum number of potential returns is half the total number of steps). Two others (Brier Creek global and Cow Creek local) showed at least one long saltatory step but were considered nondirectional because of a substan-

Table 7.4. Characteristics of trajectories depicted in figure 7.8 and assignment
of trajectory type

Site	Stress 1 in NMDS	Number of Returns toward Centroid vs. Number of Steps	Number of Outliers, Tukey, or Generalized ESD	Trajectory Type
Piney global	0.088	4 of 10	0	A
Brier global	0.103	5 of 16	2 GESD	D
Brier snorkel	0.209	8 of 22	0	A
Borrows	0.128	1 of 6	0	B
Brier local	0.067	4 of 16	0	A
Coal	0.123	6 of 24	0	A
Cow	0.133	7 of 26	3 GESD 1 Tukey	D
Garrett	0.0002	2 of 15	4 by eye	F
Little Elm	0.0009	5 of 12	3 GESD	F
Little Glasses	0.140	2 of 10	0	most like C
Roanoke local	0.120	2 of 6	0	A

Note: Trajectory types are as follows: A, gradual, nondirectional; B, gradual, directional; C, gradual, directional with return; D, saltatory, nondirectional; F, saltatory, directional with return. From Matthews et al. (2013, table 1).

tial number of returns after displacement (fig. 7.8B) and were classified as trajectory type D (saltatory, nondirectional). Trajectory types A and D are both consistent with expectations of loose equilibrium, so these seven communities are consistent with the LEC of community dynamics (Matthews and Marsh-Matthews 2016). Two local communities (Garrett and Little Elm Creeks) had long saltatory steps (fig. 7.8B) that included directional movement away from earlier structure but then exhibited return toward the earlier or more typical structure. They were thus classified as trajectory type F (saltatory, directional with return). Little Glasses Creek alone was considered type C (gradual, directional with return). The NMDS biplot for Little Glasses Creek (fig. 7.8B) showed general movement in a positive direction (to the right) on NMDS Axis 1 for six steps, but in the last three steps the community moved sharply back toward the centroid, then farther back toward the left. Overall, nine of the ten global or local communities whose trajectory patterns we examined were consistent with loose equilibrium, considered to fit hypothetical trajectory types A, C, D, or F of Matthews et al. (2013).

The community at one local site (Borrows at US 70; fig. 7.8B) was not consistent with expectations of loose equilibrium and was classified as type B (gradual, directional). The Borrows site consists of shallow backwaters along a highway north of Idabel, Oklahoma, filled during wet periods by overflow from a branch of Yanubbe Creek but greatly reduced or dried nearly completely during droughts in recent decades. In figure 7.8B it is apparent that the Borrows site has moved substantially from left to right on the first NMDS axis. The raw data showed 2 periods of sharp reduction in total species detected (from 13 in 1990 to only 4 by 2000; and from 14 species to only 3 in 2006). Western Mosquitofish numerically dominated the limited community in the most recent survey (2006). Several interesting local species that are most typical of the Southern Coastal Plain—including Pygmy Sunfish (*Elassoma zonatum*), Flier (*Centrarchus macropterus*), Bantam Sunfish (*Lepomis symmetricus*), and Western Starhead Topminnow (*Fundulus blairae*)—and that were previously common at the site had disappeared or were in very low numbers in this last sample. It appears that during drying or low-water periods the local fish community is decimated, followed by some return of lost species, but not sufficiently to return the community to structure similar to that in the earliest surveys. There seemed overall to be much "coming and going" of species, but not in any distinct pattern or with groups consistently arriving or leaving together.

Brier Creek and Piney Creek: Contrasting the Fish Community Dynamics

For decades we have studied the fish communities of Piney Creek and Brier Creek (see plates 1 and 2). These systems differ in harshness and environmental stability, and previous papers with a variety of different metrics or approaches have reported greater variation in the Brier Creek fish community than in Piney Creek. Here we revisit and compare these systems with respect to the long-term dynamics of their fish communities using the same analyses for both systems. Our analyses suggest that the dynamics of both systems fit the loose equilibrium model (Matthews et al. 2013; Matthews and Marsh-Matthews 2016; and analyses in this chapter), although Piney Creek fits a "gradual" model while Brier Creek fits a "saltatory" model. The systems also differ in the degree to which they vary around their "average" condition. Comparison of MH values of consecutive collections (fig. 7.9) shows that the Piney Creek fish community tends to change less from time to time, as indicated by a higher median MH and smaller interquartile range than those for Brier Creek.

As another comparison of variation in the fish communities of Piney and Brier Creeks, we performed a common-platform NMDS of MH values for both creeks (fig. 7.9), including all species found in either site. Over all surveys, there were 63 species in both creeks combined, 15 of which were shared by the 2 creeks, but with 15 unique to Brier Creek and 33 unique to Piney Creek. In the resulting NMDS plot (final stress = 0.097), the convex hull enclosing Piney Creek collections was smaller than that enclosing Brier Creek collections (fig. 7.9). The fact that both of these systems display dynamics consistent with loose equilibrium, despite differences in the overall variability of their fish communities, suggests that the LEC may serve as an appropriate null hypothesis for many community types.

Summary

Over the past century, views of community dynamics changed from an assumption of equilibrium or "climax" communities in the early years, to community variation in

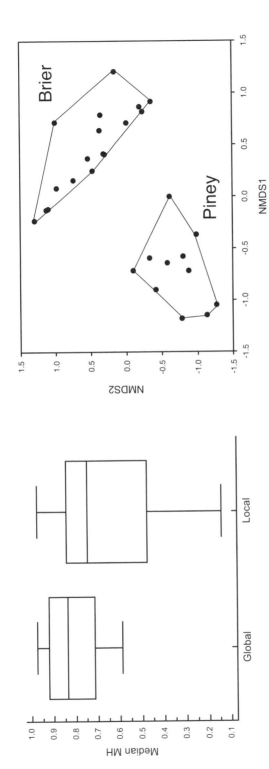

FIG. 7.9 *Left,* Boxplots of Morisita-Horn values for consecutive surveys for Brier Creek and Piney Creek fish communities. *Right,* Common-platform NMDS of Brier and Piney Creeks.

response to disturbance since the mid-1970s, then to more recent models of communities that vary around an average condition in a "loose equilibrium." Recently, numerous studies using a wide variety of metrics and analytical tools have addressed temporal dynamics of fish communities, and many found evidence of persistence but did not specifically attempt to classify community dynamics with respect to the fit to a loose equilibrium model.

Our own studies of long-term dynamics of fish communities span more than 40 years, with most emphasis on 2 systems: Piney Creek in Arkansas and Brier Creek in Oklahoma. These two systems vary in environmental characteristics, with the spring-fed Piney Creek system having less variation in flow and water temperature and having continuous flow throughout the year. In contrast, Brier Creek is subject to seasonal variation in flow, temperature, and dissolved oxygen and may be reduced to isolated pools during summer. Both systems have been subject to catastrophic floods over the timeline of our studies, and Brier Creek has experienced several periods of severe drought. Several of our publications (or those with colleagues) have addressed responses to floods and droughts in these systems and compared their dynamics. Overall, our studies suggest that the fish community of Brier Creek is more variable than that of Piney Creek.

In this chapter we have assessed temporal dynamics of the fish communities in multiple study systems at short- to long-term scales (including reassessment of Brier Creek and Piney Creek). We asked whether individual species were lost or gained in a given system (based on analyses of presence-absence) and examined transitions in the numerically dominant species from time to time. In addition to exploring community changes at the level of individual species, we used the Morisita-Horn (MH) similarity index to compare the structure of the entire community from time to time. We chose the MH index because it is density invariant (i.e., it is based on relative abundances of species in the community and is not sensitive to overall sample size). We also used MH as the similarity metric in nonmetric multidimensional scaling (NMDS), an ordination technique with which we can visualize proximity of samples in multivariate space and connect consecutive samples to depict a trajectory of community change over time. Finally, we examined the trajectories for fit to the predictions of the loose equilibrium concept (LEC).

We used two criteria to evaluate fit to the LEC. The number of "returns" of the trajectory toward the centroid of the NMDS plot allowed us to assess whether the trajectory was "nondirectional," "directional," or "directional with return." A second criterion examined the length of the arrows between successive collections and asked if there were any outliers. If it had no exceptionally long "steps" the trajectory was scored as "gradual," but with outliers it was scored as "saltatory." Using these combined criteria, trajectories were assigned to one of six types as described in Matthews et al. (2013). Four of these types are consistent with loose equilibrium.

Our examination of changes in presence-absence of individual species between 1978 and 1995 at 61 sites across the lower Midwest showed that on average, 77% of species captured in the first survey were recaptured 17 years later. For our nine global systems and for the Brier Creek snorkel survey, we asked if species had been lost or gained based on presence-absence in multiple surveys. In both the Brier Creek seine collections (global) and the Brier Creek snorkel surveys in a 1 km reach of pools, 3 minnow species declined or were lost. Sabine Shiner was the only species identified as lost in Piney Creek. Although this species was only common at one site, dozens to hundreds of individuals were collected at times, but none were collected in the last two surveys of Piney Creek. The only species that gained in our study systems was Western Mosquitofish, in Piney Creek.

Minnows (Family Cyprinidae) numerically dominated almost all of our study systems. One (sometimes two) minnow species dominated most surveys in a given system except in Piney Creek, which has a more complex minnow community than other sites and where several species switched dominance among surveys.

Community similarity between consecutive surveys (as measured by MH) varied widely among both global and local systems. Using the median value and interquartile range of MH values, we examined potential factors related to the magnitude to community similarity from time to time. For the global systems, there was no relationship between either median MH or IQR and factors related to schedule of sampling (number of surveys, time span of study), environment (intermittency, mean temperature, elevation, stream width, stream gradient), or community structure (number of species detected). For the 31 local communities, there was a negative relationship between median MH and time span and a positive relationship between IQR and number of surveys, both of which suggest that longer-term surveys are more likely to detect community change (possibly associated with rare events). The only environmental factor related to community similarity from time to time was stream gradient, which was positively related to median MH and negatively related to IQR, suggesting that fish communities in higher-gradient streams are less variable than those in lower-gradient streams. Overall, there was a difference in the median values of MH for the global and local systems, with global systems having higher MH and less variation in median MH.

Examination of temporal trajectories for 11 systems with at least 6 surveys showed that 5 were best classified as type A, 1 as type B, 1 as type C, 2 as type D, and 2 as type F. The only one of these that was not consistent with the LEC was type B. This fish community was a local system in the Borrow ditches of far southeastern Oklahoma that had changed dramatically as the habitat dried during extreme droughts in the area.

Our reanalyses of Brier Creek and Piney Creek using comparable metrics confirmed the findings of earlier papers: the fish community in Brier Creek was more variable over time and showed a saltatory, nondirectional trajectory, compared to the gradual nondirectional trajectory for Piney Creek. The magnitude of sample-to-sample variation was higher for Brier Creek, and the spread of collections for Brier Creek was greater than that for Piney Creek when both were plotted on the same, common-platform NMDS biplot.

Over all of the systems we have studied, at both local and global spatial scales, and on temporal scales from months to decades, we have found little change in qualitative composition, few changes in the more abundant species, and a tendency for loose equilibrium in community structure in most systems.

References

Allee, W. C., A. E. Emerson, O. Park, T. Park, and K. P. Schmidt. 1949. Principles of animal ecology. W. B. Saunders, Philadelphia, PA.

Blair, A. P. 1959. Distribution of the darters (Percidae, Etheostominae) of northeastern Oklahoma. Southwestern Naturalist 4:1–13.

Bormann, F. H., and G. E. Likens. 1979. Catastrophic disturbance and the steady state in northern hardwood forests. American Scientist 67:660–669.

Cheek, C. A., and C. M. Taylor. 2015. Salinity and geomorphology drive long-term changes to local and regional fish assemblage attributes in the lower Pecos River, Texas. Ecology of Freshwater Fish. doi:10.1111/eff.12214.

Collie, J. S., A. D. Wood, and H. P. Jeffries. 2008. Long-term shifts in the species composition of a coastal fish community. Canadian Journal of Fisheries and Aquatic Sciences 65:1352–1365.

Collins, S. L. 2000. Disturbance frequency and community stability in native tallgrass prairie. American Naturalist 155:311–325.

Connell, J. H. 1983. On the prevalence and relative importance of interspecific competition: evidence from field experiments. American Naturalist 122:661–696.

Cowles, H. C. 1899. The ecological relations of the vegetation on the sand dunes of Lake Michigan: I. Geographical Relations of the Dune Floras. Botanical Gazette 27:95–117.

Davis, M. B. 1986. Climatic instability, time lags, and community disequilibrium. Pages 269–284 in J. Diamond and T. J. Case, eds. Community ecology. Harper and Row, New York, NY.

DeAngelis, D. L., J. C. Waterhouse, W. M. Post, and R. V. O'Neill. 1985. Ecological modelling and disturbance evaluation. Ecological Modelling 29:399–419.

Echelle, A. A., and P. J. Connor. 1989. Rapid, geographically extensive genetic introgression after secondary contact between two pupfish species (Cyprinodon, Cyprinodontidae). Evolution 43:717–727.

Elton, C. 1946. Competition and the structure of ecological communities. Journal of Animal Ecology 15:54–68.

Gauch, H. G., Jr. 1982. Multivariate analysis in community ecology. Cambridge University Press, Cambridge, UK.

Gause, G. F. 1937. Experimental populations of microscopic organisms. Ecology 18:173–179.

Gido, K. B., and D. A. Jackson, eds. 2010. Community ecology of stream fishes: concepts approaches, and techniques. American Fisheries Society Symposium 73. American Fisheries Society, Bethesda, MD.

Gido, K. B., K. N. Bertrand, J. N. Murdock, W. K. Dodds, and M. R. Whiles. 2010. Disturbance-mediated effects of fishes on stream ecosystem processes: concepts and results from highly variable prairie streams. Pages 593–617 in K. B. Gido and D. A. Jackson, eds. Community ecology of stream fishes: concepts, approaches, and techniques. American Fisheries Society Symposium 73. American Fisheries Society, Bethesda, MD.

Gillette, D. P., A. M. Fortner, N. R. Franssen, et al. 2012. Patterns of change over time in darter (Teleostei: Percidae) assemblages of the Arkansas River basin, northeastern Oklahoma, USA. Ecography 35:855–864.

Grossman, G. D., and J. L. Sabo. 2010. Incorporating environmental variation into models of community stability: examples from stream fish. Pages 407–426 in K. B. Gido and

D. A. Jackson, eds. Community ecology of stream fishes: concepts, approaches, and techniques. American Fisheries Society Symposium 73. American Fisheries Society, Bethesda, MD.

Grossman, G. D., P. B. Moyle, and J. O. Whittaker Jr. 1982. Stochasticity in structural and functional characteristics of an Indiana stream fish assemblage: a test of community theory. American Naturalist 120:423–454.

Hardin, G. 1960. The competitive exclusion principle. Science 131:1292–1297.

Hubbs, C. L., and A. I. Ortenburger. 1929a. Further notes on the fishes of Oklahoma with descriptions of new species of Cyprinidae. Publications of the University of Oklahoma Biological Survey 1:17–43.

Hubbs, C. L., and A. I. Ortenburger. 1929b. Fishes collected in Oklahoma and Arkansas in 1927. Publications of the University of Oklahoma Biological Survey 1:47–112.

Hutchinson, G. E. 1957. Concluding remarks. Cold Spring Harbor Symposia on Quantitative Biology 22:415–427.

Jost, L., A. Chao, and R. L. Chazdon. 2011. Compositional similarity and beta diversity. Pages 66–84 in A. E. Magurran and B. J. McGill, eds. Biological diversity—frontiers in measurement and assessment. Oxford University Press, Oxford, UK.

Keast, A., and D. Webb. 1966. Mouth and body form relative to feeding ecology in the fish fauna of a small lake, Lake Opinicon, Ontario. Journal of the Fisheries Research Board of Canada 23:1845–1874.

Lake, P. S. 2011. Drought and aquatic ecosystems: effects and responses. Wiley-Blackwell, West Sussex, UK.

Lubchenco, J., and B. A Menge. 1978. Community development and persistence in a low rocky intertidal zone. Ecological Monographs 59:67–94.

Lugo, A. E., C. S. Rogers, and S. W. Nixon. 2000. Hurricanes, coral reefs and rainforests: resistance, ruin and recovery in the Caribbean. Ambio 29:106–114.

MacArthur, R. H. 1958. Population ecology of some warblers of northeastern coniferous forests. Ecology 39:599–619.

Magurran, A. E., and P. A. Henderson. 2010. Temporal turnover and the maintenance of diversity in ecological assemblages. Philosophical Transactions of the Royal Society B: Biological Sciences 365:3611–3620.

Marsh-Matthews, E., and W. J. Matthews. 2002. Temporal stability of minnow species co-occurrence in streams of the central United States. Transactions of the Kansas Academy of Science 105:162–177.

Marsh-Matthews, E., and W. J. Matthews. 2010. Proximate and residual effects of exposure to simulated drought on prairie stream fishes. Pages 461–486 in K. B. Gido and D. A. Jackson, eds. Community ecology of stream fishes: concepts, approaches, and techniques. American Fisheries Society Symposium 73. American Fisheries Society, Bethesda, MD.

Marsh-Matthews, E., W. J. Matthews, and N. R. Franssen. 2011. Can a highly invasive species re-invade it native community? the paradox of the Red Shiner. Biological Invasions 13: 2911–2924.

Matthews, W. J. 1986. Fish faunal structure in an Ozark stream: stability, persistence, and a catastrophic flood. Copeia 1986:388–397.

Matthews, W. J. 1998. Patterns in freshwater fish ecology. Chapman and Hall, New York, NY.

Matthews, W. J., and L. G. Hill. 1980. Habitat partitioning in the fish community of a southwestern river. Southwestern Naturalist 25:51–66.

Matthews, W. J., and J. D. Maness. 1979. Critical thermal maxima, oxygen tolerances and success of cyprinid fishes in a southwestern river. American Midland Naturalist 102:374–377.

Matthews, W. J., and E. Marsh-Matthews. 2003. Effects of drought on fish across axes of space, time and ecological complexity. Freshwater Biology 48:1232–1253.

Matthews, W. J., and E. Marsh-Matthews. 2006a. Temporal changes in replicated experimental stream fish assemblages: predictable or not? Freshwater Biology 51:1605–1622.

Matthews, W. J., and E. Marsh-Matthews. 2006b. Persistence of fish species associations in pools of a small stream of the southern Great Plains. Copeia 2006:696–710.

Matthews, W. J., and E. Marsh-Matthews. 2007. Extirpation of Red Shiner in direct tributaries of Lake Texoma (Oklahoma-Texas): a cautionary case history from a fragmented river-reservoir system. Transactions of the American Fisheries Society 136:1041–1062.

Matthews, W. J., and E. Marsh-Matthews. 2015. Comparison of historical and recent fish distribution patterns in Oklahoma and western Arkansas. Copeia 103:170–180.

Matthews, W. J., and E. Marsh-Matthews. 2016. Dynamics of an upland stream fish community over 40 years: trajectories and support for the loose equilibrium concept. Ecology 97:706–719.

Matthews, W. J., R. C. Cashner, and F. P. Gelwick. 1988. Stability and persistence of fish faunas and assemblages in three midwestern streams. Copeia 1988:945–955.

Matthews, W. J., B. C. Harvey, and M. E. Power. 1994. Spatial and temporal patterns in the fish assemblages of individual pools in a midwestern stream (U.S.A.). Environmental Biology of Fishes 39:381–397.

Matthews, W. J., E. Marsh-Matthews, G. L. Adams, and S. R. Adams. 2014. Two catastrophic floods: similarities and differences in effects on an Ozark stream fish community. Copeia 2014:682–693.

Matthews, W. J., E. Marsh-Matthews, R. C. Cashner, and F. Gelwick. 2013. Disturbance and trajectory of change in a stream fish community over four decades. Oecologia 173: 955–969.

May, R. M. 1973. Stability in randomly fluctuating versus deterministic environments. American Naturalist 107:621–650.

McCune, B., and J. B. Grace. 2002. Analysis of ecological communities. MjM Software Design, Gleneden Beach, CA.

Odum, E. P., with H. T. Odum. 1959. Fundamentals of ecology, 2nd ed. W. B. Saunders, Philadelphia, PA.

Ortenburger, A. I., and C. L. Hubbs. 1926. A report on the fishes of Oklahoma, with descriptions of new genera and species. Proceedings of the Oklahoma Academy of Science 6: 123–141.

Peckarsky, B. L. 1983. Biotic interactions or abiotic limitations? a model of lotic community structure. Pages 303–323 in T. D. Fontaine III and S. M. Bartell, eds. Dynamics of lotic ecosystems. Ann Arbor Science, Ann Arbor, MI.

Pickett, S. T. A., J. Kolasa, J. J. Armesto, and S. L. Collins. 1989. The ecological concept of disturbance and its expression at various hierarchical levels. Oikos 54:129–136.

Rahel, F. J. 2010. Homogenization, differentiation, and the widespread alteration of fish faunas. Pages 311–326 in K. B. Gido and D. A. Jackson, eds. Community ecology of stream fishes: concepts, approaches, and techniques. American Fisheries Society Symposium 73. American Fisheries Society, Bethesda, MD.

Resh, V. H., A. V. Brown, A. P. Covich, M. E. Gurtz, H. W. Li, G. W. Minshall, S. R. Reice, A. L. Sheldon, J. B. Wallace, and R. C. Wissmar. 1988. The role of disturbance in stream ecology. Journal of the North American Benthological Society 7:433–455.

Rinne, J. N., R. M. Hughes, and B. Calamusso, eds. 2005. Historical changes in large river fish assemblages of the Americas. American Fisheries Society Symposium 45. American Fisheries Society, Bethesda, MD.

Roberts, J. H., and N. P. Hitt. 2010. Longitudinal structure in temperate stream fish communities: evaluating conceptual models with temporal data. Pages 281–299 in K. B. Gido and

D. A. Jackson, eds. Community ecology of stream fishes: concepts, approaches, and techniques. American Fisheries Society Symposium 73. American Fisheries Society, Bethesda, MD.

Ross, S. T. 1986. Resource partitioning in fish assemblages: a review of field studies. Copeia 1986:352–388.

Ross, S. T. 2013. Ecology of North American freshwater fishes. University of California Press, Berkeley, CA.

Ross, S. T. 2015. Fish out of water: evolutionary and ecological issues in the conservation of fishes in water-altered environments: introduction to the symposium: eco-evolutionary change and the conundrum of Darwinian debt. Copeia 103:125–131.

Ross, S. T., and W. J. Matthews. 2014. Evolution and ecology of North American freshwater fish assemblages. Pages 1–49 in M. L. Warren Jr. and B. M. Burr, eds. Freshwater fishes of North America. Vol. 1. Petromyzontidae to Catostomidae. Johns Hopkins University Press, Baltimore, MD.

Ross, S. T., W. J. Matthews, and A. A. Echelle. 1985. Persistence of stream fish assemblages: effects of environmental change. American Naturalist 126:24–40.

Schoener, T. W. 1974. Resource partitioning in ecological communities. Science 185:27–39.

Schoener, T. W. 1983. Field experiments on interspecific competition. American Naturalist 122:240–285

Shelford, V. E. 1911. Ecological succession: I. stream fishes and the method of physiographic analysis. Biological Bulletin 21:9–35.

Shelford, V. E. 1913. Animal communities in temperate America as illustrated in the Chicago region. Bulletin of the Geographic Society of Chicago 5:1–368.

Smith, C. L., and C. R. Powell. 1971. The summer fish communities of Brier Creek, Marshall County, Oklahoma. American Museum Novitates 2458:1–30.

Strobeck, C. 1973. N species competition. Ecology 54:650–654.

Taylor, C. M. 2010. Covariation among plains stream fish assemblages, flow regimes, and patterns of water use. Pages 447–459 in K. B. Gido and D. A. Jackson, eds. Community ecology of stream fishes: concepts, approaches, and techniques. American Fisheries Society Symposium 73. American Fisheries Society, Bethesda, MD.

Wiens, J. A. 1977. On competition and variable environments. American Scientist 65:590–597.

Zaret, T. M., and A. S. Rand. 1971. Competition in tropical stream fishes: support for the competitive exclusion principle. Ecology 52:336–342.

CHAPTER 8

Spatiotemporal Dynamics of Stream Fish Communities

History and Our Approach

Spatial patterning of communities and temporal community dynamics have both been major foci in community ecology, and each has a long history in the discipline. One of the first comprehensive studies of spatial distribution of freshwater fish communities in North America was Forbes (1907), which was a quantitative treatise on the distributions of species of darters (Percidae) in Illinois, based on statewide surveys. Forbes introduced the idea of "association coefficients," used half a century later by Blair (1959) for darters in northeastern Oklahoma. Shelford (1911) considered spatial and temporal (in geologic time) patterns in the longitudinal distributions of fish species in small tributaries to Lake Michigan. Shelford's paper marked the beginning of interest in "longitudinal zonation," which has been a basic tenet of fish distribution ever since (e.g., McGarvey 2011). Shelford (1911) also discussed "ecological succession" in fish communities, that is, change in community structure in ecological time.

Spatiotemporal variation in communities can be approached either by (1) examining temporal variation in spatial patterning or (2) assessing spatial variation in temporal dynamics. Previous studies of spatiotemporal variation in fish communities have taken both approaches, as we do in this chapter. We examine temporal variation in spatial patterning at several spatial scales (including variation in longitudinal patterning in our nine global communities), and we also investigate spatial patterns of temporal dynamics in our study systems.

Previous Studies

Temporal Variation in Spatial Patterns

To assess temporal variation in spatial patterns of stream fish communities, it is first necessary to establish the spatial patterns within a stream, by adequate and repeated sampling, as discussed in chapter 1, across multiple surveys. Longitudinal differences in fish species, from headwaters to lower mainstems ("zonation"), has been the subject of concepts like the "fish zones" (Huet 1959) that have been much used in Europe and on some other continents (Balon and Stewart 1983). In North America the idea of longitudinal succession from up- to downstream has prevailed (e.g., Burton and Odum 1945), but with competing ideas about downstream addition versus replacement of species. Papers on longitudinal distributions of stream fishes in temperate and tropical waters through the mid-1990s were reviewed in Matthews (1998, 296–312). Matthews (1998) noted that zonation might be a realistic model for longitudinal patterns of fishes in Europe or mountainous regions (e.g., Hutchinson 1939; Rahel and Hubert 1991), where colder, turbulent, high-gradient headwaters with "trout zones" give way downstream to fish adapted for warmer, lower-gradient, slow-moving waters. Matthews (1998, 299) concluded that, "for small and large stream systems outside montane regions, transitions from one fish zone to another may be so gradual that it is better to ignore zonation and consider the fish assemblages to change in modest increments at increasing distances downstream."

As one example of a lack of distinct zones, Matthews (1998, 305–307) compared the locations of upstream and downstream distribution limits for fish in Strawberry River, a moderate-sized river in northern Arkansas that transitions from the high-gradient Ozark uplands to the low-gradient Mississippi Delta. This assessment showed a distinct transition in fauna between the upland and lowland part of this river system but no finer division of the river into distinct fish zones (fig. 8.1).

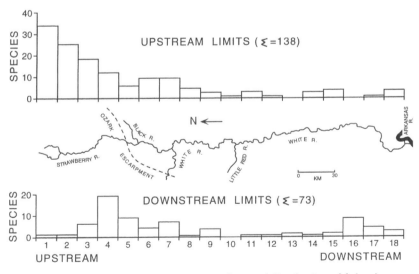

FIG. 8.1 Location of upstream and downstream limits of distribution of fishes known from the Strawberry River, Arkansas. From Matthews (1998). With permission of Springer Science+Business Media.

Based on pooled seasonal samples in Piney Creek in 1972–1973 from WJM's MS thesis, Matthews and Harp (1974) noted upstream-to-downstream differences in distributions of some minnows, madtom catfishes, and sculpins but also that recurrence of habitat types precluded sharp longitudinal zonation, as headwater species occurred well downstream in appropriate habitats. Matthews et al. (1978) also documented noteworthy spatial patterns in abundant Piney Creek minnow species, with differences corresponding roughly to stream order.

Matthews (1986) tested evidence for distinct faunal "breaks" from site to site on the mainstems of 23 warm-water streams in the eastern and central United States, including sites on Piney and Brier Creeks and the Kiamichi River sampled to that time. There had been substantial use of the Horton-Strahler stream-order system to conceptually organize stream ecology or predict longitudinal distributions of fish (e.g., Kuehne 1962), so Matthews (1986) also asked whether breaks occurred where a stream transitioned from a lower to a higher stream order (i.e., two streams of similar order joined to form a next-higher-order stream). There were qualitative longitudinal breaks in presence-absence of species in only 8 of the 23 streams and quantitative breaks based on abundances in 11 of the 14 streams for which abundances were available. However, the locations of qualitative and quantitative breaks in community structure were not associated with changes in stream order. Matthews (1986, tables 2–4) also can be interpreted by eye to suggest the degree of change (beta diversity; nestedness) in fish communities within the Roanoke River, Brier Creek, or Kiamichi River. Matthews (1986, table 1), for the Roanoke River, suggested relatively similar communities up- and downstream, based on samples pooled across months in a year and a lack of nestedness. In contrast, his table 2 for Brier Creek and table 3 for Kiamichi River both showed a strong decrease in presence of species from downstream to headwaters, likely related to substantial nestedness (as determined formally below, in the discussion of beta diversity).

What was lacking in historical assessments of longitudinal distribution of stream fishes was a temporal component, but several recent papers have addressed this. Hoeinghaus et al. (2003) showed differences in distribution of large-bodied fishes within tropical streams in two different years. Hitt and Roberts (2012) reassessed the distributions of fish in the streams studied by Burton and Odum (1945) in mountains in Virginia after a 69-year hiatus, finding that longitudinal patterns in 2 of the three streams had changed markedly in the interim. Taylor and Warren (2001) evaluated spatiotemporal variation seasonally for 2 years in local stream fish communities in the Ouachita Mountains, emphasizing colonization, extinction, and nestedness, finding less stability of local communities and more nestedness as a result of extinctions in smaller and more variable stream sites.

Cashner et al. (1994) evaluated spatiotemporal dynamics of local fish communities at three sites on Bayou LaBranche and three on its smaller tributary, Bayou Trepagnier, in southern Louisiana, based on electrofishing we did in May 1989 and June 1990. The fish community of this bayou system was dynamic, changing as salinity gradients ebbed or flowed in the system near Lake Pontchartrain. But a detrended correspondence analysis (DCA) showed that five of the six sites were relatively consistent in their locations in multivariate space, with the headwater sites on Bayou Trepagnier clustering together in both years, two downstream sites on Bayou LaBranche clustering together, and the most downstream site on Bayou Trepagnier consistently intermediate between the others. Only one upstream site on Bayou LaBranche showed substantial difference from one year to the next in the DCA biplot. Thus the local fish communities in this bayou system were well separated spatially, and the spatial patterning was largely consistent between the two years (with one site excepted).

Taylor et al. (1996), in WJM's lab, sampled 10 sites on mainstem Red River or its tributaries monthly for a year. They found that the greatest month-to-month change was in the springtime, coinciding with floods, but also related to population dynamics. DCA biplots showed that most sites were consistently within distinct subsets of multivariate space, indicating the importance of spatial differences among sites. Tributary and mainstem sites did not differ in their average month-to-month change, but the timing of changes differed, as tributary sites were particularly affected by springtime floods. In his MS thesis with WJM, Chad Hargrave resampled nine of the Taylor et al. sites monthly for a year a decade later, finding that many of the patterns remained the same but that there was more seasonality in the drainage as a whole and in local fish communities in the latter survey, due to differences in flow between the two survey years (Hargrave 2000).

Many of the papers above or in chapter 7, on temporal change, were based on similarity indices like the original Morisita index or the percent similarity index (PSI), accompanied in some cases by multivariate cluster analyses for the communities. Matthews (1990) took a different approach in a spatiotemporal analysis of monthly collections of riffle fishes in the Roanoke River, using a two-way (space-time) ANOVA. For species with significant overall variance, variance components were then combined in ratios that showed the relative importance of "temporal," "fixed spatial," and "ephemeral spatial." For four abundant riffle species, spatial variation exceeded temporal variation. For two of these species, fixed spatial was more important, and for the other two species, ephemeral spatial variation was greater. For these latter two, the interpretation was that spatial differences in abundance, while important, differed significantly from month to month.

Using a similar approach for all of the common species in monthly samples in Sister Grove Creek, Texas, Meador and Matthews (1992) showed that ephemeral spatial variation (i.e., a site x month interaction) accounted for more than 60% of the variation in abundance of all but 1 species. They concluded that although both spatial and temporal differences were important in fish assemblage structure, spatial differences were most important, but particularly for site-level differences in abundance that did change over time (i.e., ephemeral spatial variation).

It is also possible to consider temporal dynamics of spatial fish distributions at very small scales. In the 14 consecutive pools in Brier Creek that WJM studied in snorkel surveys in 1982–1983, Power et al. (1985) showed that distributions of Central Stoneroller (*Campostoma anomalum*) minnows and the predatory Largemouth Bass (*Micropterus salmoides*) and Spotted Bass (*Micropterus punctulatus*) usually were complementary, but the pattern broke down in 2 of 7 surveys, after large floods. Following up to assess spatiotemporal relationships among all the common fish species in those pools in eight surveys from 1982 to 1984, Matthews et al. (1994, fig. 6) found that across all surveys most individual pools occupied relatively small, distinct portions of multivariate space in a principal components biplot based on species abundances. They also found that within individual pools there typically was strong similarity in the fishes in consecutive surveys, with a median Morisita (original formula) similarity of 0.78. These results all suggested concordance in spatial patterns of fish distribution among these pools over time.

Matthews and Marsh-Matthews (2016) detailed temporal variation in global and local fish communities in Piney Creek, as described in chapter 7. But we also assessed spatiotemporal patterns related to differences (turnover) between longitudinally adjacent local fish communities in Piney Creek and the way that the turnover differed from time to time. To represent differences between spatially consecutive fish communities in Piney Creek, we calculated beta diversity for eight pairs of sites from downstream to headwaters (e.g., P-1 to P-2, P-2 to P-4, P-4 to P-5, etc.; see fig. 2.1) in five summers for which we had complete surveys of the entire drainage. We followed Baselga (2010), using presence-absence data (methods are summarized in the appendix to this chapter) to calculate within each summer survey, separately, one dissimilarity index representing total beta diversity, one representing pure spatial turnover, and one representing nestedness (fig. 8.2). We then tested for similarity in the spatial patterns among surveys by Kendall's W, which showed no concordance among summers in total beta diversity, pure spatial turnover, or nestedness. Overall, pure spatial turnover across all summers and sites accounted for approximately 77% of total beta diversity, leaving only 23% of the difference between spatially adjacent sites from nestedness (fig. 8.2). We (Matthews and Marsh-Matthews 2016) concluded that the lack of nestedness in Piney Creek was related to strong differential distribution in some common species within the watershed, with some "headwater" and "downstream" species having minimal overlap. Additionally, the location within the watershed with greatest total beta diversity varied among surveys, likely owing to increases or decreases in breadth of distributions of other species.

Spatial Variation in Temporal Dynamics

Fewer studies have assessed spatial patterning of temporal dynamics in stream fishes, probably because documenting temporal dynamics for any given site requires longer-term studies, which are still relatively rare for stream fish communities (Matthews 2015). Schlosser (1982) found more temporal change, seasonally over two years, in fish communities in headwaters than farther downstream in a small, environmentally unstable Illinois stream. Those observations informed his highly influential general

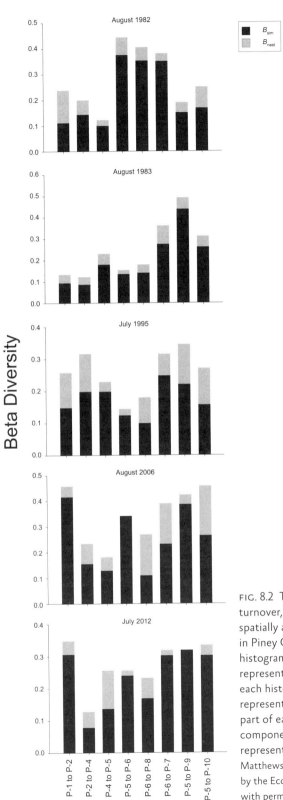

FIG. 8.2 Total beta diversity, pure spatial turnover, and nestedness for eight pairs of spatially adjacent sites over five summers in Piney Creek, Arkansas. Within each histogram, spatial pairs from left to right represent increasing distances upstream. In each histogram, the total height of the bar represents total beta diversity, the upper part of each bar represents the nestedness component, and the lower part of each bar represents pure spatial turnover. From Matthews and Marsh-Matthews (2016). Copyright by the Ecological Society of America. Reprinted with permission.

model of small stream fish community dynamics (Schlosser 1987). At a small local scale, seasonally across three study sites in Flint Creek, northeastern Oklahoma, Bart (1989) found substantial temporal consistency for pool-associated fishes but not for riffle or "inlet" fishes. In monthly samples in Battle Branch, a small tributary to the stream that Bart studied, Gelwick (1990) found more longitudinal increase in numbers of species from headwaters to downstream in pools than in riffles and that riffles showed more temporal variation than pools with respect to fish assemblage composition.

Ross et al. (1985) found less temporal stability at upstream than at downstream sites across four years of summer surveys at five Brier Creek sites. A cluster analysis of the fish communities over time at five of our Brier Creek sites (excluding the extreme headwater site BR-1) showed that in summer 1981, after extreme drought and unusually hot weather the previous summer, longitudinal relationships among the local fish communities had changed, such that the fish community at midreach site BR-4 was more like the community that had existed at headwater sites BR-2 and BR-3. Ross et al. (1985) suggested that in the postdrought year of 1981 the fish fauna throughout the creek was more similar to the previously headwater-associated taxa, implying movement of some species from headwaters to farther downstream. Red Shiners (*Cyprinella lutrensis*), for example, which had been most abundant upstream (see plates 22, 23, and 25) in some earlier surveys, made up a substantial part of the community at downstream sites BR-5 and BR-6 (see plate 2) in 1981. Overall, Ross et al. (1985) suggests that there can be temporal changes in spatial relationships among the local communities of a watershed, especially during or after unusually harsh environmental events.

Temporal Variation in Beta Diversity and Beta Components in Nine Global Streams

Beta diversity, introduced by Whittaker (1960), is a tricky concept (Baselga 2010), with at least two dozen formulations for its calculation or interpretation, and disagreement about its meaning (Koleff et al. 2003). See Koleff et al. (2003) for a review and comparison of 24 beta diversity measures that were in the literature to that date. In Matthews and Marsh-Matthews (2016) and in this chapter we followed the approach of Baselga (2010), using dissimilarity measures to calculate total beta diversity and then decompose the total beta into "pure spatial turnover" and "nestedness." Total beta diversity represents the degree to which samples at adjacent sites in a watershed differed in qualitative species composition; thus high total beta indicated large differences between the local fish communities and low total beta indicated little change from one site to another. (It is important to keep that distinction in mind when viewing the figures below; the small bars in histograms represent small differences from site to site in the species that are present.)

It is also important to think of the pure spatial turnover and the nestedness components for stream fishes in light of historical concepts about "longitudinal zonation" for stream fishes. Much of the literature for warm-water North American streams has viewed changes in composition of fish communities from upstream to downstream, starting with fewer species in headwaters, then assessing the way species richness increases downstream. The competing models have been "addition" (simply adding new species while headwater species remain in the community) versus "replacement," such that the headwater species drop out and are replaced downstream by other species.

Examining stream fish species distributions for the nestedness component of beta diversity implies the opposite question, of whether successive communities in an upstream direction have fewer species, consisting, if strongly nested, of reduced number of species from the more speciose downstream pool. Hypothetically, strong nestedness could occur if increasingly harsh environments farther upstream (such as more drying or greater temperature fluctuation), changes in availability of habitats, or competition among species excludes some of the downstream species. Beta diversity and nestedness in particular represents an opposing question from the more traditional "what happens from headwaters down to the mainstem?"

We now address beta diversity and beta component partitioning for our nine global streams. Readers familiar with beta partitioning or methods in Baselga (2010) should read straight ahead, but readers wanting a quick tutorial should refer to the appendix to this chapter. Table 8.1 summarizes beta diversity for our nine global streams. First, we used the grand mean of all cases (comparisons of adjacent sites in all surveys) as a rough approximation of how much difference there was in presence of species between adjacent mainstem sites. There was no correlation between the number of species and total beta diversity (table 8.2). Table 8.1, in which sites are ranked by the total number of species, showed that there were no more differences in the presence of species between adjacent stations in speciose streams (e.g., Kiamichi, Piney) than in those with fewer species (Alum, Bread, Blaylock, Crooked).

Table 8.1. Beta diversity for adjacent sites on the mainstems of nine global streams

Stream	Total S	Grand Mean B_{sor}	Percentage Spatial > Nestedness	Percentage Nestedness > Spatial	Kw for B_{sor}	Kw for B_{sim}	Kw for B_{nest}
Kiamichi	48	0.392	88	12	0.200	0.088	0.184
Piney	45	0.235	92	8	0.248	0.055	0.224
Roanoke	34	0.314	58	17	0.359	0.353	**0.408**
Brier	30	0.310	60	20	0.275	0.247	0.350
Crutcho	24	0.354	70	23	0.098	0.111	0.118
Alum	19	0.530	50	39	0.519	**0.700**	0.442
Blaylock	16	0.247	40	50	**0.519**	0.200	0.129
Bread	14	0.360	61	33	0.422	0.517	0.200
Crooked	9	0.212	10	50	0.406	0.400	0.077

Note: Data are ranked by number of species across sites and times (Total S) and include: grand mean B_{sor}, total beta diversity; percentage of all cases in which pure spatial turnover exceeded nestedness; percentage of all cases in which nestedness exceeded pure spatial turnover; and Kendall's W (measure of concordance in the patterns of values from site to site, across time; cases in bold are significant at $p = 0.05$) for total beta diversity (B_{sor}), pure spatial turnover (B_{sim}), and nestedness (B_{nest}).

Table 8.2.   Product moment correlations among beta components

	S	%Spatial	%Nest	YY B_{sor}	Kw B_{sor}	Kw B_{sim}	Kw B_{nest}
S	1.000						
%Spatial	**0.847**	1.000					
%Nest	**−0.910**	**−0.886**	1.000				
YY B_{sor}	0.036	0.236	−0.077	1.000			
Kw B_{sor}	−0.606	**−0.628**	**0.737**	0.097	1.000		
Kw B_{sim}	**−0.624**	−0.528	0.536	0.528	**0.724**	1.000	
Kw B_{nest}	0.238	0.177	−0.295	0.562	0.243	0.476	1.000

Note: Boldface indicates significance at $p = 0.05$.

The total number of species in the mainstems was strongly and positively correlated (table 8.2) with the percentage of cases in each stream in which pure spatial turnover was greater than nestedness ($r = 0.847$; $p = 0.002$), and total species was negatively correlated ($r = −0.910$; $p = 0.0003$) with the percentage of cases in which nestedness was greater than pure spatial turnover. In streams with more species, pure spatial turnover typically exceeded nestedness. In the more speciose streams, low nestedness implied that from downstream to upstream, the more upstream site was not simply a subset of the downstream species. This is apparently related to the tendency in speciose streams for spatial differences between downstream and headwater species, as Matthews and Marsh-Matthews (2016) and earlier chapters in this book pointed out for Piney Creek. In Piney Creek some minnow species are essentially ubiquitous, but numerous other minnows have a marked tendency to occur or be more abundant either downstream or upstream (Matthews and Harp 1974; Matthews et al. 1978). In Piney Creek there are also pairs of sculpins and of darters that differentially occur either upstream or downstream. Such longitudinal patterns of individual species resulted in low nestedness (B_{nest}) and thus more pure spatial turnover (B_{sim}) between adjacent sites.

In contrast, streams with low numbers of species, including Crooked, Bread, Blaylock, and Alum Creeks, all high-gradient habitats in the Ouachita Mountains of southwest Arkansas, had the highest percentage of cases, from 33% to 50 % (table 8.1), in which nestedness (B_{nest}) was greater than pure spatial turnover (B_{sim}). These small streams with simple faunal composition tended more toward a reduction of species from downstream to upstream than did the larger streams. But overall, across eight of our nine global streams, there was more pure spatial turnover than nestedness.

Finally, this chapter is about spatiotemporal dynamics of stream fishes, so let's address the degree to which spatial patterns in beta diversity or its components may change over time within a stream. Kendall's W, which is a measure of the overall concordance, showed only 3 instances of significant (at an unadjusted $p = 0.05$) concordance of any of the aspects of beta diversity across time in any streams. In all other

24 possible test cases, Kendall's W was nonsignificant, suggesting that spatial patterns in total beta diversity, pure spatial turnover, or nestedness from site to site at a given time may be different in another survey. This implies a lack of any consistent spatial location in these streams where there is a strong break in community composition (sensu Matthews 1986). At one time there may be a substantial difference in presence of species among two adjacent sites (high total beta diversity), but the total beta between those two sites might be much lower at another time if, say, a vagile minnow species moved from upstream to downstream or vice versa. The lack of consistent patterns in the location of greatest beta diversity between sites in each stream implies that, although many species tend toward upstream or downstream distributions, there is considerable movement of many species up or downstream among sites over time. This finding actually helps to substantiate the conclusion that we implied early in this book in addressing global communities as an important level at which to consider fish dynamics.

Calculation of beta diversity or an approach such as partitioning beta into its components reduces a lot of biology into just a few numbers. To really understand the biology underlying the numbers, we need to dig a bit deeper. As an example, consider the distribution of species in the Kiamichi River mainstem (table 8.1), which had the highest grand mean total beta diversity and which, like Piney Creek, had a high percentage of cases in which pure spatial turnover exceeded nestedness (fig. 8.3).

Matthews (1986, table 4), showing the fishes collected in summer 1981 in the Kiamichi River (see plate 3), suggested that from the headwaters many species are added to the more downstream communities. But it also appeared that some of the species in the headwaters, such as Highland Stoneroller (formerly *Campostoma anomalum*, now *C. spadiceum*) and Bigeye Shiner (*Notropis boops*), dropped out at one or more of the lower mainstem sites. In other words, there was some indication in the Kiamichi River that fish diversity increased by replacements as well as additions of species. Our beta partitioning for the Kiamichi River data for 1981 indicated that between the most upstream site (K-8) and the next site downstream (K-7), $B_{sim} = 0.167$ and $B_{nest} = 0.119$, representing 58.3% of beta diversity due to pure spatial turnover but 41.7% due to nestedness, with 5 species shared between the 2 sites, 3 at K-7 but not at K-8, and 1 at K-8 not present downstream in K-7. There is a slight disparity between this analysis and the data of Matthews (1986, Table 4), because in all new analyses we combined Blackstripe Topminnow (*Fundulus notatus*) and Blackspotted Topminnow (*Fundulus olivaceus*) into one *Fundulus* spp. Recent molecular work by Jacob Schaefer and colleagues (Duvernell et al. 2013), showing substantial hybridization and some convergence in their body forms in the Ouachita Highlands (which includes the Kiamichi River), means that it is now impossible to determine which species was taken at sites in the older Kiamichi surveys.

A high proportion of pure spatial turnover in the 1981 Kiamichi River data was also evident between sites K-2 (downstream) and K-3 (about 35 km farther upstream). Between these 2 sites in 1981 only 5 species were shared, while 9 species were found

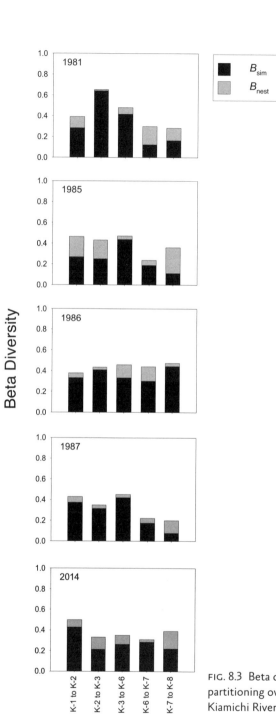

FIG. 8.3 Beta diversity and beta partitioning over time between sites on the Kiamichi River.

at K-2 but not at K-3, and 10 species were found at K-3 and not K-2. This disparity in occurrence of most species between these 2 adjacent sites resulted in high pure spatial turnover in 1981, with $B_{sim} = 0.643$, compared to B_{nest} of only 0.012. But the point is that in the Kiamichi River we have evidence both of additions and replacements of species from headwaters to downstream, making local communities substantially different from site to site; hence the high grand mean total beta diversity for consecutive pairs of sites in that river. Additions and replacements apparently result in beta partitioning being dominated by a high proportion of pure spatial turnover and low nestedness. From downstream to upstream in the Kiamichi River there was little evidence in 1981 or at any other time that upstream fish communities were mere subsets of more speciose downstream sites, as would be expected from models that invoke competition or habitat stress to successively winnow downstream taxa from more upstream sites. These patterns were not fixed across time, and variations in the pattern were evident in the histograms showing beta partitioning for the Kiamichi River from 1981 to 2014. Kendall's W tests of concordance were quite low (table 8.1) for all beta components in this system; greatest beta diversity was between different adjacent sites in the Kiamichi River in different years (e.g., between K-2 and K-3 in 1981; K-3 and K-6 in 1985; K-7 and K-8 in 1986; K-3 and K-6 in 1987; K-1 and K-2 in 2014). While there was much more spatial turnover than nestedness in the Kiamichi River overall, a spatiotemporal view of the actual locations of differences between sites showed substantial variation from time to time.

In Crooked Creek, which had a low number of species ($S = 9$) in contrast to the Kiamichi River ($S = 48$) and low overall beta diversity (Crooked Creek grand mean for $B_{sor} = 0.212$; table 8.1), pure spatial turnover exceeded nestedness in only 10% of the cases, whereas nestedness exceeded pure spatial turnover in 50% of the cases. (For the remaining cases there was no beta diversity between adjacent sites.) The histograms of beta partitioning for Crooked Creek (fig. 8.4) showed some simple patterns, with turnover of zero for three cases in October 1989, two cases each in May 1990 and May 1991, and one case in October 1990.

In each of the cases with beta diversity equal to zero, all the same species occurred at consecutive sites, usually with only three species shared, including Highland Stoneroller, Creek Chub (Semotilus atromaculatus), and Orangebelly Darter (Etheostoma radiosum). In these cases turnover and beta diversity is zero, as all species are shared. In 10 other cases, beta diversity in Crooked Creek was entirely due to nestedness, as upstream communities were subsets of those downstream. The most extreme situation was between Crooked Creek sites CK-5 and CK-6. CK-6, the most upstream site, is above an almost vertical rock-face waterfall that is 4–5 m high, precluding upstream movement of fishes (see plate 5). Just below the falls at CK-5, Creek Chub and Orangebelly Darter were always present and abundant, and Highland Stoneroller was usually present, but above the falls only Orangebelly Darter was ever found, and it was typically abundant there. It might be argued that for a stream with a simple fauna like Crooked Creek and with at least one obvious barrier to exchange of species, so-

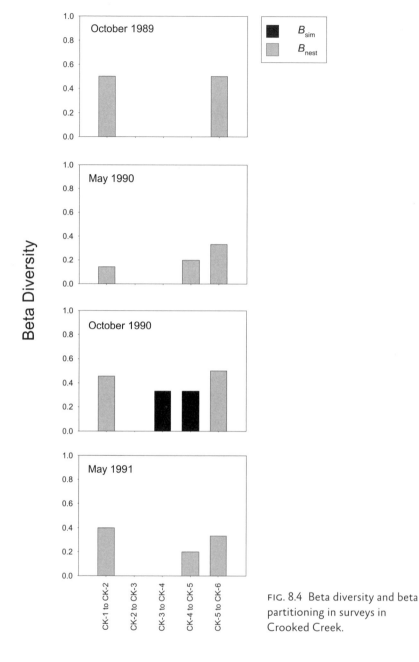

FIG. 8.4 Beta diversity and beta partitioning in surveys in Crooked Creek.

phisticated calculations of beta diversity might not be needed, as much as it is "obvious by inspection," but we offer Crooked Creek in contrast to the Kiamichi River to suggest the wide range of possibilities for beta diversity from stream to stream.

The other seven global streams showed a wide range of beta diversity and beta partitioning intermediate between the extremes represented by the Kiamichi River and

Crooked Creek. But most streams showed more pure turnover between adjacent sites (B_{sim}) than nestedness (B_{nest}). In many streams there are headwater specialists (like Southern Redbelly Dace, *Chrosomus erythrogaster*, associated with small spring habitats, and some darters or sculpins), in contrast to many species that occur downstream but are rare or absent in small headwaters. In chapter 5 we provided examples of species in the Kiamichi River, Piney Creek, and Blaylock Creek that show strong propensity for either headwaters or mainstems, and the beta diversity results suggest that such habitat specialization may be the norm, rather than the exception, for many fishes. And some species seem to actually have a propensity for midreach habitats, as illustrated by the distribution of Bigeye Shiner in Piney Creek (table 8.3). From 1972 to about 1983, Bigeye Shiners were most abundant at midreach sites and often absent in headwaters or in lower mainstem sites. But table 8.3 also illustrates that spatial relationships can change over time, because from 1994 to 2012, Bigeye Shiner was much more widely distributed in the drainage and at least moderately abundant at many sites where it had not been detected in earlier surveys. The spatiotemporal dynamics of stream fish communities do indeed depend on the dynamics of their constituent species.

But while our beta diversity analyses show that for most streams we evaluated there was a strong spatial turnover component, the Kendall's *W* concordance tests showed little concordance in the actual locations of, or spatial patterns in, turnover (or nestedness) from survey to survey. We suspect that there is a strong temporal component to the spatiotemporal dynamics of the fish communities in almost any warm-water stream. Discordance in spatial patterns from time to time was evident, as distributions of species in a watershed can change. There may be substantial movement either up- or downstream as environmental conditions change in a stream (e.g., Ross et al. 1985; Grossman et al., 1998).

Temporal Variation in Spatial Patterns among Midwest Samples in 1978 and 1995: Dynamics for Two Points in Time

For a total of 59 matched sites throughout the Midwest, from Nebraska and Iowa to Texas coastal drainages, all collected in the month of June in 1978 (Matthews 1985) and 1995 (Marsh-Matthews and Matthews 2000), we calculated Morisita-Horn (MH) similarity on the basis of abundances of species, among all possible pairs of sites separately for the two surveys. A Mantel test with 10,000 randomizations was used to test for congruence, between 1978 and 1995, of the intersite similarities in the triangular MH matrices, based on a total of 92 fish species. The pattern of similarities among all sites was highly congruent between the 2 surveys (Mantel $Z = 0.634$; $p = 0.0001$). At this scale of spatial analysis, however, we would have expected congruence between surveys, because there are differences in distributions of many of the species across a north-to-south gradient, between river basins in the Great Plains (Marsh-Matthews and Matthews 2000), or among Texas coastal river drainages. Across the Texas drainages, for example, many species are shared, but others such

Table 8.3.   Abundance of Bigeye Shiner at 12 sites in Piney Creek, from the headwater sites to most downstream sites, in surveys from summer 1972 to summer 2012

	S72	W72	A73	S82	W83	A83	S83	W94	A95	S95	S06	SO12
Headwater sites												
M-3	0	0	0	0	1	1	0	426	637	106	140	14
P-10	1	0	0	3	0	0	3	4	0	0	0	0
P-9	0	0	0	7	0	0	7	0	1	0	2	17
P-7	13	1	0	67	203	60	67	136	55	59	69	42
P-8	11	15	19	92	53	30	92	278	290	57	40	25
P-6	38	28	23	34	39	98	34	431	426	172	54	53
Downstream sites												
M-2	0	0	0	26	3	13	26	62	32	116	45	40
P-5	19	2	11	0	4	3	0	89	155	48	3	12
P-4	0	0	2	14	4	1	14	203	84	52	2	15
M-1	0	0	2	0	1	1	0	8	12	1	4	11
P-2	0	0	0	0	1	0	0	11	19	24	2	21
P-1	0	1	1	0	1	2	0	11	5	24	11	36

as Plateau Shiner (*Cyprinella lepida*), Texas Shiner (*Notropis amabilis*), Greenthroat Darter (*Etheostoma lepidum*), or several localized *Gambusia* species are restricted to one or more of the basins (Connor and Suttkus 1996). At extreme ends of this latitudinal gradient, few species are shared, with the exception of widespread species like Red Shiner (Matthews 1985). It would have been rather surprising if the general spatial patterns among local fish communities throughout the Midwest were not concordant over time.

In light of bias from the north–south differentiation or basin-specific range limits of many midwestern species, we conducted MH and Mantel analyses at smaller spatial scales, limiting comparisons of similarities to sites (numbers in parentheses) within the Kansas River Basin (10) and the Arkansas River Basin (12). Within the Arkansas River Basin, spatial relationships among all sites were highly congruent between 1978 and 1995 (Mantel $Z = 0.666$; $p = 0.0004$).

What was surprising is that among our 10 sites in the Kansas River basin there was absolutely no congruence in the similarities of sites based on their fish communities in 1978 and 1995! In fact, the lack of congruence was so strong (Mantel $Z = 0.067$; $p = 0.384$) that we initially thought there must have been a mistake in the analysis. But a check of the raw data showed marked differences in presence or abundances of numerous species from 1978 to 1995, such as our samples in Poop Creek (yes, that is what the landowner called it), where we caught no Channel Catfish (*Ictalurus punctatus*) in 1978 but 23 juveniles in 1995, or in "No Name" Creek, where we caught only 13 Red Shiners and 4 Sand Shiners (*Notropis stramineus*) in 1978, compared to 728 Red Shiners and 539 Sand Shiners in 1995. Our field notes from 1978 and 1995 suggested that both of these creeks had changed some in the 17 years between surveys. Field notes from 1978 (WJM 295) indicated that Poop Creek was a "channelized ditch," whereas by 1995 (WJM 2812) it was still described as channelized, but small trees had grown in the riparian, and tall weeds and some fringing vegetation were present. Its width in 1978 was recorded as "2–5 ft" (about 1.5 m wide), with collections made to a depth of only 6 in (about 15 cm). In 1995 it was as much as 6 m wide, with depth to 100 cm. It appears that in 1995 there was substantially more water, and our field notes and photos of the site from 1995 suggest considerable increases in habitat complexity, so perhaps it is not surprising that more fish were present in the latter survey.

For nearby "No Name" Creek, field note WJM 298 indicated that in 1978 it was 5 m wide, compared to 8 m in 1995 (WJM 2815)—not really a big difference, but in 1995 it was 100 cm deep compared to only about 45 cm deep in 1978. These are only two examples of changes that may have taken place in our sites in the Kansas River Basin, where sites ranged from small streams with unstable banks to substantially larger streams like the Big Blue River at Beatrice, Nebraska, or the Smoky Hill River south of Russell, Kansas, both of which were about 40 m wide at the time of our collections in 1995. For whatever reason, MH similarity values between the possible pairs of these 10 sites were quite low in some cases (table 8.4).

Table 8.4. Morisita-Horn similarity values comparing all possible pairs of 10 sites in the Kansas River basin of Kansas and Nebraska in 1995

	POOP95	BLUE95	TURK95	REP95	GNT95	ELMCK95	SOL95	SAL95	SMK95	NON95
POOP95	1.000									
BLUE95	**0.482**	1.000								
TURK95	0.461	0.970	1.000							
REP95	**0.455**	0.972	0.970	1.000						
GNT95	0.376	**0.558**	0.520	0.598	1.000					
ELMCK95	**0.450**	0.953	0.957	0.968	0.593	1.000				
SOL95	**0.445**	0.805	0.747	0.859	0.808	0.815	1.000			
SAL95	0.425	**0.780**	**0.856**	**0.726**	0.345	**0.740**	0.459	1.000		
SMK95	**0.401**	0.837	0.914	0.807	0.317	**0.820**	**0.454**	**0.901**	1.000	
NON95	0.505	**0.920**	**0.963**	**0.894**	**0.446**	**0.932**	0.618	0.889	**0.953**	1.000

Note: Boldface indicates sites that were markedly changed in MH value (either increased or decreased) compared to MH values for the same pairs of sites in 1978.

Temporal Variation in Spatial Patterns from Survey to Survey for Nine Global Streams: Multiple Points in Time

Here we tested whether spatial similarities among local fish communities within watersheds were consistent across time. For our nine global streams, we used the MH index as above, to make a triangular matrix of similarity of sites within the stream each time it was surveyed. We then tested congruence of each triangular MH matrix to the MH matrix for the next survey by a Mantel test with 10,000 randomizations. Because of making multiple Mantel comparisons (time to time to time, etc.) in each stream, we used Rice's (1989) sequential Bonferroni approach to adjust alpha from $p = 0.05$ within each stream (with exceptions for Piney and Brier Creeks, below). Moran (2003) argued against using sequential Bonferroni tests, because if used in a harsh manner (i.e., overly adjusting alpha when there are a large number of comparisons), Type II error is introduced. But for most of our global streams the adjustment was not severe, except for having a large number of tests in Piney and Brier Creeks, and for those we didn't use it.

As a simple example, consider Alum Creek, with four surveys at each of six sites over a two-year period. There are 3 comparisons, so following Rice (1989) we made the first test at a level of alpha $0.05/3 = 0.017$. Time 2 to Time 3 on Alum Creek had a normalized Mantel statistic $(Z) = 0.542$, and $p = 0.006$, indicating significant congruence in the spatial distribution of communities in that watershed between those 2 surveys. The first step having been significant, we adjusted alpha for the second test to $0.05/2 = 0.025$, and comparing Time 3 to Time 4 had $Z = 0.418$ and $p = 0.018$, also significant by adjusted Bonferroni. The third test was simply at $p = 0.05$, and congruence between Time 1 and Time 2 was significant, with $Z = 0.750$ and $p = 0.025$. Thus the spatial similarities among sites in Alum Creek were congruent from time to time.

Blaylock, Bread, and Crooked Creeks in the Ouachita National Forest, Arkansas, were surveyed like Alum Creek, with two surveys in spring and two in autumn from 1989 to 1991. For Blaylock Creek, all three comparisons of the four consecutive surveys were significantly congruent by Mantel tests. For Bread Creek, however, none of the Mantel comparisons showed spatial community structure to be congruent from time to time. For Bread Creek we additionally tested for congruence between the seasonal samples (spring to spring; fall to fall), but none of the comparisons was significant. Bread Creek was subject to extreme drying in the headwaters, reduced to isolated pools in two surveys, and once had no fish at all in the most upper site, so it may be that the fishes were sufficiently rearranged in the watershed during harsh conditions to cause breakdown in spatial structuring. For Crooked Creek, with the simplest fish fauna of all the four Ouachita National Forest streams, spatial patterning was never congruent from time to time. The only potential congruence of distribution patterns for fishes in Crooked Creek was between autumn 1989 and autumn 1990, with $Z = 0.883$ and unadjusted $p = 0.026$. Thus two of the four streams in the Ouachita National Forest had consistent spatial patterning in the fish communities among local sites, and two did not. This was not readily explained by any measured environmental feature, with the exception that headwater drying in Bread

Creek may have contributed to the lack of consistency of spatial patterning in that stream. Additionally, the two streams with strong consistency of patterning (Alum and Blaylock Creeks) are in two different drainages (Saline River vs. Little Missouri River), and the two lacking consistency of patterning also were in those two different river drainages. Consistency of spatial patterning, or lack thereof, could not be explained by any simple drainage basin differences.

For Kiamichi River and Roanoke River (mainstem sites 1–7 only in the Roanoke drainage, excluding some small spring "runs" that were direct tributaries to the mainstem with extremely different fishes), there were five surveys each. In Kiamichi River the spatial similarities of fish communities among six sites were significantly congruent for all comparisons from survey to survey (1981–1985, 1985–1986, 1986–1987, and 1987–2014). Because of the long hiatus of 24 years between the last 2 surveys, we also tested congruence of the spatial patterns in local communities between the first survey, in 1981, and the last, in 2014. This comparison was congruent, with $Z = 0.839$ and $p = 0.006$. The distribution of fishes in local communities up- and downstream in the Kiamichi River was highly congruent between all consecutive surveys and across the greatest span of time for which we did surveys.

In the Roanoke River, results were exactly the opposite: spatial patterns among the local communities in the river mainstem were never significantly congruent from time to time! A review of the raw MH similarities for individual sites on the Roanoke River suggested great differences within local communities among surveys. For example, comparison of sites 1 and 2 (the most downstream) at different times resulted in MH values as high as 0.729 and as low as 0.262. Comparison of sites 2 and 3 across different times showed MH values as high as 0.823 and as low as 0.080 (almost no similarity in the structural differences between these two sites at some times). And a similar dichotomy existed for comparisons between sites 5 and 6 (farther upstream), with MH values ranging from 0.880 to 0.065. Throughout the Roanoke River mainstem there was little consistency in similarities of spatially adjacent communities from time to time, possibly related to seasonal changes, episodes of wet versus dry weather, and so on, but we have no specific explanation for why the mainstem fishes in one river (Kiamichi) were consistently patterned spatially over four decades, whereas in another river (Roanoke) there was no consistency to the spatial patterns among local communities. Because all Kiamichi surveys were in summer and the Roanoke surveys were at approximately monthly intervals across seasons, maybe seasonal variation in distributions outweighs long-term consistency within a single season. But to synthesize these outcomes critically, we have to admit that we simply do not know the factors producing these results.

We also had five surveys over approximately 2 years at 10 sites in the Crutcho Creek watershed (see plate 6), on Crutcho Creek mainstem and its large tributary, Kuhlman Creek. For the first three comparisons by Mantel (Times 1–2, 2–3, 3–4), spatial patterns among the local fish communities were highly congruent. The last comparison, from Time 4 to Time 5, was nonsignificant. However, a comparison of the first and last surveys by Mantel test had $Z = 0.621$ and $p = 0.0001$, indicating strong congruence in the spatial patterning within this watershed, with the exception of 1 time when spatial distributions of fish were apparently altered. Inspection of the raw data did not reveal any marked changes in abundances of most fish species at Time 4, but at Time 4 at the lowest site (CR-1, near the confluence of the creek with the North Canadian River) we found a large influx ($N = 209$) of juvenile Gizzard Shad (*Dorosoma cepedianum*), which might have been sufficient to make comparisons of Time 4 to other times, lacking such large numbers of this species, noncongruent.

With Brier and Piney Creek fish communities we had more challenge in deciding on an appropriate level for testing significance. In Brier Creek there were 14 intervals (table 8.5) between consecutive surveys, so a sequential Bonferroni alpha would begin with $p = 0.05/14 = 0.0036$, which would be a much more rigorous test than in any of the previous streams, potentially introducing a serious potential for Type II error, that is, overlooking biologically interesting relationships that are real (Moran 2003). For Brier Creek we omitted Bonferroni adjustment and focused on the magnitude of the Z scores and on unadjusted p values at 0.05 (table 8.5).

There was marked variation in Brier Creek in congruence between MH matrices for different intervals between surveys (Z scores; table 8.5). But at $p < 0.05$, 12 of the 14 comparisons had congruent intersite similarity patterns from 1 survey to the next. The temporal sequence of Mantel Z values was interesting. For the first 8 comparisons between surveys (1976 to 1996) the p value was <0.05, indicating congruence in the spatial patterns in Brier Creek fishes throughout the first 20 years of study. The

Table 8.5. Comparisons of triangular Morisita-Horn matrices by Mantel tests for congruence, across the intervals indicated, for six sites on Brier Creek, Oklahoma

Survey Years	Mantel Z	Unadjusted p Value	Comments
1976–1981A	0.597	0.018	big drought / heat wave summer 1980
1981A–1981B	0.600	0.018	drier in July than in June
1981B–1985	0.558	0.011	
1985–1986	0.822	0.003	
1986–1991	0.858	0.005	1988 sample omitted; upper sites dry
1991–1994	0.436	0.047	
1994–1995	0.629	0.044	
1995–1996	0.452	0.005	1996 sample at low-water conditions
1996–1999	0.179	0.238 n.s.	spans extreme drought in 1998
1999–2001	0.723	0.002	spans extreme drought in 2000
2001–2002	0.365	0.137 n.s.	
2002–2004	0.503	0.008	
2004–2008	0.554	0.004	spans severe drought in 2006
2008–2012	0.461	0.056	follows severe drought 2011–2012
1996–2012	0.494	0.041	spans severe/extreme droughts 1998, 2000, 2006, and 2011–2012
1976–2012	0.590	0.032	spans all years and events in study

spatial pattern persisted in spite of numerous moderate floods, a flood in October 1981 that was the most extreme on record for Brier Creek, and several droughts. The first time that congruence of spatial patterns was weak was between 1996 and 1999, which included what was then the "drought of record" for south Oklahoma. But there was an even worse drought in 2000, and spatial patterns were congruent between collections in 1999 and 2001. The other interval lacking congruence was from 2001 to 2002, during which there were three moderate floods (a complete summary of all flood and drought events in Brier Creek from 1976 to 2008 is in ESM1— "Study Area and Events"—in the online supplementary material from Matthews et al. 2013).

Overall, it appeared that for most intervals between surveys, the spatial patterns in similarity among local Brier Creek fish communities were persistent, and only for two intervals was there lack of congruence. Two other comparisons in table 8.5 are also of interest. Spatial structure was congruent in 1996 and 2012, from before the first of the severe droughts that started in 1998 until the end of the surveys in 2012, suggesting that even though there were intervals of interruption of the inter-site patterns at times in the later years of the study, the spatial patterns had become reestablished by our last survey. Additionally, spatial patterns in the first survey (1976) to the last survey (2012) were congruent. So, spatial relationships among fish communities at our permanent local sites on Brier Creek were generally congruent across time, in spite of many episodes of flood or drought in this physically highly dynamic system.

For Piney Creek there were 11 intervals between consecutive surveys from July 1972 to July 2014 (table 8.6). In this comparison we did include July 1972 although fish in riffles were undersampled, because it appeared that the spatial patterns, which mostly depended on pool fishes, were adequately documented. Results for Piney Creek (table 8.6) are much simpler than for Brier Creek, as all comparisons of the spatial similarities among local communities by Mantel tests between consecutive surveys showed strong congruence. The spatial structure of fish communities across the 12 long-term survey sites was preserved throughout the 40 years spanned by our surveys, in spite of 2 incredibly large "floods of the century" (Matthews et al. 2014). Spatial patterns in the first survey (July 1972) were strongly congruent with patterns in the last survey 40 years later (July 2012). In Brier Creek, where there was a lack of congruence between some survey pairs, all samples were in summer, without any seasonal component. In Piney Creek many of the samples were in summer but others were in winter or spring, introducing a seasonal component. But even with that potential source of greater variation, the spatial patterns of fish communities throughout Piney Creek were persistent across all comparisons. This difference in spatial persistence between the two creeks reinforces the concept of Ross et al. (1985) and much of our subsequent work, all suggesting that fish communities or their distribution in space tends to be more stable or persistent in the benign environment of Piney Creek than in the more environmentally harsh Brier Creek.

Table 8.6. Mantel Z and unadjusted *p* values for comparisons of Morisita-Horn matrices between consecutive surveys of 12 sites in Piney Creek

Survey Times	Mantel Z	Raw *p* Value	Comments
July 1972 to Dec. 1972	0.563	0.001	
Dec. 1972 to Apr. 1973	0.442	0.007	
Apr. 1973 to Aug. 1982	0.511	0.0004	
Aug. 1982 to Jan. 1983	0.474	0.0004	100-year flood in Dec. 1982
Jan. 1983 to Apr. 1983	0.524	0.0002	
Apr. 1983 to Aug. 1983	0.502	0.0005	
Aug. 1983 to Dec. 1994	0.596	0.0001	
Dec. 1994 to Apr. 1995	0.729	0.0001	
Apr. 1995 to July 1995	0.705	0.0001	
July 1995 to Aug. 2006	0.508	0.0006	
Aug. 2006 to July 2012	0.726	0.0001	
July 1972 to July 2012	0.505	0.0095	100-year flood in Mar.–Apr. 2008

Spatial Variation in Morisita-Horn: Similarity within Individual Sites for Nine Global Streams

We should have said this a lot earlier: the cool thing about creating a book like this is that you get to find out things you didn't already know! It seems to us that most people who write scientific books already know the results, conveying the answers in a way that readers can learn what the authors have already learned. But by new analyses of our 9 global stream communities and the communities at 31 local sites, we continued to learn new things while we were writing this book. If you had asked us before we started this book where in most watersheds local fish communities would be most consistent or stable, we probably would have said (based on experience in Piney and Brier Creeks and general observations elsewhere) that local communities would show more temporal change in the headwaters than downstream, because the headwaters are more environmentally variable (as also noted by Schlosser 1982, 1987) and because lower mainstems are larger, with relatively more stable conditions. In one summer in Brier Creek, for example, long-term temperature loggers regularly showed daily fluctuations of as much as 10°C in the headwaters compared to only a few degrees at sites farther downstream, and oxygen concentrations in some headwater pools were measured at 1 ppm on numerous occasions (Matthews 1987). And in Brier Creek we have seen one or more headwater sites entirely dry in 1982, 1983, and 1988, as well as during the big droughts of 1998 and 2000. Detailed studies in 1982–1983 (Matthews 1987) followed recolonization and successful reproduction by tolerant species at headwater site BR-2. In the more environmentally stable Piney Creek, none of our sites have ever been completely dry during our surveys, and only P-8, the uppermost site on the Piney Creek mainstem, has been reduced to isolated pools lacking any connecting flow (in three of our six summer samples at that site). Our intuition would have been that in warm-water streams there would be more similarity within sites over time at larger, permanently watered downstream sites with persistent flow and more temporal variation for local fish communities in the headwaters. But by actually examining patterns in similarity within sites over time in our nine global systems, we got some surprises.

For Piney Creek, we would have been right. In figure 8.5, Piney Creek has generally high intersurvey MH values in the five lowermost sites in the watershed (Piney Creek and its large tributary, Mill Creek; see fig. 2.1), including P-1, P-2, M-1, P-4, and P-5. These are wide sites with deep pools, large and swift riffles (see plate 1), and generally stable oxygen and temperatures. Farther toward headwaters the medians and ranges of MH values were variable but trended downward toward the headwaters. The least consistent, by far, was the most upstream site, M-3, on Mill Creek. This site has really changed over the decades and has been subject to essentially total rearrangement of the stream channel and riparian vegetation by floods (see plate 21), with substantial reshaping of most habitat features over the years. But overall, we would have been right about Piney Creek: its intrasite similarity over a span of 40 years was greatest downstream and lowest at some upstream sites.

For Brier Creek (fig. 8.5), an a priori prediction of more variability downstream and less upstream was poorly supported. The median for the three most downstream sites (BR-6, BR-5, BR-4) was higher than the median for the two more upstream sites (BR-2 on the mainstem and BR-3, a small headwater tributary), but the boxplot shows great overlap in MH values for all five sites. BR-1, the most extreme headwater site, appears in figure 8.5 to be extremely stable. Although that initially seems counterintuitive, in recent decades there usually has been only one species (Green Sunfish, *Lepomis cyanellus*) present in this extremely harsh and variable site. Either seven or eight species were detected in BR-1 in all surveys from 1976 to 1986, but the site went completely dry in 1988. After that, in 10 surveys from 1991 to 2012, the only species we collected at that site was Green Sunfish (often as hundreds of juveniles; many counted and released), with the exception that we found four Central Stonerollers there in 2004. If there is only one species and it is present in a series of surveys, the MH values between surveys are by definition 1.00 and there is no variation. Thus a depauperate local community can appear stable, but only because only one or maybe a few very hardy species persist.

Kiamichi River, Blaylock Creek, and Crooked Creek (fig. 8.5) each showed a downstream-to-upstream pattern in MH similarity that was reversed from the pattern in Piney Creek and would not have been predicted by us a priori. In the Kiamichi River (fig. 8.5), the lowest median MH similarity was at the most downstream site (K-1), and the highest (most similar community) across time was the most upstream site (K-8). Between those two, the pattern in median MH similarity fluctuated. For the Kiamichi River, some of the explanation for this pattern may be in a more detailed consideration of the physical or environmental setting of the lowermost (K-1) and most upstream site (K-8). Site K-1 is a big-water site, well downstream in the drainage, and subject to drastic fluctuations in water level. It is also within the reach of the

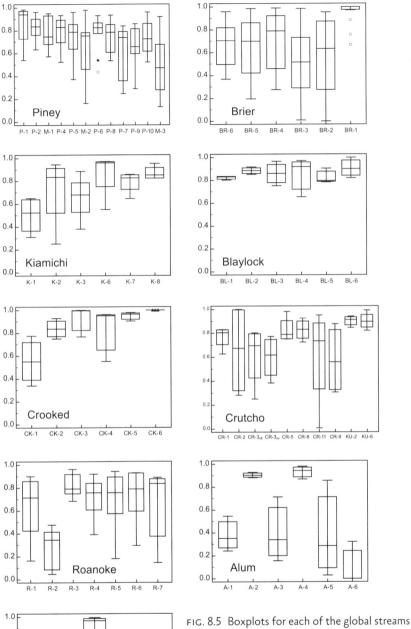

FIG. 8.5 Boxplots for each of the global streams of Morisita-Horn similarities between consecutive surveys. In each boxplot the horizontal bar represents the median, the upper and lower bounds of the "box" contain 25% or 75% of cases, and the total length of the vertical bar represents the range. Within each stream, sites are arranged from downstream on the left to most upstream on the right.

river that can back-flood from a downstream reservoir, Lake Hugo, after extreme rainfall. This most downstream site has had quite different conditions at various times in our surveys, from times like 1985, when there was easy sampling around gravel bars and stands of water willows, to times like 2014, when the water was relatively deep bank to bank and most sampling was in water waist deep or deeper. Some of the lower temporal similarity within K-1 could be related to sampling efficiency. But under high-water conditions in 2014, WJM sampled this site with a crew of five graduate students in ichthyology or other well-trained persons, and the sample was certainly adequate to detect species and relative abundances. In fact, in 2014 we caught more raw numbers of several minnow species (Redfin Shiner, *Lythrurus umbratilis*; Emerald Shiner, *Notropis atherinoides*; Rocky Shiner, *Notropis suttkusi*) than ever before at that site.

The headwater site K-8 had a much higher median MH and a small range of values (fig. 8.5). It was characterized by a relatively simple community, consistently dominated numerically by Redfin Shiners and Bigeye Shiners, which were always more abundant than all other species combined. Brook Silversides (*Labidesthes sicculus*) and Longear Sunfish (*Lepomis megalotis*) also were taken consistently at this site. The site has varied little in environmental conditions in all our surveys, consisting upstream of long rocky or boulder strewn riffles and one large deep pool under a highway bridge. In Kiamichi River we had a pattern from upstream to downstream that was opposite to that in Piney Creek but probably accounted for by a reversal in stability of the environment.

Blaylock and Crooked Creeks, both relatively small streams in the Ouachita National Forest, had trends for lower MH values (less temporal consistency) at their most downstream sites, with upstream sites in both creeks with highest median MH values (fig. 8.5). In Blaylock Creek the trend was weak, and all median MH values were relatively high, indicating that there was relatively high consistency of local communities within all sites. The lowermost site on Blaylock Creek, with low MH compared to some upper sites, consisted of one long swift riffle flowing into a large, deep pool, and the local community included 10 species. Catches of several species in that reach were variable over time. We consistently caught substantial numbers of Bigeye Shiners, but numbers of Highland Stonerollers, Redfin Shiners, Northern Studfish (*Fundulus catenatus*), and Greenside Darters (*Etheostoma blennioides*) varied from absent or scarce to relatively abundant at that site. In contrast, the uppermost site BL-6 had no deep pools and consisted instead of shallow riffles with modest flow and a series of rock-bound "step pools." We never found more than three species at BL-6, and Creek Chub and Orangebelly Darters strongly and consistently dominated the local community. Here again, much as for Brier Creek headwaters, we found an extremely simple community consistently dominated by only one or two species, thus having a high intersurvey median and range of MH similarity.

In Crooked Creek the downstream-to-upstream pattern in MH values was even stronger (fig. 8.5), with the lowermost site having the most variable local community and the two uppermost sites the most consistent. Downstream we found nine species overall, with Orangebelly Darters always abundant in riffles. But in downstream pools the dominant species was inconsistent, as Creek Chub was most abundant minnow in the first two surveys, but Striped Shiner dominated thereafter. Upstream, Creek Chubs and Orangebelly Darters were most numerous, and the only other species we ever found there was a few Highland Stonerollers. That site mimicked the headwater community in Blaylock Creek, both of which are tributaries to the Little Missouri River. Again, we find that small headwaters can be dominated by one or a few species that are consistently abundant, thus giving strong similarity in the local community from one time to the next. Finally, at the uppermost site on Crooked Creek, above the nearly vertical Crooked Creek Falls (see plate 5), we found the most extreme possible similarity across surveys, and all MH values were 1.00, because only a single species (Orangebelly Darter) was ever collected there.

Crutcho Creek, a tributary to the North Canadian River in central Oklahoma, also showed the highest consistency across time at two sites in the headwaters (actually on a tributary, Kuhlman Creek; see two rightmost boxplots for Crutcho in fig 8.5). These two sites had a total of six and eight species, respectively, detected across all surveys, and were dominated by hardy or environmentally tolerant species, including Red Shiner, Golden Shiner (*Notemigonus crysoleucas*), Sand Shiner, Fathead Minnow (*Pimephales promelas*), Western Mosquitofish (*Gambusia affinis*), Green Sunfish, and Longear Sunfish. In every survey at these Kuhlman Creek sites, Sand Shiners, Green Sunfish, and Longear Sunfish were in substantial numbers, and Fathead Minnows were present in 9 of 10 samples. All of these species are well known for their tolerance or success in potentially stressful habitats (Matthews 1987; Jester et al. 1992), and their persistence resulted in strong consistency, with median MH similarities >0.90 for these headwater fish communities.

The three other global streams (Roanoke River, Bread Creek, and Alum Creek) showed no distinct upstream-to-downstream pattern in MH similarities between consecutive surveys. In the Roanoke River, most sites were rather similar in their median MH similarities (fig. 8.5), with the exception that intersurvey MH values were consistently low at site R-2. This was a relatively wide site on the mainstream, just downstream from the small community of Lafayette, Virginia, with apparently good riffle and pool habitat, but in September 1978 we found few fish there for reasons that were never clear. Bread and Alum Creeks, both in the Saline River drainage in the Ouachita National Forest, showed relatively low intersurvey MH values at their lowermost and uppermost sites, with sites through the middle of these watersheds varying from high to low MH values (fig. 8.5). In the most upstream sites there were only five (A-6) and three (BD-6) species ever detected and very few individuals. In both of these headwater sites there were some intersurvey MH values of zero, because, twice in A-6 and once in BD-6, we caught no fish although water was available and seining was thorough. At A-5 and A-6 in October 1990, conductivity was extremely high for no apparent reason (measured reliably at 4600 μmhos each), and there were few fish at A-5 and none at all taken at A-6, even though riffles were flowing and pools were 75 cm deep, so "something" had happened. In May 1991, no fish were taken at Alum Creek A-6, although rain the day before resulted in a 10 to 12 cm rise in the stream, with flowing riffles and substantial pools at this headwater site (WJM 2524). Likewise (WJM 2517), there were no fish taken at Bread Creek BD-6, although riffles were flowing and pools were as deep as 40 cm. Sometimes, for unexplained reasons, fish may simply be absent or at least below levels of detection.

These last three streams, with highly variable intersurvey similarities, contrasted with Kiamichi River, Blaylock Creek, and Crooked Creek, in which relatively depauperate extreme headwaters were nevertheless consistent in composition from survey to survey. But one difference was obvious: Kiamichi River and Blaylock and Crooked Creeks all had substantial flow in the headwaters during all surveys, whereas Alum and Bread Creeks headwaters were nonflowing and reduced to isolated pools during one or more surveys.

So in five of the nine streams (Kiamichi River and Brier, Blaylock, Crooked, and Crutcho Creeks) the greatest within-site MH similarity over time was at headwater sites, whereas in three streams (Piney, Bread, and Alum Creeks) the extreme headwater sites had the lowest within-site similarity. The mechanisms by which these headwater sites differ in their fish community consistency may partly be explained by consistency of flow in the headwaters, but overall they are also idiosyncratic. If we review the streams above with regard to consistency of flow at the headwater sites (F, always flowing; NF, nonflowing at least once in our surveys), the pattern is inconsistent, as follows: highest MH similarity in headwaters = Kiamichi (F), Brier (NF), Blaylock (F), Crooked (F), Crutcho (F); lowest MH similarity in headwaters = Piney (F), Bread (NF), and Alum (NF). We did find one headwater site on Piney Creek (P-8) to be pooled up on two occasions, but the most extreme headwater site in our Piney Creek surveys was M-3 on Mill Creek, which was spring fed and never ceased flow in our surveys, seen at rightmost in the Piney Creek panel of fig. 8.5. Four of the five streams with highest MH similarity in headwaters were consistently flowing at the most upstream site, whereas for two of three streams with low MH similarity in headwaters, there were periods lacking flow. Brier headwater site BR-1 had an extremely simple fauna, consisting in most surveys in recent decades of only one species, the extremely hardy Green Sunfish. For this Brier Creek headwater we can attribute high MH similarity to the fact that the environment there is so harsh (frequently drying and then being recolonized) that only one species is successful. The other exception to the flow/no-flow pattern was M-3 in the Piney Creek watershed, which never lacked flow but was the most variable site in the watershed over time. We think that the low intersurvey similarity at this site, which has a relatively complex fish fauna, can be attributed to the fact that two great floods (1982 and 2008) in the watershed resulted in extreme instream and riparian habitat destruction at this site, with rearrangement of the basic structure of the stream and substantial differences in the fish fauna from time to time after such events. And, while not directly related to either flood event, late in our surveys of Piney Creek (1990s to present), the Western Mosquitofish became abundant at M-3 where it had not been found earlier.

These two exceptions aside, the differences in consistency of flow in the headwaters may explain much of the rest of the differences among streams. Remember that the extreme headwater sites in all four Ouachita National Forest streams (Blaylock, Crooked, Alum, and Bread Creeks) had simple faunas of not more than five species. In Blaylock and Crooked Creeks, where the headwaters were always flowing, these simple faunas persisted, but in Alum and Bread Creeks, where flow ceased at times, there were differences in community dominants from time to time, and in some surveys no fish were found. For those four streams at least, we would conclude that flow or lack of flow can be a substantial factor in persistence of headwater fish communities.

Finally, for the nine streams overall, the lack of any general patterns in within-site similarities from downstream to upstream did surprise us and may be a good example of how unalike different streams and their fish communities can be and how elusive overarching patterns in stream fish community dynamics may be. For five of the streams (Piney, Brier, Blaylock, and Crooked Creeks and Kiamichi River) there was some kind of identifiable downstream-to-upstream pattern in MH similarities within sites (if you ignore the unrealistically high MH values for the extreme headwater site on Brier Creek where one species dominated). For the other four streams (Roanoke River and Crutcho, Alum, and Bread Creeks), no simple longitudinal pattern in MH values was evident. So, we asked if there was an alternate hypothesis that could explain variation in site stability. Specifically, we asked if larger (wider) sites had more consistent fish communities, regardless of where in the watershed the bigger sites were found. To test this, we graphed within each global stream the median MH similarity index by site versus stream width (fig. 8.6) and regressed the median MH value against width. There was some support for a "bigger, more consistent" hypothesis, but the results were mixed. Several of the panels suggested trends, but linear regression analyses indicated that for only two (Piney and Alum Creeks) the median MH was significantly and positively related to stream width at $p < 0.05$, and for one (Crooked Creek) they were marginally and negatively related at $p < 0.10$. With $N = 6$ for 6 of the streams, the power of regression was low, increasing the chance for Type II error. Trends, albeit nonsignificant, for a negative relationship between median MH value (lower consistency) and stream width for Blaylock Creek, Kiamichi River, and Roanoke River in figure 8.6 may suggest investigating a size-variability relationship for researchers with data on other streams sampled over multiple sites and times.

At the end of all these analyses, we had five cases for the nine global streams in which a longitudinal pattern in variation (median consecutive MH similarities) within sites was apparent and one case (Alum Creek) in which the pattern was not longitudinal but could be explained by size of the stream at individual sites. For three streams we have no explanation about why different sites had different levels of intrasite variation in the local community over time.

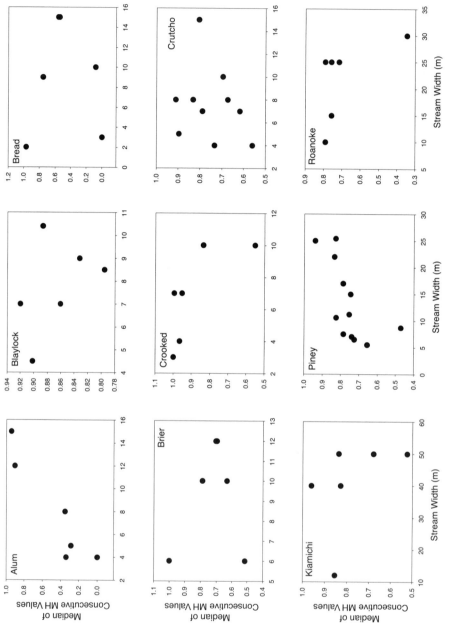

FIG. 8.6 Median Morisita-Horn similarity versus width for sites on nine global streams.

Spatial Differences among Individual Sites in Piney and Brier Creeks over Time

In this section we examined longitudinal differences in the temporal trajectories of local fish in Brier Creek and compared results for Brier Creek to similar analyses for Piney Creek (Matthews and Marsh-Matthews 2016). We followed the procedures outlined in chapter 7 for temporal trajectories, including (1) calculating a large, triangular matrix of MH similarity values among all possible pairs of 96 individual samples by sites and times; (2) performing nonmetric multidimensional scaling (NMDS) of the MH similarity matrix to produce a biplot; (3) connecting consecutive surveys across time within each site to reveal site-specific trajectories, as well as the overall amount of multivariate space occupied by each site over time; and (4) following guidelines in Matthews and Marsh-Matthews (2016) and in chapter 7 to classify the trajectory at each site as one of 6 hypothetical trajectory types in Matthews et al. (2013), to determine whether each site exhibited long-term loose equilibrium, following Collins (2000), Matthews et al. (2013), and Matthews and Marsh-Matthews (2016). We only analyzed trajectories for individual sites on Brier Creek because, following our own advice in chapter 7, with fewer than six surveys it may be less likely that long-term patterns that actually exist will be revealed. So, we did not include this kind of analysis for the other seven global streams for which we had only four or five surveys.

Repeated runs of a two-dimensional NMDS of the MH triangular matrix for all Brier Creek sites x times stabilized with similar solutions. The final NMDS run had Stress 1 = 0.192. This is a bit higher than Stress 1 < 0.15 recommended by McCune and Grace (2002). But it is well known that for very large data sets it is difficult to achieve stress below that level, as stress in NMDS typically increases with data set size. We relied on the two-dimensional solution, which can be displayed successfully with two axes, for interpretation.

The biplot from the NMDS two-dimensional solution showed lower overall variation (amount of area occupied in both dimensions, if a polygon surrounding all times is drawn by eye) for the two downstream sites BR-6 and BR-5 and the midreach site BR-4, more overall variation at a headwater tributary BR-3, greatest variation at mainstem headwater site BR-2, and least variation of all at the extreme headwater site BR-1. Note that this total variation within sites in Brier Creek was also consistent with the pattern showed by the MH values between consecutive surveys for Brier Creek, in which the median MH values were higher (less variation) for BR-6, BR-5, and BR-4; lower for the next two sites (BR-3 and BR-2); and then very high (very low variation between surveys) for the extreme headwater site (BR-1). All of this makes sense for Brier Creek, whether we consider time-to-time similarity (fig. 8.5) or overall variation in NMDS space within a site (fig. 8.7).

Matthews et al. (2013) demonstrated that temporal dynamics of the global Brier Creek fish community over a span of 40 years were consistent with expectations of loose equilibrium. Matthews and Marsh-Matthews (2016) similarly showed that the global community of Piney Creek was consistent with the loose equilibrium concept

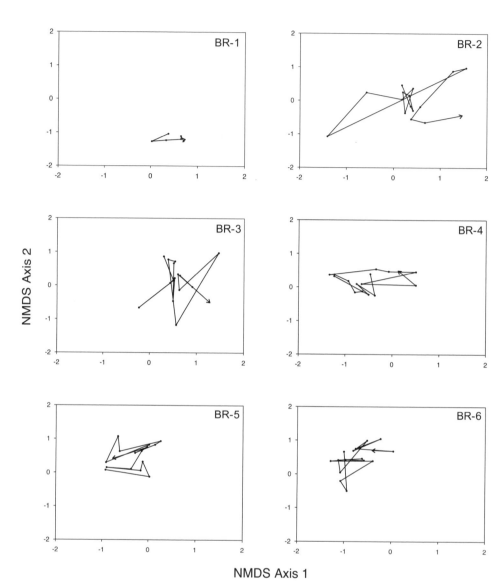

FIG. 8.7 Trajectories of six individual Brier Creek sites in two-dimensional biplots of nonmetric multidimensional scaling (NMDS) based on Morisita-Horn similarities among all possible pairs of sites and times. Each panel is scaled to the same lengths on NMDS axes 1 and 2, so distances or spaces are comparable among sites. The trajectories begin with the first survey (1976) and end with the last survey (2012), which is represented by an arrowhead. The arrowhead for BR-1 covers 10 different surveys since 1988, all with identical plotting scores.

(LEC) and that in Piney Creek dynamics of 8 of the 12 individual sites also met expectations of the LEC. Here we asked if trajectories of individual sites on Brier Creek were consistent with loose equilibrium. Based on criteria we outlined in chapter 7 (tendency of a site to return toward average community structure in a multivariate biplot after a displacement away from average, and presence or absence of long saltatory steps), the temporal dynamics of 5 of the 6 individual Brier Creek sites appeared consistent with the LEC (fig. 8.7).

The only Brier Creek site not consistent with the LEC was the headwater site BR-1. Although BR-1 occupied only a small amount of multivariate space in fig. 8.7, it had permanent movement away from its earliest community structure, from a relatively diverse community of seven or eight species to usually having only Green Sunfish. We classified site BR-1 as hypothetical trajectory type E of Matthews et al. (2013), not consistent with expectations of loose equilibrium.

Brier Creek site BR-2, also a headwater site on the main creek but downstream from BR-1, exhibited only two returns from displacement and two saltatory (long) steps. This site was one of the most physically variable in the creek, sometimes having water from bank to bank in much of the stream channel and at other times reduced to long reaches of bare bedrock with isolated pools (see plates 22 and 23). And in the early 1980s it was completely dry in two different years (Matthews 1987). This site had the greatest overall variation in composition over time, based on area occupied in figure 8.7. From 1976 to 1994 the local community actually varied little, in spite of the episodes of drying, after which it was recolonized rapidly (Matthews 1987). But in 1995 and 1996 it diverged sharply toward the bottom left in the panel for BR-2 in figure 8.7, and by 1999 it had reversed direction to lie to the extreme upper right in the BR-2 panel. But after both of these saltatory leaps in structure, the community structure reversed and moved back toward the centroid. The last two surveys (2008 and 2012) showed that BR-2 had again moved away from the centroid, but on the basis of the earlier strong tendency to return toward average, we classified BR-2 as trajectory type D (saltatory, nondirectional) and considered it consistent with the LEC.

Site BR-3 was saltatory, with one step an outlier by the Tukey test and with five returns toward the centroid of points. Based on the large number of returns and the general pattern in the BR-3 panel of figure 8.7, we also classified this site as trajectory type D (saltatory, nondirectional), consistent with the LEC.

Midstream site BR-4 and the two downstream sites BR-5 and BR-6 were all classified as hypothetical trajectory type A (gradual, nondirectional) on the basis of a lack of outliers by either a Tukey or generalized extreme studentized deviate (ESD) test and five or six returns after displacement. The panels for these sites in figure 8.7 show numerous reversals within their trajectories, and in addition, their community structures in the final survey (arrowheads) were closer to the first survey than were at least several other surveys in time.

Overall, temporal dynamics within 5 of the 6 individual Brier Creek sites were consistent with loose equilibrium, as were 8 of 12 sites in the Piney Creek watershed

(Matthews and Marsh-Matthews 2016). But patterns where LEC and non-LEC sites were located were different in the two watersheds. In Brier Creek, all of the sites were LEC with the exception of the smallest, most extreme headwater site. In Piney Creek, sites lacking LEC dynamics were at the two largest, lowermost sites (P-1 and P-2), one midreach site (P-6), and one small headwater site (P-10) (for site numbers, see fig. 2.1). In Piney Creek, four other headwater sites that were far upstream on the named creek channel (P-8) or on small tributaries (P-7, P-9, M-3) were consistent with the LEC. Comparing the two watersheds, the similarities were that the majority of individual sites were in loose equilibrium on both, as were the global communities for both creeks (Matthews et al. 2013; Matthews and Marsh-Matthews 2016). The differences were that although Brier Creek has a substantially harsher environment than Piney Creek (Ross et al. 1985), sites throughout the creek were in LEC with the exception of the uppermost site, whereas in the environmentally benign Piney Creek, two of the largest downstream sites had trajectories that represented apparent real change in the local community over time, as did one midreach and one headwater site. Thus no longitudinal pattern in sites exhibiting LEC existed in Piney Creek, whereas the pattern in LEC in Brier Creek could be interpreted as longitudinal. In both cases, the fact that at least some sites in either creek lacked LEC underscores the conclusion that there is more "noise" in the dynamics of fish communities within local stream reaches than for a watershed as a whole.

Summary

Spatial patterns of fish distribution and the resulting variation in composition of communities have been studied for more than a century. These studies resulted in the development of concepts such as longitudinal zonation in community structure or discrete fish zones within watersheds. More recently, studies have begun to address temporal variation in these patterns and thus to address community variation in a broader spatiotemporal context.

Spatiotemporal analyses can be approached in two ways. One can examine the temporal variation in spatial patterns or the spatial variation in temporal patterns. In this chapter we have taken both approaches. We examined temporal variation in spatial patterns of species presence-absence for each of the nine global communities in the context of beta diversity (partitioned into spatial turnover and nestedness components). Using abundance-based data, we used Mantel comparisons of Morisita-Horn similarity to ask whether the spatial pattern in community structure detected across 59 sites in the lower Great Plains in 1995 was concordant with the pattern observed 17 years earlier (in 1978). We also used Mantel comparisons to address concordance of spatial patterns within watersheds in order to minimize the contribution of major zoogeographic changes across space.

At a smaller spatial scale, we asked if the similarity of the fish community at sites within a watershed varied over time. To answer this question, we calculated MH similarity measures among all pairs of sites and used Mantel analyses to determine whether spatial patterns were concordant among surveys.

In other analyses, we examined spatial patterns in temporal variation in community structure for each of our nine global systems. To do this, we calculated MH similarity indices for consecutive samples at each site within the system and examined the variation detected as a function of position of the site in the watershed. For each site, we calculated MH similarity of consecutive samples and constructed boxplots to summarize the variation in the magnitude of time-to-time change at each site. Boxplots were graphed representing sites from downstream to upstream, which allowed us to look for any longitudinal pattern in temporal variation.

Finally, we constructed trajectories of change (based on nonmetric multidimensional scaling) for each site within the watershed for the fish community in Brier Creek. Trajectories were compared for total amount of multivariate space traversed (as another measure of within-site variability). Each trajectory was evaluated for fit (or not) to the loose equilibrium model.

Beta diversity (B_{sor}) analyses for the nine global streams showed that there was no correlation between total number of species in the global community and the overall beta diversity (measured as the grand mean over all sites and times). When beta diversity was partitioned into pure spatial turnover (B_{sim}) and nestedness components (B_{nest}), in eight of the nine streams B_{sim} tended to be greater than B_{nest}. But there was a strong positive correlation between number of species and percentage of cases in which pure spatial turnover was greater than nestedness. This result suggests that in the more speciose streams, communities at upstream sites are not simply subsets of those at downstream sites and that there are species that are predictably found upstream or downstream. In streams with low numbers of total species, nestedness was often greater than pure spatial turnover. In these streams, headwater communities were typically a subset of species present downstream.

Comparison of the beta diversity patterns over time showed that the spatial patterns of turnover and nestedness within streams were generally not temporally concordant. This suggests a lack of any predictable location for faunal turnover or species loss within most systems. This finding would be expected if fishes moved among sites (even if tending to remain within a given section of stream).

Analyses of MH similarity among all sites in our Midwest collections in 1978 and 1995 showed highly significant spatial concordance between surveys. This result was not surprising, given that species composition of communities at this spatial scale was highly influenced by breaks in species' distributions across several zoogeographic boundaries. When temporal comparisons were limited to sites within a river basin, results were mixed. There was strong concordance of communities between surveys within the Arkansas River basin but not within the Kansas River basin. Fish communities at several sites within the Kansas River basin changed substantially between surveys, rendering the overall spatial pattern within the basin nonconcordant.

At the spatial scale of individual streams, there was variation among our nine global communities in the concordance of spatial similarity from time to time (based on Mantel comparisons of consecutive matrices of similarity among all sites within a stream). There were four streams (Alum Creek, Blaylock Creek, Kiamichi River, and Piney Creek) for which the MH spatial similarity matrices were concordant across every interval, indicating no significant variation in spatial patterning of communities at local sites over time. In two additional streams (Crutcho Creek and Brier Creek), spatial patterns were congruent over most intervals. In contrast, there were two streams (Bread Creek and Roanoke River) that showed no congruence in spatial patterning from time to time and a third (Crooked Creek) that only showed congruence in one of four intervals. Incongruence of spatial patterns may have resulted from different causes for the different streams: Bread Creek headwaters were dry during part of the study in that system, and sampling was more seasonal in the Roanoke River than in other streams, but there was no obvious feature of sampling in Crooked Creek that might explain spatiotemporal variation in that system.

In our analyses of spatial variation in temporal patterns in our global systems, we found a longitudinal pattern in the community dynamics from time to time at a given site in four of the nine streams. In Piney Creek, downstream sites tended to have higher similarity than those upstream, but for the other three (Kiamichi River, Blaylock Creek, Crooked Creek) the headwater sites showed the greatest time-to-time similarity. Although the greater community stability in headwaters in the Kiamichi River likely reflects less environmental variation there than at more downstream sites, the apparent stability in the other two creeks exists in spite of variable and often harsh conditions in the headwaters and probably reflects the presence of only one or a few tolerant species. Although Brier Creek did not show a general trend for increasing stability from downstream to upstream, the greatest time-to-time similarity in that system was also in a harsh headwater site with only Green Sunfish present in most collections.

In five of the nine streams (Brier, Crutcho, Roanoke, Alum, and Bread) there was no pronounced longitudinal pattern in time-to-time community similarity. In Alum Creek, however, temporal similarity was related to size (stream width) of the site, such that larger sites showed greater community stability (which was also true for Piney Creek), but the relationship between stream width and temporal stability was not apparent for most of our global systems.

Trajectories of fish communities at each of six sites on Brier Creek showed that the headwater site exhibited a directional shift in community structure over the course of the study, but variation in the local fish community at all other sites was consistent with a pattern of loose equilibrium. Among sites in loose equilibrium, however, details of dynamics differed among sites. The three downstream sites displayed dynamics characteristic of gradual, nondirectional change, while two sites farther upstream (but not the headwater) showed saltatory, nondirectional dynamics.

Overall, our studies of spatiotemporal dynamics at these various spatial scales suggest that large-scale spatial patterns (among or within large basins) likely persist over time, except where systems are heavily altered. Within streams, however, spatial structuring may show some degree of temporal variation both in presence-absence of species and in patterns of relative abundance that probably reflect within-stream movements of fishes. This result supports the global community as the appropriate scale for long-term studies. Our analyses of spatial variation in temporal dynamics (at the scale of within streams) did not find any overarching pattern or factor that explained time-to-time variation in community structure at a given site. Evidence for a longitudinal pattern in stability was equivocal, and extremely high stability at various sites may result from different causes.

Appendix: Methods and Examples of Beta Diversity and Beta Partitioning

Here we provide a summary of the beta diversity partitioning following Baselga (2010) and provide examples of its application. We begin with a traditional comparison of two samples by accumulating species into the crossed, or "box," arrangement in which "a" indicates a species present in both samples, "b" indicates a species present in the first sample but not the other, and "c" indicates a species present in the second sample but not the first. (For purposes of calculation, it does not matter which sample is first and which is second. In other words, you can switch the values in the b and c cells and the results are the same). The fourth cell, "d," equals shared absence in the two samples, so it is irrelevant and not considered in calculations. By way of illustration, keep the box below in mind.

Appendix Table 8A. Cross-classification of species by presence-absence across two samples

Number of Species	Present in Sample A	Absent in Sample A
Present in Sample B	a	b
Absent in Sample B	c	d

We will consider situations in which, for a pair of samples, (A) no species are shared, comprising complete turnover; (B) there are shared species a, and the b and c cells are equal; (C) there are shared species, but the b and c cells are unequal; (D) there are shared species, and the b and c cells are unequal, with more species; and (E) with completed nestedness.

The calculating formulas are:

$$B_{sor} = \text{total beta diversity} = b + c \, / \, 2a + b + c$$
$$B_{sim} = \text{pure spatial turnover} = \min(b,c) \, / \, a + \min(b,c)$$
$$B_{nest} = B_{sor} - B_{sim}$$

In the simplest case, comparing samples A and B in table B, no species are shared between samples, and b and c are equal. For this example, $B_{sor} = (3+3) \, / \, ((2 \times 0) + 3 + 3)$ is equal to 1.00. B_{sim} equals 3/3, also equivalent to 1.00. By subtraction of B_{sim} from B_{sor}, $B_{nest} = 0$. In this simple example, total beta diversity equals pure spatial turnover, and there is no nestedness. Neither sample is a subset of the other.

Appendix Table 8B.   Hypothetical communities for calculation of beta diversity and beta partitioning

Species	A	B	C	D	E	F	G
				Sample			
1	X						
2	X						
3	X		X		X	X	X
4		X	X	X	X	X	X
5		X	X	X	X	X	X
6		X		X			
7				X			
8				X			
9					X	X	X
10					X	X	X
11					X	X	
12					X	X	
13						X	
14						X	
15						X	
16						X	

To make this a bit more complicated, consider comparison of sample B to sample C. Two species are shared, and one species in each is not shared with the other. In this case, $a = 2$, $b = 1$, and $c = 1$. So $B_{sor} = 2/6 = 0.333$, $B_{sim} = 1/3 = 0.333$, and by subtraction $B_{nest} = 0$. In this case there are shared species, but the beta diversity is again entirely due to pure spatial turnover, and there is no nestedness. Another way to look at it is that while they do share some species, neither is nested within the other in terms of one sample having fewer species than, and being nested within, a more speciose sample.

Next, consider the case of samples C and D, in which some species are shared, but the b and c cells are unequal. In this example, two species are shared ($a = 2$), $b = 1$, and $c = 3$. $B_{sor} = 4/8 = 0.500$; $B_{sim} = 1/3 = 0.333$, and by subtraction $B_{nest} = 0.167$. In this case, beta diversity is decomposed into more pure spatial turnover (0.333) than nestedness (0.167), but for the first time we do see nestedness. Actually, in this example it is sample C that is a partial subset of the more speciose sample D, but the arithmetic works the same.

In a fourth example, using samples D and E in the table, $a = 2$ shared species, $b = 3$ species (in D but not E), and $c = 5$ (in E but not D). $B_{sor} = 8/12 = 0.666$; $B_{sim} = 3/5 = 0.600$, and $B_{nest} = 0.066$. In that example there is a lot of pure spatial turnover and weak nestedness.

If we further increase the number of species in the more speciose sample, not shared with the other, we can consider a comparison between samples D and F in table B. In this example, a remains equal to two shared species; b remains three species in D but not in F, and c increases to 10 (species in F that are not in D). $B_{sor} = 13/17 = 0.765$, $B_{sim} = 3/5 = 0.600$, and by subtraction $B_{nest} = 0.165$. Comparing this example to the previous one, note that the total beta diversity has increased from 0.666 to 0.765, as the overall disparity in shared and nonshared species increased in the present example.

Finally, let's examine a case with complete nestedness. All species in sample G of the table are a subset of the species in sample F. So, $a = 5$ shared species; $b = 6$ species in F but not in G; c = 0 because G has no species not included in F. For this case, $B_{sor} = 6/16 = 0.375$, $B_{sim} = 0/5 = 0.000$, and by subtraction B_{nest} is the same as $B_{sor} = 0.375$. In this case, nestedness accounts for all of the total beta.

In the examples above we have seen a range of outcomes, from total beta being totally to spatial turnover (no species shared in common) to being due entirely to nestedness (when all species in the less speciose sample are a subset of the more speciose sample). Also, regardless of the number of shared species, nestedness is zero if b and c are identical, as neither sample is nested within the other. Finally, if the number of shared species remains the same but disparity between b and c increases, the result is more total turnover. There are other outcomes that can be envisioned, as we will see in the real-world examples from our nine global streams, but the examples above provide what is, we hope, a range of explanations. And, overall, keep in mind that in viewing bar graphs or considering raw values for beta, higher total beta (B_{sor}) means that two samples are less similar, and low B_{sor} means that two samples are similar in composition.

References

Balon, E. K., and D. E. Stewart. 1983. Fish assemblages in a river with unusual gradient (Luongo, Africa-Zaire system), reflections on river zonation, and description of another new species. Environmental Biology of Fishes 9:225–252.

Bart, H. L., Jr. 1989. Fish habitat association in an Ozark stream. Environmental Biology of Fishes 24:173–186.

Baselga, A. 2010. Partitioning the turnover and nestedness components of beta diversity. Global Ecology and Biogeography 19:134–143.

Blair, A. P. 1959. Distribution of the darters (Percidae, Etheostominae) of northeastern Oklahoma. Southwestern Naturalist 4:1–13.

Burton, G. W., and E. P. Odum. 1945. The distribution of stream fish in the vicinity of Mountain Lake, Virginia. Ecology 26:182–194.

Cashner, R. C., F. P. Gelwick, and W. J. Matthews. 1994. Spatial and temporal variation in the distribution of fishes of the LaBranche wetlands area of the Lake Pontchartrain estuary, Louisiana. Northeast Gulf Science 13:107–120.

Collins, S. L. 2000. Disturbance frequency and community stability in native tallgrass prairie. American Naturalist 155:311–325.

Conner, J. V., and R. D. Suttkus. 1996. Zoogeography of freshwater fishes of the western Gulf slope. Pages 413–456 in C. H. Hocutt and E. O. Wiley, eds. The zoogeography of North American freshwater fishes. John Wiley and Sons, New York, NY.

Duvernell, D. D., S. L. Meier, J. F. Schaefer, and B. R. Kreiser. 2013. Contrasting phylogeographic histories between broadly sympatric topminnows in the *Fundulus notatus* species complex. Molecular Phylogenetics and Evolution 69:653–663.

Forbes, S. A. 1907. On the local distribution of certain Illinois fishes: an essay in statistical ecology. Bulletin of the Illinois State Laboratory of Natural History 7:273–303.

Gelwick, F. P. 1990. Longitudinal and temporal comparisons of riffle and pool fish assemblages in a northeastern Oklahoma Ozark stream. Copeia 1990:1072–1082.

Grossman, G. D., R. E. Ratajczak Jr., M. Crawford, and M. C. Freeman. 1998. Assemblage organization in stream fishes: effects of environmental variation and interspecific interactions. Ecological Monographs 68:395–420.

Hargrave, C. W. 2000. A decade of stability in fish assemblage of a harsh, prairie river system. MS thesis, University of Oklahoma, Norman, OK.

Hitt, N. P., and J. H. Roberts. 2012. Hierarchical spatial structure of stream fish colonization and extinction. Oikos 121:127–137.

Hoeinghaus, D. J., C. A. Layman, D. A. Arrington, and K. O. Winemiller. 2003. Spatiotemporal variation in fish assemblage structure in tropical floodplain creeks. Environmental Biology of Fishes 67:379–387.

Huet, M. 1959. Profiles and biology of western European streams as related to fish management. Transactions of the American Fisheries Society 88:155–163.

Hutchinson, G. E. 1939. Ecological observations on the fishes of Kashmir and Indian Tibet. Ecological Monographs 9:146–182.

Jester, D. B., A. A. Echelle, W. J. Matthews, J. Pigg, C. M. Scott, and K. D. Collins. 1992. The fishes of Oklahoma, their gross habitats, and their tolerance of degradation in water quality and habitat. Proceedings of the Oklahoma Academy of Science 72:7–19.

Koleff, P. K. J. Gaston, and J. L. Lennon. 2003. Measuring beta diversity for presence-absence data. Journal of Animal Ecology 72:367–382.

Kuehne, R. A. 1962. A classification of streams, illustrated by fish distribution in an eastern Kentucky creek. Ecology 43:608–614.

Marsh-Matthews, E., and W. J. Matthews. 2000. Geographic, terrestrial and aquatic factors: which most influence the structure of stream fish assemblages in the midwestern United States? Ecology of Freshwater Fish 9:9–21.

Matthews, W. J. 1985. Distribution of midwestern fishes on multivariate environmental gradients, with emphasis on *Notropis lutrensis*. American Midland Naturalist 113:225–237.

Matthews, W. J. 1986. Fish faunal "breaks" and stream order in the eastern and central United States. Environmental Biology of Fishes 17:81–92.

Matthews, W. J. 1987. Physicochemical tolerance and selectivity of stream fishes as related to their geographic ranges and local distributions. Pages 111–120 in W. J. Matthews and D. C. Heins, eds. Community and evolutionary ecology of North American stream fishes. University of Oklahoma Press, Norman, OK.

Matthews, W. J. 1990. Spatial and temporal variation in fishes of riffle habitats: a comparison of analytical approaches for the Roanoke River. American Midland Naturalist 124:31–45.

Matthews, W. J. 1998. Patterns in freshwater fish ecology. Chapman and Hall, New York, NY.

Matthews, W. J. 2015. Basic biology, good field notes, and synthesizing across your career. Copeia 103:495–501.

Matthews, W. J., and G. L. Harp. 1974. Preimpoundment ichthyofaunal survey of the Piney Creek watershed, Izard County, Arkansas. Arkansas Academy of Science Proceedings 28:39–43.

Matthews, W. J., and E. Marsh-Matthews. 2016. Dynamics of an upland stream fish community over 40 years: trajectories and support for the loose equilibrium concept. Ecology 97:706–719.

Matthews, W. J., B. C. Harvey, and M. E. Power. 1994. Spatial and temporal patterns in the fish assemblages of individual pools in a midwestern stream (U.S.A.). Environmental Biology of Fishes 39:381–397.

Matthews, W. J., E. Marsh-Matthews, G. L. Adams, and S. R. Adams. 2014. Two catastrophic floods: similarities and differences in effects on an Ozark stream fish community. Copeia 2014:682–693.

Matthews, W. J., E. Marsh-Matthews, R. C. Cashner, and F. Gelwick. 2013. Disturbance and trajectory of change in a stream fish community over four decades. Oecologia 173: 955–969.

Matthews, W. J., W. D. Shepard, and L. G. Hill. 1978. Aspects of the ecology of the Dusky-stripe Shiner, *Notropis pilsbryi* (Cypriniformes, Cyprinidae) in an Ozark stream. American Midland Naturalist 100:247–252.

McCune, B., and J. B. Grace. 2002. Analysis of ecological communities. MjM Software Design, Gleneden Beach, CA.

McGarvey, D. J. 2011. Quantifying ichthyofaunal zonation and species richness along a 2800-km reach of the Rio Chama and Rio Grande (USA). Ecology of Freshwater Fish 20:231–242.

Meador, M. R., and W. J. Matthews. 1992. Spatial and temporal patterns in fish assemblage structure of an intermittent Texas stream. American Midland Naturalist 127:106–114.

Moran, M. D. 2003. Arguments for rejecting the sequential Bonferroni in ecological studies. Oikos 100:403–405.

Power, M. E., W. J. Matthews, and A. J. Stewart. 1985. Grazing minnows, piscivorous bass and stream algae: dynamics of a strong interaction. Ecology 66:1448–1456.

Rahel, F. J., and W. A. Hubert. 1991. Fish assemblages and habitat gradients in a Rocky Mountain–Great Plains stream: biotic zonation and additive patterns of community change. Transactions of the American Fisheries Society 120:319–332.

Rice, W. R. 1989. Analyzing tables of statistical test. Evolution 43:223–225.

Ross, S. T., W. J. Matthews, and A. A. Echelle. 1985. Persistence of stream fish assemblages: effects of environmental change. American Naturalist 126:24–40.

Schlosser, I. J. 1982. Fish community structure and function along two habitat gradients in a headwater stream. Ecological Monographs 52:395–414.

Schlosser, I. J. 1987. A conceptual framework for fish communities in small headwater streams. Pages 17–24 in W. J. Matthews and D. C. Heins, eds. Community and evolutionary ecology of North American stream fishes. University of Oklahoma Press, Norman, OK.

Shelford, V. E. 1911. Ecological succession: I. stream fishes and the method of physiographic analysis. Biological Bulletin 21:9–34.

Taylor, C. M., and M. L. Warren Jr. 2001. Dynamics in species composition of stream fish assemblages: environmental variability and nested subsets. Ecology 82:2320–2330.

Taylor, C. M., M. R. Winston, and W. J. Matthews. 1996. Temporal variation in tributary and mainstem fish assemblages in a Great Plains stream system. Copeia 1996:280–289.

Whittaker, R. H. 1960. Vegetation of the Siskiyou Mountains, Oregon and California. Ecological Monographs 30:279–338.

CHAPTER 9

What's It All Mean?
Ecosystem Effects

A rt Stewart asked of our *Campostoma* and algae data in reflective moments, "What's it all mean, Bill? What's it all mean?" This chapter answers this question, considering how dynamics of fish communities that we quantified in previous chapters could affect stream ecosystems.

Fish Effects in Ecosystems: Background
The roles of fishes in ecosystems received only scattered attention throughout most of the history of aquatic ecology. The importance of salmonid decomposition in northern streams was known since the 1930s (Juday et al. 1932). But fish were largely ignored for most of the twentieth century, by both limnologists (in lakes or ponds) and stream ecologists. The limnology book used most often in the 1970s was Ruttner (1963), in which "fish" or "fishes" appeared only twice, with nothing about their effects in ecosystems. G. E. Hutchinson's four-volume *A Treatise on Limnology*, published from 1957 to 1993, did not include fishes. Hynes's (1970) classic *The Ecology of Running Waters* had four descriptive chapters on fishes but nothing to suggest that they could affect stream ecosystems. Based on fish consumption of benthic macroinvertebrates, however, Hynes (1970, 432) noted that "Such results . . . illustrate very well the fallacy of treating the benthos as a whole as if it were a unit quite separate from the fishes."

Limnologists were first to appreciate that fish could alter the trophic structure of ecosystems, based on Hrbacek et al. (1961). Then Brooks and Dodson (1965; currently cited more than 3000 times) showed clearly that fish could alter zooplankton and thus food web dynamics. Hall (1972) showed that migrations of stream fish could transfer ions upstream and contribute to stream metabolism. But in Cummins's (1974) seminal paper "Structure and Function of Stream Ecosystems," fish were not mentioned. In his influential paper "The Stream and Its Valley," Hynes (1975) mentioned fish only once, briefly noting the work by Hall (1972). Even in the paradigm-changing "River Continuum Concept" (Vannote et al. 1980), little was said about roles of fishes in stream ecosystems.

By the 1970s, much of stream ecology had turned to quantifying energy budgets and carbon processing (Cummins 1974). Studies of Bear Brook (Fisher and Likens 1973) and Hubbard Brook (Likens and Bormann 1974) demonstrated the importance of inputs and outputs of organic material, processing of organic material, and nutrient dynamics. This new view of streams also emphasized roles of macroinvertebrates in the transformation of organic materials (Cummins and Klug 1979). Shredders break down allochthonous leaves; scrapers remove algae or biofilms from stones; collector-gatherers accumulate finer particles. Aquatic insect larvae were accorded the key roles, along with bacteria and fungi, in processing particulate organic matter (POM) from coarse (CPOM) to large (LPOM) to medium (MPOM) to fine (FPOM) and eventually ultrafine (UPOM), until nutrients were made available to the stream as dissolved organic matter (DOM) (Cummins 1974). But as of 1980, the importance of fish, along with bivalve mussels (Vaughn et al. 2008; Vaughn 2010; Atkinson and Vaughn 2015), in the functional dynamics of stream ecosystems was not yet recognized. Based on the only three field studies available, Allan (1983) concluded, "At present, the weight of evidence indicates that fish do not commonly play a major community structuring role in running waters" but added that "further studies would certainly be of value, as we do not yet have sufficient information to be confident of this generalization."

Effects of fish in stream ecosystems were not really appreciated until Mary Power's (1983; 1984a,b) work in Panamanian streams showing that algivorous catfishes (Loricariidae) could regulate standing crops of algae (Power 1990a) or until Power and Matthews (1983) and Power et al. (1985) showed that algae-grazing Central Stoneroller (*Campostoma anomalum*; see plate 27) could control distribution of algae in temperate streams. Minshall (1988, table 1) cited Power and Matthews (1983) and Power et al. (1985) in an overview of "historical derivation of stream ecological theory illustrated by selected references," considering them, with about a dozen other papers, to underscore the importance of biotic factors in the "era of refinement and experimentation" in stream ecology.

As stream ecologists came to understand fish to be important in ecosystems, other investigators made significant early contributions. Grimm (1988) showed that Longfin Dace (*Agosia chrysogaster*) had strong effects on algae standing crops and nutrient dynamics in an Arizona desert stream. In Venezuela, Flecker (1992, 1996) found that detritivorous fishes (*Prochilodus*) and algivorous catfish (*Chaetostoma*) had strong effects on algae and benthic invertebrates. Power (1990b) showed that California Roach (*Hesperoleucas symmetricus*) regulated attached algae in the Eel River, California, and were important in a trophic cascade. Numerous other studies in tropical and temperate streams further cemented the importance of the trophic activities of algivorous or detritivorous fishes in stream ecosystems (reviewed in Matthews 1998, 583–559). As a result, David Allan's (1995) book *Stream Ecology: Structure and Function of Running*

Waters had substantial sections on fish effects, including Power et al. (1985) and her work in California, indicating that herbivorous fish clearly could control benthic algae in streams. At least one recent textbook, *Community Ecology* (Morin 2011, 196–197), cited Power et al. (1985) as one of the best early examples of trophic cascades in streams. Within about two decades, concepts in stream ecology changed from little importance being attributed to fishes to widespread recognition that "fish matter" in stream ecosystems.

Central Stoneroller: A Key Actor in Stream Ecosystems

Much knowledge about effects of algivorous fishes in North America is based on the Central Stoneroller. Stonerollers of several species are extremely abundant in many streams in the eastern United States, often occurring in large shoals. They vigorously swipe, scrape, or shovel periphyton from stone surfaces (fig. 9.1; see plate 26) with a cartilaginous ridge on their lower jaw (see plate 27) or nip at longer filamentous algae strands or epiphytes (Matthews et al. 1987). Initial observations by Mary Power and WJM in pools of Brier Creek in November 1982 showed that stonerollers reduced the height of attached algae and were central to a trophic cascade including black bass (*Micropterus* spp.)–stoneroller–algae (Power and Matthews 1983). Then, Power and WJM, joined by Art Stewart, found experimentally (Power et al. 1985, 1988) and by snorkeling surveys in Brier Creek and other streams in Oklahoma and

FIG. 9.1 Group of stonerollers grazing algae. Photo courtesy of Brandon Brown.

western Arkansas that stonerollers consistently had strong effects on attached algae (Matthews et al. 1987).

A trial in a divided pool in Brier Creek (Power et al. 1985) revealed distinct differences in composition and amount of algae in areas with and without stonerollers (see plate 28). There also were dramatic effects from introducing a Largemouth Bass (*Micropterus salmoides*) to a stream pool with a large shoal of stonerollers, as some of the minnows were apparently eaten, but others emigrated from the pool or refuged in shallows, resulting in rapid regrowth of filamentous algae in parts of the pool patrolled by the bass (Power et al. 1985).

Much research on stoneroller effects was also performed in Baron Fork of the Illinois River (see plate 29), a clear, slate- and chert-bottomed Ozark stream. The stoneroller research group stayed for several field seasons east of Tahlequah, Oklahoma, at Camp Egan, which owned a reach of Baron Fork ideal for experimental studies, and where stonerollers (mostly Central Stonerollers with a few Largescale Stonerollers, *Campostoma oligolepis*) and the algivorous Ozark Minnow (*Notropis nubilus*) made up about 50% and 10%, respectively, of all fishes. We observed that in many Ozark streams, natural stones of streambeds often had grazing scars (fig. 9.2; see plate 26) made by stonerollers and were dominated by a low-growth form of cyanobacterial "felts" dominated by *Calothrix* sp. We also observed characteristic grazing scars left by stonerollers on submerged logs where attached periphyton grew (fig. 9.2). In troughs in Baron Fork placed to exclude grazing stonerollers, low-growth form cyanobacterial "felts" on clay tiles and natural cobbles were rapidly overgrown by filamentous diatoms several centimeters long (Power et al. 1988). When clay tiles with these diatom growths were exposed to stonerollers, the fish swarmed to the new resource and rapidly stripped tiles bare, and cyanobacterial felts were reestablished within days (Power et al. 1988).

Power, Stewart, and WJM also showed in stream surveys throughout eastern Oklahoma and northwest Arkansas that densities of bass were important, and identity of the bass also was important (Matthews et al. 1987). Smallmouth Bass (*Micropterus dolomieu*), which eat minnows but actually consume more invertebrates (Coble 1975), were common in Baron Fork but had little influence on stoneroller distribution within or among pools (Matthews et al. 1987).

The clear water of Baron Fork and a nearby tributary, Tyner Creek, allowed extensive observations of stoneroller behavior. Matthews et al. (1986, 1987) found distinctive size-sorting of stonerollers by depth, with larger individuals generally deeper in midchannel and juveniles in large numbers near stream edges. In warm weather, shoals of adults, estimated at hundreds to thousands of individuals, typically grazed on the streambed, heads upstream against the current. A shoal of busily feeding stonerollers would slowly drift downstream with the current until, nearing the end of the pool, they would make a burst upstream as a group (acting almost like a coordinated "school"), then settle down to resume feeding and drifting slowly downstream as a group until the cycle was repeated. They appeared to effectively cover most of

FIG. 9.2 Grazing scars (indicated by arrows) made by stonerollers on a large cobble in an Ozark stream (*above*) and on a submerged log in Baron Fork (*below*).

the streambed in this fashion, giving rise to the widespread occurrence of the dark, low-growth form of "felts" on flatter stones, described in Power et al. (1988), or to bright green and highly productive "new" algae on surfaces of cobbles. Stonerollers sometimes remained in one place on large boulders, however, and continued to scrape or swipe at an exposed grazing edge in the algae on the stone surface. Feeding intensively in one place may help keep periphyton free of inorganic silt, making it more available as food, reminiscent of Power's (1984b) observation that armored catfish kept grazed areas free of unusable organic sediments.

After the findings about the impacts of stonerollers on algae, the University of Oklahoma Biological Station (UOBS) research group, including at various times Mary Power, Art Stewart, Bret Harvey, Beth Goldowitz, David Partridge, Bob Cashner, Fran Gelwick, Tom Gardner, Marsha Stock, Caryn Vaughn, and WJM, found stonerollers important to a variety of ecosystem properties. Matthews et al. (1987) summarized our findings through about 1985, and Matthews (1998, 583–589, 611–616) reviewed findings through the mid-1990s. In Brier Creek, Stewart (1987) found that nutrients could be limiting to growth of benthic algae on stream substrates, but that even when N-P-K fertilizer was added, grazing by stonerollers at natural densities could "outrun" periphyton growth. Vaughn et al. (1993) showed experimentally that stonerollers had negative effects on crayfish (*Orconectes virilis*) production, presumably through resource monopolization by the fish, but that stonerollers indirectly enhanced snail (*Physella virgata*) production. Vaughn et al. (1993) also discovered that snails in artificial streams with stonerollers delayed reproduction and grew larger, that is, displayed a shift in life-history traits. Gardner (1993) showed in experimental streams that grazing by stonerollers resulted in significantly more FPOM than in control streams and less of the UPOM fraction, thus finding that stonerollers had the potential to alter particle size distribution in streams. Gardner also showed in unpublished data that grazing by stonerollers enhanced uptake of dissolved organic carbon (DOC) by algae.

From March to June 1988, Gelwick and Matthews (1992) conducted a two-part experiment on broader ecosystem effects of stonerollers in whole pools in Brier Creek. Plastic mesh fences across shallow, natural riffles isolated eight natural stream pools in the reach previously studied by Power and Matthews (1983) and Power et al. (1985). Four pools with large schools of stonerollers were "grazed," and four with few to no stonerollers were "nongrazed." Pools grazed by stonerollers developed lower ash-free dry mass (AFDM, a measure of algae biomass), lower net primary productivity (NPPR) per unit area but higher NPPR per algal biomass, more blue-green algae, less green algae, lower bacterial density, smaller fractions of POM, and higher invertebrate densities (Gelwick and Matthews 1992). Nongrazed pools grew more long strands of filamentous algae, forming columns or floating mats. The reduction of algae by stonerollers was analogous to a "lawn mower" effect, whereby the productivity of attached algae grazed by stonerollers is like a mowed lawn that may be more productive per unit biomass than an unmowed lawn that has a large standing crop of senescent material.

After the main experiment, a changeover showed that effects of stonerollers were rapid when newly introduced to pools, but that in pools from which they were removed they can have a substantial legacy effect (Gelwick and Matthews 1992). The significance of the study by Gelwick and Matthews (1992) was that it showed for the first time that stonerollers not only had strong effects on standing crops or kinds of algae, but also that these differences had indirect effects through many of the approximately 30 measured variables all related to ecosystem function in Brier Creek (see the appendix to this chapter).

Gelwick et al. (1997) did a conceptually similar experiment in Baron Fork to ask if patterns detected in the smaller Brier Creek would apply in the very different environment of a larger, stony-bottomed, clear-water Ozark stream. The size of Baron Fork precluded fencing off whole pools, so an inclusion/exclusion experiment was set up using an arrowhead pen design that Mary Power and WJM had piloted earlier. In July 1988, with help from Scott and Amy Matthews and Irene Camargo, 12 pens were built, each with one side open to fish and one side closed (fig. 9.3; see plate 29). Unglazed clay tiles were scattered throughout pens to provide surfaces for measuring algae growth, AFDM, and primary productivity. We could not control which fish entered and grazed in the open sides of the pens but attributed fish effects to stonerollers and Ozark Minnows, which made up the majority of fish in that reach of Baron Fork.

During the experiment, sides of pens with fish excluded grew large standing crops of filamentous algae, which also accumulated a heavy silt load and senesced and sloughed away after about 15 days (fig. 9.3; see plate 30). In contrast, grazed sides of all pens consistently had only a low growth of bright-green algae (fig. 9.3; see plate 30) that was highly productive per unit biomass. In this experiment (Gelwick et al. 1997, fig. 2), AFDM per area and NPPR per area were significantly lower, but NPPR per biomass was higher in grazed sides of pens. Grazing by fishes resulted in lower standing crop of algae and lower net primary productivity per square meter but also resulted in grazed algae that was more productive per unit of biomass, attributed to the removal of the older overstory of algae by fishes, leaving low-growth forms of young, highly productive (and very green) attached algae. Closed pens had a much higher percentage of long-strand diatoms, whereas in grazed pens the percentages of low-growth form of cyanobacteria and green algae were higher. There was more total benthic POM (BPOM) and a greater MPOM fraction in closed pens but more UPOM in open pens, suggesting that grazing by fish indirectly helped to move POM through stages of breakdown. Overall, for 19 of 23 ecosystem properties measured in the experiment, there was a significant "fish" effect, for 8 there was a "time" effect, and for 6 there was a fish x time effect (Gelwick et al. 1997, table 1).

At the end of the experiment, pens were opened so that fish (87% were stonerollers) could access the sides with accumulated algae growth. Newly arrived fish consumed algae with an average removal rate of AFDM of 4.4 $g/m^2/hr$, which, like Stewart (1987) found in Brier Creek, slightly exceeded the accumulation rate of algae on protected tiles, calculated at 3.7 $g/m^2/hr$. In both Brier Creek and Baron Fork it appeared that stonerollers could consume filamentous algae as fast as it could grow, leaving only the very low-growth form of taxa adherent to stone surfaces (Gelwick et al. 1997). Grazing by stonerollers does not scrape algae bare, and even within grazing scars there were cropped diatoms in a microscopic "grazing lawn" (see scanning electron microscope micrographs in Gelwick et al. 1997, fig. 6).

FIG. 9.3 *Above*, Arrowhead-shaped pens used by Gelwick et al. (1997) in Baron Fork, looking upstream, with access by grazing fishes allowed on the left-hand side. *Below*, Effects of grazing fishes (stonerollers and Ozark Minnows) in one pen on day 16 (30 July 1988) of an experiment in which fish were excluded from the left side of pen, where there is a heavy accumulation of silt and algae, and algae on some tiles has begun sloughing, whereas the tiles and natural substrates on the right side are "clean." See also plates 29 and 30.

During the Baron Fork experiment (Gelwick et al. 1997), we estimated from snorkel surveys and videotaping that densities of adult stonerollers ranged from 7 to 20/m² and juveniles were about 60/m², averaged over whole pools (Gelwick et al. 1997). A series of video recordings (Matthews and Gelwick, unpub. data) also showed stonerollers removing algae from tiles that had been protected from grazing and allowed to accumulate algae. Adults and juveniles, depending on where tiles were placed in the stream, would swarm to the newly offered tiles and strip them bare within minutes. Calculations based on timed removal rates and density of fishes on tiles indicated that juvenile stonerollers, because of their large numbers, consumed more of the algae from tiles than did adults. On freshly exposed clay tiles we found a mean density of 12 adults compared to 40 juveniles per tile, actively feeding on newly available algae. We calculated a removal rate of 0.15 g/hr/adult, compared to only 0.08 g/hr/juvenile, but when multiplied by densities and extrapolated to potential rate of removal of AFDM per square meter of streambed, the results suggested that on average, juveniles at natural densities could remove 111 g/m²/hr, compared to 58 g/m²/hr for adults.

Now, as a speculation, do stonerollers actively "farm" the algae like some reef fishes, keeping silt from accumulating, preventing accumulation of a senescent overstory, and leaving highly productive "grazing lawns" (Gelwick et al. 1997)? The algae that is evident in stoneroller-grazed microhabitats almost always appears bright green, and measurement both in Brier Creek and Baron Fork indicated that this grazed algae is more productive per unit biomass than older, senescent algae that accumulates if there is no grazing pressure. Some reef fishes, particularly damselfishes, are well known to actively "farm" algae (Hata and Kato 2004). Damselfish can, through a variety of activities, arrest development of the algae community in reef habitat so it remains dominated by palatable algae (Ceccarelli et al. 2011). It is tempting to think (albeit with no proof) that where stonerollers are dense and grazing is active, the low-growth forms of early-stage, typically more productive algae that is maintained by their activity could be either (1) better for stonerollers nutritionally or (2) more easily consumed by the scraping or swiping behavior that they most often exhibit (Matthews et al. 1986).

Other Fishes and Complex Ecosystem Effects

Algivorous Fishes

Other than stonerollers, there are not many primarily algivorous fish species in the central United States. About the only two are Southern Redbelly Dace (*Chrosomus erythrogaster*; *Phoxinus erythrogaster* in previous literature) and Ozark Minnow, both primarily eating diatoms or other periphyton (Miller and Robison 2004). Ozark Minnows were implicated to have grazing effects in the Baron Fork by Gelwick et al. (1997) but were only a fraction of the fish community, and their specific effects in ecosystems have not been studied by us. WJM and Gelwick attempted one experiment to compare grazing effects of stonerollers, Southern Redbelly Dace, crayfish, and snails, but many dace died during the experiment, leaving us unsure of the reliability of results, so we never published those data.

However, Keith Gido, whose PhD under WJM's direction was a study of nutrient effects of benthic fishes in Lake Texoma (Gido 2002), joined the faculty at Kansas State University, where he built a set of experimental streams at the Konza Prairie Biological Station (KPBS) matching those at the UOBS (Matthews et al. 2006). Gido and his students have studied effects of Southern Redbelly Dace (Gido et al. 2010) in the KPBS mesocosms and in nearby Kings Creek on the KPBS. Bertrand and Gido (2007) showed that moderate densities of dace in Kings Creek pens or in mesocosms transiently reduced algae filament length and mean size of POM, but without affecting algal biomass or NPPR. Also in the Gido lab, Bengston et al. (2008) showed that dace lowered algal height and chironomid density in autumn, but not in springtime. Murdock et al. (2010) found in Gido's experimental streams that dace had only minimal effect on algal assemblages or ecosystem functional rates during recovery from an experimental drought. And one other experiment in the Gido stream system (Kohler et al. 2011) confirmed the earlier finding by Bertrand and Gido (2007) that dace density had little effect on algae biomass. Based on all available information, it would appear that the ecosystem effects of dace are less than the effects of stonerollers.

Insectivorous Fishes

In the artificial streams (Matthews et al. 2006) at the UOBS, insectivorous Red Shiner (*Cyprinella lutrensis*) density caused a linear, threefold increase in benthic primary production during the first 35 days of an experiment, presumably by feeding on winged chironomids that alighted on the water surface, and transferring nutrients to benthic algae by excretion (Gido and Matthews 2001). But later, up to 203 days, Red Shiner effects disappeared, likely as a result of autumnal input of leaf litter or other seasonal effects, suggesting that the effects of water-column minnows might be of importance only when nutrients are limiting. Hargrave (2005) compared ecosystem effects of several insectivorous species, as described below in the section on complex pathways.

Piscivorous Fishes

We have previously described the importance of piscivorous Largemouth Bass and Spotted Bass (*Micropterus punctulatus*) in a trophic cascade in Brier Creek (Power and Matthews 1983; chapter appendix), in which presence of bass in stream pools excluded Central Stonerollers, thereby enhancing growth of attached algae in bass pools (Power and Matthews 1983; Power et al. 1985). Harvey (1991) showed that Largemouth Bass within stream pools can strongly regulate microhabitat and feeding activity of small sunfish, reducing their negative effects on larval fishes. We also have documented the effects of intermediate-sized "mesopredators" of several sunfish (*Lepomis* spp.) species as potentially responsible for the demise of Red Shiners in Brier Creek (Matthews and Marsh-Matthews 2007; Marsh-Matthews et al. 2011, 2013) and the decline or loss of three other minnow species (raw data in Matthews et al. 2013). One exception to the thinking that "sunfish reduce minnows" in Brier Creek has been the continued success of Bigeye Shiner (*Notropis boops*) in the stream in spite of increases in sunfish mesopredators. We attribute this to the microhabitats most used by Bigeye Shiners, generally not present in parts of pools where sunfish or bass hunt (Marsh-Matthews et al. 2013), and possibly to their ability to escape predators by a fast-start response (James Cureton, pers. comm. of unpub. data).

More Complex Pathways and Interactions among Species

Chad Hargrave's (2005) doctoral dissertation with WJM tested direct and indirect pathways by which algivores, surface- or water-column-feeding insectivores, and benthic insectivores could affect ecosystems (chapter appendix). First, in the UOBS mesocosms he tested effects of six fish species with different feeding modes, including Central Stoneroller, Orangebelly Darter (*Etheostoma radiosum*), Brook Silverside (*Labidesthes sicculus*), Golden Redhorse (*Moxostoma erythrurum*), Striped Shiner (*Luxilus chrysocephalus*), and Rocky Shiner (*Notropis suttkusi*), on periphyton biomass, benthic invertebrate density, and BPOM, finding up to 13-fold differences among species. Hargrave then calculated the relative effects of 12 species on productivity of benthic algae, as the slope of a regression line from a density of zero to the density at which each species had maximum effect in trials in the UOBS stream mesocosms (Hargrave 2005, chap. 3, table 1). We use these relative effects in the second part of this chapter to estimate how important these particular species might actually be in stream ecosystems in our region.

Hargrave (2006) next compared pathways or effects on primary productivity for a surface-feeding insectivore (Western Mosquitofish, *Gambusia affinis*), a benthic and water-column omnivore-insectivore (Bullhead Minnow, *Pimephales vigilax*), and a benthic insectivore (Orangethroat Darter, *Etheostoma spectabile*). Orangethroat Darters enhanced algal NPPR in a trophic cascade by reducing algae-grazing invertebrates, releasing algae from grazing pressure; Western Mosquitofish feeding at the surface enhanced growth of algae by increasing nutrient translocation to the benthos; and the benthic feeding of Bullhead Minnows disturbed substrates and enhanced algal production by releasing nutrients (Hargrave 2006). Hargrave (2009) also showed, in randomly selected combinations of one to six fish species, drawn from a pool of twelve potential species, that both species richness and species identity had strong effects on benthic algal PPR. There was a significant linear effect of number of fish species on PPR, but identity was also important, as randomly selected combinations of species that included Sand Shiner (*Notropis stramineus*), Green Sunfish (*Lepomis cyanellus*), Brook Silverside, and Common Carp (*Cyprinus carpio*) resulted in more enhancement of PPR. One important additional idea to come from Hargrave's (2005) dissertation was the calculation of synergistic effects of various combinations of species on attached algal biomass, relative to their potential additive effects predicted from single species experiments, finding apparent synergistic enhancement of algal biomass in 90% of the mesocosms having two or more species. Finally, Hargrave (2005) did an experiment in enclosures in Brier Creek with Blackstripe Topminnow (*Fundulus notatus*), Longear Sunfish (*Lepomis megalotis*), or Orangethroat Darter as fish treatments. All three species treatments resulted in enhanced PPR relative to fishless controls. This study provided evidence that ecosystem effects demonstrated for fishes in experimental mesocosms may translate into effects within natural streams.

A group of our graduate students, known as the FishLab, carried out an additional experiment that underscores the importance of indirect effects that fish of one trophic group can have on another. In the UOBS experimental streams, consumption of grazing invertebrates by Orangethroat Darters allowed increased growth of benthic algae, which was then consumed by Central Stonerollers (Hargrave et al. 2006). After 83 days in this experiment, stonerollers with darters had greater gut fullness, condition, and growth than stonerollers without darters.

How Far Downstream Are Fish Effects Important?

By the late 1980s the research group at the UOBS had shown that stonerollers could have significant ecosystem effects where the fish were grazing. But we did not know how far downstream a "zone of influence" of algivorous or insectivorous fishes might extend. If a large group of minnows is feeding in one pool, do the effects extend to the next pool downstream or elsewhere in the system? As one example of fish-related transport of materials, Gelwick and Matthews (1997) observed that in Brier Creek pools without stonerollers, attached algae grew long, then senesced and sloughed from substrates, resulting in large mats of detached algae drifting downstream to other pools or stranding in riffles.

David Partridge was the first in the UOBS group to quantify downstream transport of organic matter caused by stonerollers, first in a natural stream, then by experiments in flowing troughs (Partridge 1991). Partridge (1991) showed that POM caught in drift nets in Baron Fork was correlated with numbers of stonerollers grazing within 1 m upstream. But POM transport in his fieldwork also was related to depth or current speed, so it was hard to tease fish effects from abiotic effects. His follow-up laboratory trough experiment, over six different weekly trials, suggested that stoneroller feeding could result in an overall increase in export of CPOM (consisting mostly of strands of filamentous green algae, *Rhizoclonium*) and in export of MPOM at times (Partridge 1991). Stonerollers are sloppy eaters! The work by Partridge suggested that grazing fishes could have effects downstream, as well as in the vicinity of feeding shoals.

A trial in autumn 2000 in the UOBS experimental streams (Matthews et al. 2006) by Keith Gido and WJM was designed to compare the downstream effects of stonerollers to those of the insectivorous Blacktail Shiner (*Cyprinella venusta*). Fish were confined to the uppermost pool in each three-pool unit. By day 30, the most downstream pools in two of the three stoneroller units had much taller algae than in any of the downstream shiner or control pools. But snails, apparently stocked along with algae from Brier Creek, reached high densities during the experiment in the Blacktail Shiner units, eliminating our ability to determine direct shiner effects. Blacktail Shiners might have indirectly enhanced snail growth or densities if they did enhance production of algae as food for snails (Keith Gido, pers. comm.), but we had no data to test that idea directly.

In light of the trends shown by Partridge (1991) and by Gido and Matthews (2001; unpub. data), WJM did a pilot study in the UOBS mesocosms from Autumn 2002 to Spring 2003 to evaluate downstream effects of a more complex fish community (chapter appendix). The overall question was whether, in any particular reach of stream, output of particulate organic matter would be greater with than without fishes present. Two parallel rows of experimental stream units, each 22 m long, had six pool units connected by five riffle units. In each row, water was pumped from a downstream footbox to an upstream headbox, where water passed through coarse and fine filter materials and a 363-micron mesh drift before reentering the upstream pool. We seeded the streams with a slurry of filamentous algae from Brier Creek, added fertilizer, and allowed a month for colonization by flying insects and algal growth before introducing fish. Before fish were added, export of POM was sampled from the drift nets at the lower end of each stream on three dates. On 22 November, we seined all possible fish from a 25 m reach of Brier Creek and added all fish to one of the two streams, randomly chosen, including: 271 Bigeye Shiners, 149 Blackstripe Topminnows, 79 Orangethroat Darters, 63 Central Stonerollers including large adults, 36 juvenile Longear Sunfish, 5 juvenile Green Sunfish, 1 juvenile Largemouth Bass, and 1 adult Bluegill (*Lepomis macrochirus*). WJM often observed that actively feeding stonerollers disturbed and suspended sediments and that Orangethroat Darters suspended sediments into the water column by flicking their pectoral fins on the substrate as they "darted" about in the units. In four of the five samples during this part of the experiment, there was more export of organic, inorganic, and total materials in the "fish" stream than in the "no-fish" stream. Then a changeover was done by moving fish between streams, and after the changeover the pattern of export was reversed. Both the main experiment and the changeover showed that substantially more total materials, including POM and feces and inorganic silt, were transported downstream out of a stream reach with than without fish.

Recent work by Martin et al. (2016) in Keith Gido's laboratory at Kansas State University further reinforces the idea that algae-grazing fish can have downstream influence. In the KPBS outdoor stream mesocosms, Martin et al. found in units stocked with Central Stonerollers as the only fish species, greater algae filament length and more benthic chlorophyll-a in downstream riffles than in the pools where the fish grazed. Martin et al. also found fewer floating mats of filamentous algae and lower floating chlorophyll-a, in mid- and downstream pools of their experimental streams when stonerollers were alone or in combination with dace. Greater filament lengths in riffles below stoneroller pools and more floating mats of algae in downstream pools both suggested that the effects of grazing fishes can be transferred through a "zone of effect" substantially downstream of the actual site of the fish activity.

Another potential pathway for downstream effects of fishes is through transport of feces, which store organic matter (Wotton and Malmqvist 2001). Stonerollers create copious amounts of characteristically C- or S-shaped fecal particles readily identifiable on the streambed. Our observations in Brier Creek and in the UOBS experimental streams suggest that at current speeds <7 cm/s, stoneroller feces accumulate in large quantities in deeper and slower-flowing parts of the pools where they are produced. At higher current speeds the feces are likely to be transported downstream. They may contain live cells and could serve as potential propagules for establishing algae in new areas. Art Stewart (pers. comm.) indicated that his microscopic examination of stoneroller feces suggested that some cells passed through guts alive, based on what appeared to be intact cell walls and chloroplasts. But he also emphasized that a definitive finding of live cells with growth potential would require more detailed trials than he performed, for example, using an improved protocol in an attempt to culture pure cells from stoneroller feces under controlled conditions. But if the notion of live cells is correct, it is possible that stonerollers in a pool could contribute algae to downstream pools under flow conditions sufficient to entrain feces, then deposit them in slow-flow microhabitats. Mary Power (pers. comm.) noted that this could select for the establishment of less digestible algae, perhaps favoring cyanobacteria over diatoms.

Finally, there is evidence from work in Brier Creek that fishes other than algivores can have downstream effects of importance to other fishes. David Gillette, in his PhD dissertation under WJM's direction, carried out a field study that showed that physical properties of upstream riffles and densities of benthic, insectivorous Orangethroat Darters on those riffles had substantial influence on density or foraging success of drift-feeding Bigeye Shiners in pools below the riffles (Gillette 2007, chap. 1). Specifically, foraging success and body condition of Bigeye Shiners were positively related to densities of some of the important drifting prey taxa, particularly amphipods, which were greater from riffles with higher densities of Orangethroat Darters (Gillette 2007). Insectivorous fishes as well as algivorous fishes have potential downstream effects in stream ecosystems.

Export of Energy into Terrestrial Ecosystems

There is increasing emphasis in ecology on linkages across ecosystems (e.g., terrestrial and aquatic). Jeff Wesner (2010a), in his doctoral dissertation under WJM, showed the importance of emerging aquatic insects to terrestrial food webs. Wesner (2010b) showed in field surveys that emergent aquatic insects were an average of 41% of the abundance and 34% of the biomass of winged insects in terrestrial environments near three natural streams. He then conducted two mesocosm experiments to test effects of fish predation on insect emergence. Red Shiners and Orangethroat Darters in monoculture decreased aquatic insect emergence more than 50%, mostly by their reduction in emergence of dragonflies (Wesner 2010c). In a second experiment in the UOBS mesocosms, Wesner (2012) found in trials with zero, one, two, or three insectivorous fish species, including Orangethroat Darter (benthic insectivore), Red Shiner (water column and surface feeder), and Western Mosquitofish (surface feeder) that the three-species treatment resulted in reduction in emergence of predatory dragonflies by 67% and chironomid emergence by 55%. The reduction in chironomid emergence had a cascading negative effect on the abundance of tetragnathid spiders, which feed primarily on emerging chironomids, with 75% lower spider abundance over pools that contained all three species of fish (Wesner 2012). The combination of Wesner's three studies confirmed the potential for insectivorous fish to control emergence of some of the insect taxa, and thus densities of a terrestrial predator.

In a final experiment in whole pools of Brier Creek, Wesner (2013) redistributed fish (mostly four species of sunfish and Bigeye Shiners) to create a 13× density gradient of insectivorous fishes. Enhanced fish densities resulted in lower standing crops of aquatic insect larvae in the stream, but not in any less insect emergence. During this whole-pool experiment there was substantial drying and shrinking of isolated pools, and the study was performed later than the peak of summer insect emergence. Wesner (2013) concluded that fish effects on aquatic-to-terrestrial subsidies might only occur during brief periods, depending on seasonality and variable conditions in a stream.

Recently, Wesner (2016) showed in a meta-analysis of numerous studies that although fish may have only weak effects on standing crops of benthic aquatic invertebrates, they can have strong effects on emergence of adults, presumably by focusing predation on insects that are in vulnerable states of emergence, effectively "running a gauntlet" of predators, in Wesner's terminology. In our own research we have what may be one example of this effect. Matthews et al. (1978) found that Duskystripe Shiners (*Luxilus pilsbryi*) in Piney Creek often fed heavily on chironomid pupae, which sometimes were the most frequently occurring item in stomachs. In contrast to the longer larval stage of chironomids, the pupal stage lasts only a few hours to a few days, typically with the pupa rising from the benthos to the surface to complete adult eclosion, and it is during this time that they are most vulnerable to predation (Oliver 1971). Thus the conclusion from Wesner's meta-analysis and the finding by Matthews et al. (1978), of a preponderance of chironomid pupae in the diet of a sight-feeding minnow, seem consistent with the life history of aquatic insects.

Ecosystem Effects of Fishes in Real Streams?

In this section we extrapolate from what is known about potential effects of fishes from field or lab observations, mesocosm experiments, and a few trials in natural streams to consider the likelihood that fish actually do affect real stream ecosystems. We address the ways that the distribution and abundance of individual species, temporal change in species or functional groups, or interactions among groups can affect their roles in natural streams. We then address questions about where or when fish are likely to have substantial effects.

Occurrence and Abundance versus Potential Ecosystem Effects

Density of fish and their per-capita effects are important determinants of potential impact in ecosystems (Gido and Matthews 2001; Hargrave 2005). As a general predictor, we can match the potential effects measured in mesocosms for individual species with their likelihood to be present or abundant. Hargrave (2005, chap. 3, table 1) provided estimates of the per-capita enhancement of benthic chlorophyll-a (surrogate for algal PPR) for 12 common midwestern fishes. His estimates were based on single-species trials under varying conditions, so the numbers expressing relative effects should only be considered approximations (Hargrave, pers. comm.). But even if only approximate, his estimates provide a template for thinking about which species could have greater or lesser impact. If a species has high potential impact as measured in experimental mesocosms but is rarely encountered or in low abundance in natural streams, its real-world impacts may be of limited importance.

For the species tested in mesocosms (Hargrave 2005), we asked which were most often encountered or abundant in our large spatial data set for the upper Red-Washita River basin in Oklahoma. First we ranked seven of Hargrave's water-column omnivores and insectivores for their potential effect on benthic algae, including, from greatest to least effects per capita: Brook Silverside, Striped Shiner, Sand Shiner, Green Sunfish, Western Mosquitofish, Rocky Shiner, and Red Shiner (table 9.1). The effects of all these water-column/surface-feeding species were hypothesized to occur through consumption of terrestrial insects at the water's surface, translocating nutrients by excretion into the water column, and hence to attached algae. The strongly surface-feeding Brook Silverside had the greatest effect in mesocosms (relative score = 12.0), and Red Shiner, although shown as potentially important to algal NPPR by Gido and Matthews (2001), had a relative score of only 1.0, so there was a wide range in potential effects among these species (table 9.1).

Some of the insectivorous species ranked highly for potential effects by Hargrave (table 9.1) were not widespread or typically abundant in streams of our region. In our 143 quantitative collections at separate sites in the Red-Washita basin, the number of occurrences and total abundances (in parentheses) of the seven water-column insectivore/omnivores in Hargrave (2005) were: Brook Silverside (30–610), Striped Shiner (11–165), Sand Shiner (30–1173), Green Sunfish (71–940), Western Mosquitofish (95–4730), Rocky Shiner (16–285), and Red Shiner (53–16,282) (table 9.1). All comprised a third or more of the community in at least one sample (maximum ecosystem effect, EE; table 9.1), so they could be relatively abundant and have substantial effects in some locations. For all except Red Shiner, however, the median ecosystem effect (table 9.1) was based on relative abundances of <7%, suggesting they typically were at too low a density to have major impacts. In contrast, although the potential per-capita impact for Red Shiner was the lowest for any of the water-column insectivores tested by Hargrave (2005), they were the most widespread and abundant, comprising more than 30% of the community in half of the samples where they occurred (median EE; table 9.1). The collective impact (per capita × abundance) of Red Shiners could be greater than that of some of the other water-column insectivores with higher per-capita effects but that are scarce or in low abundance. As another suggestion of their potential importance in ecosystems, Red Shiners were present in 50 of 65 collections we

Table 9.1. Potential ecosystem effects in real streams for fish species in Hargrave (2005), based on their relative effect on algae growth in his mesocosm studies, multiplied by their relative abundance in our 143 collections in the Red River drainage. "Score" is the slope of a regression line relating per-capita algal biomass to fish density (Hargrave 2005). "Sites" and "Total" are the number of sites where a species occurred and their total numbers in our 143 field samples. Ecosystem Effect (EE) values are the relative abundance multiplied by "Score" and provide an estimate of the range of potential effects of each species in the region, at field sites with the highest ("Maximum EE"), lowest ("Minimum EE"), and median ("Median EE") relative abundance for each species.

Scientific Name	Common Name	Score	Sites	Total	Maximum EE	Minimum EE	Median EE
Labidesthes sicculus	Brook Silverside	12	30	610	0.411	0.004	0.039
Cyprinus carpio	Common Carp	8	10	32	0.024	0.001	0.003
Etheostoma spectabile	Orangethroat Darter	4	43	1470	0.714	0.001	0.058
Pimephales vigilax	Bullhead Minnow	4	45	1743	0.333	0.001	0.031
Luxilus chrysocephalus	Striped Shiner	3.5	11	165	0.326	0.002	0.019
Lepomis cyanellus	Green Sunfish	3	71	940	0.642	0.000	0.019
Notropis stramineus	Sand Shiner	3	30	1173	0.486	0.001	0.018
Gambusia affinis	Western Mosquitofish	2.5	95	4730	0.860	0.002	0.065
Notropis suttkusi	Rocky Shiner	2.2	16	285	0.483	0.004	0.040
Etheostoma radiosum	Orangebelly Darter	1.8	37	1072	0.746	0.003	0.038
Campostoma anomalum	Central Stoneroller	1	55	5039	0.960	0.001	0.143
Cyprinella lutrensis	Red Shiner	1	53	16,282	0.952	0.002	0.314

made throughout the Midwest from Iowa to south Texas in June 1995 and ranked highest in abundance in half of the collections where they occurred (Marsh-Matthews and Matthews 2000a).

As another example, Striped Shiner and Green Sunfish had similar potential impacts on algal productivity (3.5 and 3.0, respectively), and their median ecosystem effects were the same (table 9.1). But Green Sunfish were more often encountered in the upper Red-Washita basin collections (71 vs. 11 occurrences; table 9.1). If we take a crude estimate of potential "regional impact" on primary productivity, Green Sunfish ($3.0 \times 71 = 213$) would have much more potential to affect stream ecosystems across the region than Striped Shiner ($3.5 \times 11 = 38.5$).

From a similar perspective, Hargrave (2005) showed that two benthic omnivores fed by disturbing sediments (Bullhead Minnow and Common Carp had relatively strong effects on algal PPR, with index values of 4.0 and 8.0, respectively [table 9.1]).

But in the data set for the Red-Washita River basin, occurrence and total abundance were for Bullhead Minnow (45–1743) and for Common Carp (10–32) (table 9.1). Common Carp, the species with higher potential effect on algae production by releasing nutrients from the sediments (Hargrave 2005), was much less common than the more widespread and potentially abundant Bullhead Minnow.

The above rough estimates of their potential regional importance in ecosystems gloss over many details about the biology of these species. But to the extent that frequency of occurrence, abundance, or relative densities are important, future research on nutrient translocation effects by insectivorous water-column fishes or benthic omnivores might best be focused on widespread or highly abundant species like Red Shiner, Green Sunfish, Western Mosquitofish, or Bullhead Minnow. And one other perspective needs consideration. There is a growing body of theory (McCann et al. 1998; McCann 2000; Rooney and McCann 2012) that suggests that instead of being relatively unimportant, weak interactions can have strong stabilizing effects in food webs and ecosystems. Even species for which Hargrave (2005) calculated comparatively weak effects might be important, individually or in aggregate, in ecosystem dynamics.

Consistency in Potential Ecosystem Effects of Species and Functional Groups

The appendix to this chapter shows the 22 midwestern fish species for which one or more ecosystem effects have been identified. If changes in local abundances or densities do relate to changes in their ecosystem effects, then it is important to consider long-term dynamics of individual species. Species with persistent high densities might have consistent effects in stream ecosystems (e.g., persistence of stoneroller-grazed algal "felts" in Ozark streams; Power et al. 1985), whereas stream ecosystems might be only transiently affected by fish species with flashy densities or boom-and-bust years (like Western Mosquitofish; Matthews and Marsh-Matthews 2011). From that perspective, we examined the variation in abundances of species listed in the chapter appendix for our three longest data sets: global Piney and Brier Creeks seining, and snorkeling data summed for the 14 pools in Brier Creek.

Piney Creek was classified in chapter 3 as a Cyprinid-Fundulid stream. Ten insectivorous or algivorous species listed in the chapter appendix, mostly minnows and topminnows, are moderately to highly abundant in Piney Creek. *Micropterus* basses and Green Sunfish (chapter appendix) also occur but are scarce in Piney Creek, so we did not consider them here. Table 9.2 below shows variation in occurrences and raw abundances, summed over all 12 sites, for the species, listed in the chapter appendix, in our collections from August 1982 to July 2012.

For Piney Creek algivores (table 9.2), abundances of Central Stonerollers (CAMANO) and Ozark Minnows (NOTNUB) have varied widely, whereas abundance of Southern Redbelly Dace (CHRERY) has increased in recent surveys, but mostly at one site (Matthews and Marsh-Matthews 2016). The fewest algivores occurred in April 1983, only months after a devastating flood. But after that, numbers of algivores

Table 9.2. Number of occurrences in seining surveys of Piney Creek and Brier Creek, mean abundance, minimum abundance, maximum abundance, and coefficient of variation (CV) for fish in chapter appendix

	Occurrence	Mean	Minimum	Maximum	CV
Piney Creek					
ALGIVOR					
CAMANO	9	849.0	306	1974	0.644
NOTNUB	9	649.7	150	1299	0.718
CHRERY	8	47.2	0	241	1.607
SF&WC INV					
LEPMEG	9	210.9	57	505	0.753
LUXCHR	9	89.7	7	258	0.891
NOTBOO	9	625.3	174	1716	0.989
LABSIC	8	12.1	0	54	1.446
GAMAFF	5	96.3	0	415	1.496
BENTHIC INV					
ETHSPE	9	358.1	35	688	0.627
MOXERY	7	12.1	0	55	1.398
Brier Creek					
ALGIVOR					
CAMANO	17	631.4	75	2194	1.003
SF&WC INV					
NOTBOO	17	536.5	48	1399	0.713
CYPVEN	14	33.9	0	91	0.848
PIMVIG	13	8.0	0	24	1.074
CYPLUT	16	65.4	0	232	1.123
FUNNOT	17	107.1	5	703	1.574
NOTSTR	12	15.4	0	119	1.938
GAMAFF	12	30.2	0	248	2.059
BENTHIC INV					
ETHSPE	17	283.1	12	1193	1.038
MOXERY	12	26.9	0	291	2.651
PISCIVORE					
MICPUN	17	189.8	44	451	0.650
MICSAL	17	268.0	34	665	0.776
LEPCYA	17	141.9	2	471	1.165
LEPMEG	13	24.8	0	122	1.398

went up dramatically in summer 1983 and were relatively high in winter 1994, spring 1995, and again in summer 2012. Overall, after recovery from the great flood of 1982, most surveys in Piney Creek showed algivores to be abundant, likely exerting strong grazing pressure in the watershed (Matthews and Marsh-Matthews 2016).

Insectivorous water-column or surface-feeding fishes in Piney Creek that are included in the chapter appendix fluctuated greatly in abundance over four decades (table 9.2), mostly as a result of large but transient increases in Bigeye Shiner (NOTBOO). We included in table 9.2 abundances of Longear Sunfish (LEPMEG), because most of the individuals we collected on Piney Creek were juveniles or small adults, more likely to eat insects than fish. Western Mosquitofish (GAMAFF) greatly increased in abundance in surveys after 2006, but only at one site (Matthews and Marsh-Matthews 2016), and they did not comprise a large proportion of the insectivores in the watershed as a whole.

Benthic insectivores in Piney Creek included in the chapter appendix were Golden Redhorse (MOXERY), a large-bodied sucker, and Orangethroat Darter (ETHSPE). These species are quite different ecologically, with the redhorse disturbing sediments to winnow invertebrates from the gravels in pools and the darter picking aquatic insects from stony substrates in riffles. But both feed on benthic insects, so they were combined. The much more abundant darter makes up most of the total (table 9.2), but one large redhorse feeding actively in a pool might consume as many macroinvertebrates as would be eaten by many darters.

Two measures in table 9.2 are useful indicators of the potential of species to have consistent or long-term impacts in the Piney Creek ecosystem. Six of the Piney Creek species occurred in all surveys from 1982 to 2012 and thus seem most likely to affect the system consistently across time. But another important indicator of consistency of potential impact may be the coefficient of variation (CV) in raw abundances of the species. In table 9.2, species are ranked from lowest to highest CV within functional groups. From that perspective, the most consistent species in the functional groups were Central Stoneroller (CAMANO, algivore), Longear Sunfish (LEPMEG, water-column insectivore), and Orangethroat Darter (ETHSPE, benthic insectivore). Other species worth noting include the algivorous Ozark Minnow (NOTNUB), which was almost as abundant, on average, as stonerollers, and also had a low CV. Note that although Bigeye Shiner (NOTBOO) was the most abundant surface-feeding or water-column insectivore on average, it had a tenfold range in raw abundance, from 174 to 1716 individuals; thus any impact this species has in Piney Creek would probably be highly variable over time. Raw data in Matthews and Marsh-Matthews (2016) show that Bigeye Shiners peaked in abundance in Piney Creek in the 1990s, whereas Longear Sunfish increased in Piney Creek in the last two decades, so examining the CV does not reveal temporal patterns but may nevertheless be a useful indicator of overall variation in abundances.

In chapter 3 we classified Brier Creek as a Cyprinid-Centrarchid community, dominated by minnows, black bass, and sunfish. In Brier Creek seine surveys, 14 of the species listed in the chapter appendix have been common, at least occasionally. The only algivorous fish in Brier Creek is the well-studied Central Stoneroller, which in seine samples varied wildly in abundance in the watershed (mean = 631, range = 75 to 2194 individuals, with CV = 1.00; table 9.2). We will detail its variability in an overview figure with all Brier Creek functional groups variation later in the chapter.

Insectivorous water-column or surface-feeding species in Brier Creek included in the chapter appendix have changed substantially in abundance over time (table 9.2). We did not include Longear Sunfish in the insectivore category for Brier Creek because substantial numbers of the individuals we have collected by seining in Brier Creek have been large enough to be piscivorous. These insectivores in Brier Creek were dominated numerically by Bigeye Shiner (NOTBOO; table 9.2), which occurred in all 17 of our surveys and had the lowest CV (0.713) of any species in this functional group. Thus, to the extent that Bigeye Shiners do have substantial per-capita roles in translocation of nutrients (which has not been studied for this species), it would be a present and consistently abundant component in the Brier Creek ecosystem. The other species in this group that was abundant occasionally was the surface-feeding Blackstripe Topminnow (FUNNOT), but its abundance was highly variable, and its high CV (1.574) suggests that it is approximately twice as variable as Bigeye Shiner in the creek. Other insectivorous species in Brier Creek known from the chapter appendix to have potential ecosystem effects have been relatively minor numerical components of this functional group (table 9.2)

Benthic insectivores in Brier Creek whose ecosystem effects have been studied (see chapter appendix) include only the Golden Redhorse and Orangethroat Darter. Their abundances also varied substantially over time in Brier Creek (table 9.2). Benthic invertivores in Brier Creek are strongly dominated by the riffle-dwelling Orangethroat Darter, but in summer 2001, after a severe drought the previous year, there was a large increase in pool-dwelling Golden Redhorse juveniles (Matthews et al. 2013), for reasons that remain unclear. In that one summer, macroinvertebrates in pools as well as riffles would have had more predator pressure than at most other times, suggesting again that variation in abundances of individual species can result in potential differences in ecosystem effects from fishes.

In Brier Creek, potential piscivores (as adults) are more abundant than in Piney Creek, and the effects of some of these species are well known in trophic cascades affecting algae (Power and Matthews 1983; Power et al. 1985; Matthews et al. 1987) or selectively reducing some water-column minnows (Marsh-Matthews et al. 2011, 2013). We included in this group Green Sunfish and Longear Sunfish, which can be piscivorous (Marsh-Matthews et al. 2013), and many we found in Brier Creek were adults. Abundances of the piscivores listed in the chapter appendix have varied widely in Brier Creek (table 9.2), and their numbers have been dominated at various times by Green Sunfish, Longear Sunfish, or Largemouth Bass (Matthews et al. 2013). But the two species with highest mean abundance (Spotted Bass and Largemouth Bass) are probably the most piscivorous and offer the greatest threat to prey species like minnows, and were less variable in abundance (CV = 0.650 and 0.77, respectively) than the potentially piscivorous sunfish species (table 9.2). We conclude that in spite of variation in raw abundances, there has been a relatively consistent presence of some piscivorous species in the Brier Creek watershed as a whole.

To expand the view of variation in potential ecosystem effects of fishes in Piney and Brier Creeks, we considered all species (not just those in the chapter appendix) in the major functional groups in each watershed (fig. 9.4). Surface-feeding and water-column insectivores (SF&WC Insect) greatly outnumbered algivores and benthic insectivores (Benthic Insect) in Piney Creek (fig. 9.4).

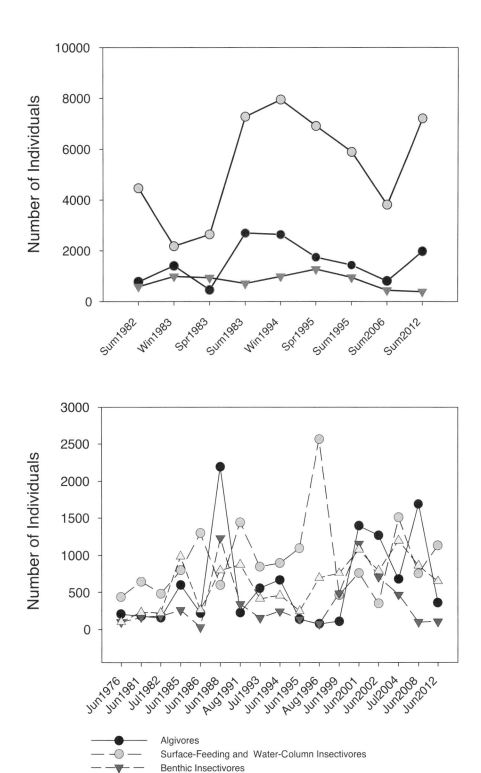

FIG. 9.4 Raw abundance of major functional groups with all species included for Piney Creek (*above*) and Brier Creek (*below*).

What does this all mean for Piney Creek? One argument would be that if inputs of nutrients by excretion from surface invertivory is important, then we might expect to see substantial accumulation of filamentous algae in Piney Creek, as predicted by Gido and Matthews (2001) and Hargrave (2005). But filamentous algae has been found in abundance only twice in our surveys (2006 and 2012) and only at one of the 12 sites (P-6). Alternatively, an indirect pathway might make surface insectivores responsible for increases in abundance of algivorous minnows by production of benthic algae. This is speculative, but beginning in April 1983, when the creek began to recover from a catastrophic flood the previous December (Matthews 1986), through all surveys to July 2012, the numbers of algivorous fishes closely tracked abundances of the surface-feeding and water-column insectivores (fig. 9.4). Patterns in the two trophic groups are so similar that a product moment correlation of their abundances from April 1983 to 2012 was highly significant ($N = 7$; $r = 0.945$; $P = 0.0013$). Although the posited indirect pathway cannot be tested without experimental evidence, empirical evidence spanning 30 years in Piney Creek strongly suggests that this indirect pathway is worth considering. If true, then temporal change in one group (the insectivores) may be closely linked to temporal change in the other (algivores) via translocation of nutrients by the former and enhanced production of food for the latter. (Alternatively, similarities in chronology of these two groups might merely reflect some extrinsic abiotic factor that we are not accounting for, causing abundances of both trophic groups to change concordantly. We have no suggestion about what such a factor might be but cannot rule it out.)

Across all Brier Creek seining surveys from 1976 to 2012, total numbers of individuals in four functional groups, with all species included, varied widely (fig. 9.4). This is in contrast to Piney Creek, where surface or water-column insectivores were the most abundant functional group in every survey. In 11 of 17 surveys in Brier Creek, surface-feeding and water-column insectivores were the most abundant (fig. 9.4). But in four surveys, algivores (Central Stonerollers) outnumbered all other groups, and in two, piscivorous were most abundant. The picture that emerges is one of marked variation from year to year in total abundances for all functional groups, suggesting that any ecosystem effects of these groups could vary substantially over time in Brier Creek.

Because designation of some sunfish and bass as either insectivores or piscivores was ambiguous for seine surveys in Brier Creek, we next considered the distribution of functional groups over time in 23 snorkel surveys of the 14 pools in Brier Creek from 1982 to 2012 in which centrarchids were recorded separately as adults or juveniles (fig. 9.5). The y axis in figure 9.5 is on a Log_{10} scale, because in two surveys there were large numbers of young-of-year or juvenile Central Stonerollers, so that an arithmetic scale for abundance made it difficult to assess patterns in variation in the other groups. In 18 of the surveys, the water-column and surface-feeding insectivores were numerically dominant; algivores (stonerollers) were numerically dominant in four surveys; and piscivores dominated in only one survey. Whether sunfish and bass are

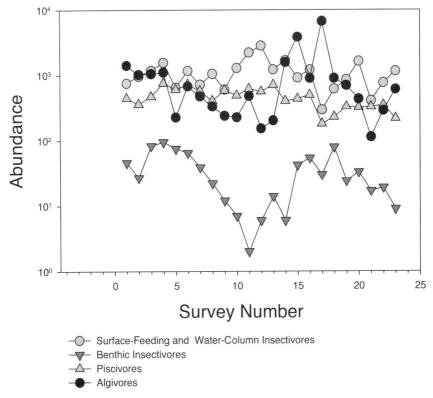

FIG. 9.5 Raw abundance of four functional groups, summed across 14 pools, from snorkel surveys in Brier Creek.

arbitrarily placed in one functional group or another, as they were for Brier Creek seining data, or assigned by size to either insectivore or piscivore groups, the picture that emerges is the same: over time, the relative abundance of different functional groups in Brier Creek varies substantially; thus any effect these different groups have in ecosystem structure or function may also vary widely over time.

Evidence for Effects of Fish or Interactions among Species in Real Streams

The calculations or speculations in the two sections above are for heuristic purposes, as ways of estimating how much the fish species listed in the chapter appendix might affect real stream ecosystems. Among algivores, the ecosystem effects of Central Stonerollers in natural streams are not in doubt. All of our observations (Power and Matthews 1983; Matthews et al. 1987) and experimental work in small or large mesocosms (Partridge 1991; Vaughn et al. 1993; Gelwick and Matthews 1997) and our experiments in real streams in troughs and pens (Power et al. 1988; Gelwick et al. 1997) or whole pools (Power et al. 1985; Gelwick and Matthews 1992) have confirmed a substantial number of effects of stonerollers. Stonerollers (both Central Stoneroller and the closely related Highland Stoneroller, *Campostoma spadiceum*) are highly abundant in many upland streams in our region, and we (Marsh-Matthews and Matthews 2000b) also have found them abundant in some low-gradient prairie streams like the North Fork of the Solomon River, Kansas. So stonerollers, of several ecologically similar species, are probably the taxa with greatest potential to affect stream ecosystems in much of the eastern United States.

Another candidate for importance as an algivore, Southern Redbelly Dace, has been studied in detail by Keith Gido's group at Kansas State University. Some of the observed dace effects in their mesocosms have been seasonal (Bengtson et al. 2008) or transitory, or effects were evident during only part of a flood recovery period (Murdock et al. 2011). But Bertrand and Gido (2007) tested effects of Southern Redbelly Dace in an experiment in four whole pools in Kings Creek to match experiments in mesocosms. At the end of 32 days, chlorophyll-a was 38% greater in the two Kings Creek pools with lower dace density, and after 39 days, algal filaments were shorter in the low-density dace pools. Overall, Bertrand and Gido (2007) considered the ecosystem effects of the dace both in Kings Creek and in their mesocosms to be less than effects that have been attributed to Central Stonerollers. But their comparison of dace effects in mesocosms and in the natural creek suggested similar findings for algal biomass and filament length, so some of the effects for this species in mesocosms may indeed translate to their having effects in real streams.

Even if their local ecosystem effects were strong, Southern Redbelly Dace would have only limited potential for widespread effects. While they can be abundant in small streams, especially spring-fed headwaters, they are rarely found in larger streams. In a total of 509 quantitative samples by WJM, with EMM or others, since 1975 in Oklahoma, western Arkansas, and southern Kansas (with repeated samples at some sites), Southern Redbelly Dace occurred in only 28 collections. And more than 100 dace were collected in only 5 samples (ranging from 129 to 249 individuals each), all at small, spring-fed headwater sites. It would seem that although Southern Redbelly Dace can affect some ecosystem properties, their potential to do so is limited to relatively few sites where they do occur in large numbers.

Among insectivorous species, Gido and Matthews (2001) first suggested the potential for surface-feeding minnows to enhance benthic algae by nutrient translocation from a mesocosm study of Red Shiners. But evidence that Red Shiners actually have measurable effects in real streams by nutrient translocation per Gido and Matthews (2001) is lacking, as we detail below. There is evidence of ecosystem effects of other insectivorous fishes in natural pools of Brier Creek in the doctoral dissertations by Chad Hargrave (2005), David Gillette (2007), and Jeff Wesner (2010a), described earlier in this chapter. Research by Hargrave (2005), Gillette (2007), and Wesner (2010a,b,c) showed for several species of insectivores of different functional feeding groups that at least some effects found in mesocosm experiments could be repeated in a natural stream, but that fish effects in a variable stream like Brier Creek might be strongly context dependent, particularly for flow or time of year. For the other sur-face or water-column invertivores listed in the chapter appendix, there have been no studies of ecosystem effects in natural streams.

What about insectivores and algivores together? Power and Matthews (1983, table 2) showed that stonerollers reduced height of filamentous algae in Brier Creek pools. In 7 of the pools in 2 surveys, there were from 10 to 50 red shiners in the same pools as stonerollers. In five of the seven pools with both species, algae height was low, apparently due to consumption of algae by stonerollers (Power and Matthews, 1983, fig. 2). Algae height was enhanced in only one pool containing both Red Shin-ers and stonerollers (WJM snorkeling data). Thus there was little evidence that algal enhancement by Red Shiners (e.g., Gido and Matthews 2001) overrode effects of stoneroller grazing, which kept algae heights low. In this example, densities of Red Shiner were always lower than densities of Central Stonerollers, so we cannot make a per-capita comparison from these field data. It would be interesting to design an ex-periment in mesocosms matched by work in actual streams with densities of the two species equal to ask if one or the other has greater direct effects on attached algae.

Bertrand et al. (2009) did complementary experiments in the KPBS mesocosms and in small whole pools of the nearby Kings Creek, with Southern Redbelly Dace and Red Shiners stocked separately. In the artificial streams they found that in the first 24 days after experimental floods, the algivorous dace had more direct effects on benthic algae, by direct consumption, than shiners did by any indirect pathway. But later in the experiment, both dace and shiner treatments resulted in more or longer algae fila-ments than in control units, as nutrient effects apparently became more important than the direct consumption of algae by dace. Bertrand et al. (2009) postulated that the switch at about day 30 from dace reducing algae to enhancing algal growth oc-curred because, over time, nutrient remineralization became more important than di-rect algivory. But in contrast to results in the experimental streams, in the whole pools of Kings Creek there was no "fish effect" for either dace or Red Shiners, which Ber-trand et al. (2009) attributed to the likelihood that nutrient dynamics were different between the mesocosms and the natural stream due to dilution by groundwater or the more complex invertebrate and algae assemblages in the natural stream. The contemporaneous experiments by Bertrand et al. (2009) underscore the challenges in carrying out controlled experiments in natural streams to test ecosystem effects that can be detected in artificial mesocosms. This does not mean that results in me-socosms are "wrong," but simply that fair tests of mesocosm results for applicability to the real world can be highly context dependent and subject to vagaries of condi-tions in real streams that may not be easy to anticipate during experimental design.

In contrast to the work on flood recovery in Kings Creek, Murdock et al. (2010) followed recovery of Kings Creek from a drought, using pens that allowed access or excluded stonerollers and dace, and found that algal biomass did increase faster if the fish were excluded. Keith Gido (pers. comm.) suggested that the difference between results in Bertrand et al. (2009) and Murdock et al. (2010) "had to do with the rapid buildup of periphyton after drought but a long delay in periphyton buildup after the flood." All of this seems to underscore the strong context-specific nature of the effects of algivorous minnows in stream ecosystems.

In spite of the experimental work above, important questions remain about whether, or how much, insectivores actually enhance algivores in natural streams. A meta-analysis by Gido et al. (2010), based on all of their experiments over a decade in experimental mesocosms at KPBS or in Kings Creek, suggested, as did Hargrave (2006), that both direct and indirect effects can be important. Gido et al. (2010, fig. 1) suggested a variety of pathways by which water-column minnows like Southern Redbelly Dace or Red Shiner can interact with algivorous fishes. The insectivores hypothetically could enhance growth of benthic algae by eating macroinvertebrate grazers of algae, releasing the algae from that grazing pressure, or by the nutrient translocation from surface insects, as proposed by Gido and Matthews (2001). Either pathway could be important at different times and needs more tests in well-designed experiments in real streams.

Other Unanswered Questions about Fish Effects in Ecosystems

In most streams, fish are the largest, most visible organisms, with the most animal biomass. Stream fish feed actively all day during most of the year (unless they are busy reproducing), with benthic algivores or detritivores vigorously grazing, swiping, or scooping attached algae from stones or biofilms from soft substrates; topminnows or mosquitofish taking winged insects at the surface; shiners nipping at particles at the surface or on the streambed or dashing into the current below riffles to catch drifting macroinvertebrates; nest-builders moving stones or guarding nests; or suckers disturbing substrates as they winnow invertebrate prey. Overlay on all this an occasional piscivore like a Largemouth Bass or an adult Green Sunfish cruising through the pool, transiently disturbing the small fishes that may perform a predator-avoidance maneuver, then resume active feeding. A careful observer will see fish producing feces (stonerollers often have long strands of feces hanging from their anus) or suspending sediments as "ecosystem engineers." Now try to imagine this underwater scene without fishes. It seems counterintuitive that with all the activity of ingestion, egestion, and physical interaction with the streambed that fish wouldn't have significant effects in stream ecosystems!

Where and when does all of this fish activity translate into significant effects in stream ecosystem function? Our knowledge about effects of fish in midwestern streams comes from observations or experiments on about 22 species (see chapter appendix) in relatively few systems, including Brier Creek, Baron Fork, and Kings Creek, and snorkeling censuses of stonerollers and bass in four other streams in Oklahoma and Arkansas (Matthews et al. 1987). (Art Stewart, Mary Power, and WJM censused stonerollers in another stream—White River, at West Fork, Arkansas—over most of a day during our surveys in the 1980s, but back at our motel I scrubbed the dive slates clean before I realized that nobody had copied the data! Now I photocopy all dive slates immediately after completing any day's observations. Lesson learned.) What is still lacking is an overview of where algivorous, insectivorous, or piscivorous fishes have significant ecosystem effects and where they might not. Most knowledge about fish effects is from small- to medium-sized streams, typically clear, with stony substrates, where fish are at high densities, and where we can actually observe them by snorkeling or streamside observation. Thirty years ago, Clark Hubbs (1987) pointed out that knowing what fish do in muddy water is difficult, and that remains correct today. We still do not know much about fish activity or behavior in shallow, sandy rivers or small turbid streams where visual observation is difficult or impossible.

How much does density matter? Gido and Matthews (2001) found effects of Red Shiner on benthic algae to be linear in artificial streams, across a range of densities similar to those in natural streams. Hargrave (2005) based his estimates of impacts on PPR for a dozen species using an assumption of linear effects. But while Partridge (1991) also found density effects for export of algae by Central Stoneroller grazing, he also found that at very low densities the fish became inactive and spent most time sheltered under stones. In our most recent snorkel survey of pools in Brier Creek in May 2012, we found fewer stonerollers overall than in surveys from the 1980s, and in 5 of the 14 pools with 20, 25, 38, 140, and 360 stonerollers, there was moderate to strong growth of attached algae, with dense, floating algae mats in some. In the 1980s, when stonerollers obviously controlled algae in these pools (Power and Matthews 1983), pools often contained 200 to 500 stonerollers. An unanswered question for essentially all of the species whose ecosystem effects have been studied is whether most effects are linear or whether, conversely, there is a threshold density below which their effects are negligible. How does it depend on the behavior of each species, for example, whether they are strongly shoaling or schooling (might enhance a density effect?) or whether they are typically active as single individuals or in small pods of a few individuals (as some of the large-bodied benthic-feeding suckers in the genus *Moxostoma* appear to be)? Behaviors and thresholds of effects need to be integrated in order to really understand effects of fish in ecosystems.

Fish density may be lower on average when they are spread out across a wide expanse of a shallow, sand bed stream, like the Canadian, Washita, or Red Rivers in Oklahoma or Texas, or in sand bed rivers like the Chikaskia and Ninnescah in Kansas. In those streams most of what we know about fish distribution or density comes from individual seine hauls, which may contain only one or two stray minnows or have hundreds or even thousands of Red Shiners, Sand Shiners, or other small minnows (Matthews and Hill 1980). Even in these expansive habitats, densities of fish in some microhabitats may be high.

"When?" is another interesting question about fish effects in ecosystems. It seems obvious that trophic effects of fish would be greatest when fish are actively feeding. In autumn (Matthews and Hill 1980) or winter (stonerollers along the Baron Fork; minnows in Roanoke River), fish can be extremely dense in small, deeper pools, so that even if their activity or metabolism is minimal, they are still using oxygen, releasing CO_2, and excreting metabolites. How does the effect of fish in any stream ecosystem differ from summer to winter simply because of changes in activity levels, metabolism, or densities in microhabitats?

Finally, how do all of our findings about dynamics of fish communities, with regard to temporal (chap. 7) or spatiotemporal (chap. 8) variation in distribution or abundance, inform our thinking about effects of fish in ecosystems? The fundamental questions for any particular species are how much it affects ecosystem function per capita and what happens if densities of that species change in a stream reach because of migration, perturbations, or population dynamics. Most of our long-term data sets suggest that the fish communities we have studied are in loose equilibrium (chap. 8; Matthews and Marsh-Matthews 2016). But some individual species do change over time, such as the declines of some minnow species in Brier Creek or increases in Longear Sunfish and Western Mosquitofish in recent decades in Piney Creek. One body of theory does suggest that temporal changes in densities of individual species are to be expected and that persistence of stability of complex communities actually depends on fluctuations in populations of individual species (McCann 2000), so there is nothing inconsistent between our previous findings that Brier, Piney, or other creeks we have studied are in loose equilibrium and the fact that some individual species have changed in abundance over time. And how would our interpretation of per-capita effects of various species change depending on whether or not effects are linear with density or whether there are thresholds to effects of individual species?

In addition to the problem of temporal variation in populations, little or nothing is known about effects of stream fish in their ecosystems as a function of time of year or temperature, or how the effects of a species may vary with location. Changes in temperature are well known to strongly affect trophic interactions and alter complex networks of feedbacks in ecosystems (Gilbert et al. 2014). Fish species are well known to have inter- or intraspecific, latitudinal differences in their ability to cope with winter conditions (Shuter et al. 2012), but most warm-water species greatly restrict or cease feeding at the coldest temperatures in their environment (e.g., Keast 1968). Comparing locations in the Midwest, does a widespread fish like Red Shiner (Matthews 1985) have ecosystem effects during less of the year in streams of blizzard-prone Nebraska and Kansas than in streams of south Texas? And how do fish effects compare seasonally to effects of macroinvertebrates? Fish may take shelter in protected, low-velocity habitats (Ross and Matthews 2014) and be less active in cold weather, but many invertebrates are active year-round in streams, with some groups, like shredders, having "autumn-winter" adapted taxa (Cummins et al. 1989). In the real world, are fish maximally important in warm weather and minimally so in cold weather, whereas macroinvertebrates are important in stream ecosystems year round?

After reviewing everything in this chapter about effects of fish in ecosystems and the potential for changes in abundances of fish species to affect ecosystem structure or function, we think the message is simple. In spite of a substantial number of studies demonstrating what fish *could* do in ecosystems, relatively few studies demonstrate what fish *actually* do in real streams. There is now a fairly rich background of "could have effects" that can be a fruitful springboard to test for "do have effects" in well-designed experiments in mesocosms, but especially in real streams. For now, in spite of the extensive published research and numerous other pilot studies, fish ecologists are far from being able to accurately predict the ways or the extent to which fish play roles in real stream ecosystems, taking into account the many as-yet unstudied individual species, the vagaries of the stream environments, and all the other complex factors that make streams fascinating ecosystems. And essentially nothing is known about how greater numbers of species in a trophic group, such as high numbers of insectivorous minnows in Piney Creek, might in aggregate have strong ecosystem effects. All of this is fertile ground for a next generation of intrepid investigators who are equally adept at experimental design and in lugging massive quantities of t-posts, netting, wire, PVC, shovels, post drivers, or whatever else is needed for replicated experiments, into streams of all kinds everywhere!

Summary

This chapter reviewed our research and that of colleagues, graduate students, or former students that relates to the ways that algivorous, insectivorous, or piscivorous fishes can affect structural or functional properties of stream ecosystems, with emphasis on small- to medium-sized streams in the Midwest. We summarized the history of awareness of fish in limnology and stream ecology, from ignoring fish for most of the twentieth century to now recognizing that fish can be extremely important in stream ecosystems. The Central Stoneroller is the species for which ecosystem effects are best known and which first showed the potential importance of algivorous fishes on ecosystem structure and in trophic cascades. Stoneroller ecosystem effects have been documented thoroughly in a substantial number of observational or experimental studies and in numerous streams.

The other algivorous fish whose effects are best known is Southern Redbelly Dace, investigated in detail by the Gido group at Kansas State University. Results by the Gido group have shown mixed and highly context-dependent effects for these dace, and in general they do not seem to have as strong effects as stonerollers. But under the right circumstances, these dace also can affect ecosystem processes.

Surface-feeding insectivorous minnows, topminnows, or other small fishes are suspected from experimental work in mesocosms to have important effects on benthic algae, by consuming surface-alighting winged insects, then translocating nutrients from the insects into the water column and to benthic algae, releasing them from potential nutrient limitation, and enhancing growth. But evidence for the importance of surface insectivores remains mostly from work in artificial systems, and much more research on their roles in real stream ecosystems would be desirable.

Piscivorous fishes clearly can have strong effects in stream ecosystems by regulating distribution of smaller fishes among microhabitats, and both observational and experimental evidence in streams has shown them to be apex predators in trophic cascades. Piscivorous fishes, especially intermediately sized "mesopredators" like sunfish, also can have differential effects on different prey taxa and may be responsible for differences in the long-term presence or abundance of minnow species in real streams, potentially influencing the magnitude of insectivory by minnows in these systems.

One issue that needs more work is how much or how far downstream the effects of fishes may be; that is, what is their "zone of influence" in a stream? Numerous studies have shown that algivorous fish, or whole fish assemblages, can increase downstream transport of particulate organic matter, but this has been difficult to test in real streams; testing in mesocosms has encountered unanticipated complications in experiments. Fish may also be important because of the production of large quantities of feces, some of which may pass through guts of algivores with cells intact and be transported by currents to serve as propagules in downstream reaches. And fish also have the potential to reduce the numbers of benthic invertebrates on the streambed or the magnitude of emergence of adult insects from the stream into surrounding terrestrial ecosystems.

Potential ecosystem effects for most fish species in midwestern streams, or elsewhere in North America, have never been studied. For those that have, best evidence for strong effects in ecosystems is from experiments in artificial systems, and evidence from trials in real streams has generally lagged behind. For numerous species for which substantial effects have been shown, it is important to take into consideration their occurrence or abundance in streams of the region. Some species with strongest demonstrated potential importance in stream processes are also some of the less widely distributed or abundant species; thus their overall potential for affecting streams in the region is limited. Conversely, some species like Red Shiner, which was less important than some other insectivores in single-species trails, are widely distributed and highly abundant in many streams, so their total effects in stream ecosystems may be important. The importance of a species or functional group in a stream ecosystem also depends on their persistence, and data from our longest data sets in Piney and Brier Creeks suggest that some are quite persistent, whereas other species, such as Western Mosquitofish, with high potential to affect systems, may have flashy or boom-and-bust abundances and thus be of lesser importance in the big picture of some stream ecosystems.

Finally, we considered the evidence that fish have substantial effects in real streams and offered thoughts about where or when fish might be important in ecosystems and where or when they are perhaps less so. There is good evidence from some field studies that fish can indeed be highly important in real streams. But one overarching conclusion from numerous studies is that ecosystem effects of fishes can be strongly context dependent, related to factors like frequency of floods or drought, temperature, or spatial or temporal variation in fish densities. A given species can have quite differing effects under different conditions, and finding any simple generality such as "species X has effect Y," under a range of real-world conditions, may be elusive. Suggestions for future research include trials with more species, under a variety of conditions, with emphasis on testing effects of fishes in real streams of many different kinds.

Appendix: Species with Known Effects in Ecosystems

Species	Effect	References
Central Stoneroller, *Campostoma anomalum*	Strong effects on attached algae standing crop, NPPR, heterogeneity or spatial distribution, taxa or growth forms; BPOM fractions; macroinvertebrates; bacteria; carbon:nitrogen ratios; alter snail life history.	Power and Matthews 1983; Power et al. 1985, 1988; Matthews et al. 1987; Stewart 1987; Partridge 1991; Gelwick and Matthews 1992, 1997; Gardner 1993; Vaughn et al. 1993; Gelwick et al. 1997; Matthews 1998; Martin et al. 2016
Ozark Minnow, *Notropis nubilus*	Probably some in reduction of algae, promote NPPR and stability of algae in Ozark stream. Role compared to Central Stoneroller probably minor, as lesser percentage of grazers.	Gelwick et al. 2007
Southern Redbelly Dace, *Chrosomus erythrogaster*	Lower algal filaments and mean size of POM, but no effect on NPPR; decrease chironomids seasonally; alter carbon:nitrogen ratios; reduce macroinvertebrates; weak effect on NPPR.	Bertrand and Gido 2007; Bengston et al. 2008; Murdock et al. 2010, 2011; Kohler et al. 2011
Red Shiner, *Cyprinella lutrensis*	Increase benthic NPPR by translocation of terrestrial nutrients from surface insects. Decrease emergence of dragonflies.	Gido and Matthews 2001; Hargrave 2005; Wesner 2010b
Orangethroat Darter, *Etheostoma spectabile*	Indirectly increase stoneroller growth by remove grazing invertebrates, enhance algae; increase total BPOM. Decrease emergence of dragonflies.	Hargrave 2005, 2006; Hargrave et al. 2006; Wesner 2010b
Orangebelly Darter, *Etheostoma radiosum*	Increase periphyton biomass.	Hargrave 2005
Brook Silverside, *Labidethes sicculus*	Increase periphyton biomass.	Hargrave 2005
Golden Redhorse, *Moxostoma erythrurum*	Increase periphyton biomass; increase total BPOM.	Hargrave 2005
Striped shiner, *Luxilus chrysocephalus*	Increase periphyton biomass.	Hargrave 2005
Western Mosquitofish, *Gambusia affinis*	Increase benthic PPR by translocation of terrestrial nutrients from surface insects.	Hargrave 2005, 2006
Sand Shiner, *Notropis stramineus*	Increase periphyton biomass.	Hargrave 2005
Rocky Shiner, *Notropis suttkusi*	Increase periphyton biomass.	Hargrave 2005

(continued)

Species	Effect	References
Common Carp, *Cyprinus carpio*	Increase periphyton biomass.	Hargrave 2005
Bullhead minnow, *Pimephales vigilax*	Increase benthic PPR by translocation of terrestrial nutrients from surface insects, and release nutrients from sediments.	Hargrave 2006
Diversity of 1–6 species, drawn at random from pool of 12 candidate species representing 5 trophic groups	More species equals synergistic enhancement of stream algae; identity of species may have been important—sampling effect?	Hargrave 2009
Blackstripe Topminnow, *Fundulus notatus*	Translocate terrestrial nutrients equals bottom–up increase in decomposers (fungi and bacteria).	Hargrave et al. 2010
Blacktail Shiner, *Cyprinella venusta*	Translocate terrestrial nutrients equals bottom–up increase in decomposers (fungi and bacteria).	Hargrave et al. 2010
Largemouth Bass, *Micropterus salmoides*	Major player in controlling trophic cascade to grazing fish and algae; regulate minnow distribution.	Power and Matthews 1983; Power et al. 1985
Spotted Bass, *Micropterus punctulatus*	Spotted bass less abundant means it is a minor player in grazing fish–algae trophic cascade.	Power and Matthews 1983; Power et al. 1985
Green Sunfish, *Lepomis cyanellus*	Reduce minnow density and reproduction in experimental streams.	Marsh-Matthews et al. 2011
Longear Sunfish, *Lepomis megalotis*	Reduce minnow density and reproduction in experimental streams.	Marsh-Matthews et al. 2011
Native stream community including: Bigeye Shiner, *Notropis boops*; Central Stoneroller; Blackstripe Topminnow; Orange-throat Darter; Green Sunfish; Longear Sunfish, *Lepomis megalotis*; Bluegill, *Lepomis macrochirus*; and Largemouth Bass	Increase export of CPOM in experimental streams.	Matthews, unpub. 2003 experiment
Red Shiner, Western Mosquitofish, Orangethroat Darter all together	Reduce emergence of dragonflies and create trophic cascade reduction of web-weaving spiders that feed on insects.	Wesner 2012

References

Allan, J. D. 1983. Predator-prey relationships in streams. Pages 191–229 in J. R. Barnes and G. W. Minshall, eds. Stream ecology—application and testing of general ecological theory. Plenum Press, New York, NY.

Allan, J. D. 1995. Stream ecology: structure and function of running waters. Chapman and Hall, London, UK.

Atkinson, C. L., and C. C. Vaughn. 2015. Biogeochemical hotspots: temporal and spatial scaling of the impact of freshwater mussels on ecosystem function. Freshwater Biology 60:563–574.

Bengston, J. R., M. A. Evans-White, and K. B. Gido. 2008. Effects of grazing minnows and crayfish on stream ecosystem structure and function. Journal of the North American Benthological Society 27:772–782.

Bertrand, K. N., and K. B. Gido. 2007. Effects of the herbivorous minnow, Southern Redbelly Dace (*Phoxinus erythrogaster*), on stream productivity and ecosystem structure. Oecologia 151:69–81.

Bertrand, K. N., K. B. Gido, W. K. Dodds, J. N. Murdock, and M. R. Whiles. 2009. Disturbance frequency and functional identity mediate ecosystem processes in prairie streams. Oikos 118:917–933.

Brooks, J. L., and S. I. Dodson. 1965. Predation, body size, and composition of plankton. Science 150:28–35.

Ceccarelli, D. M., G. P. Jones, and L. J. McCook. 2011. Interactions between herbivorous fish guilds and their influence on algal succession on a coastal coral reef. Journal of Experimental Marine Biology and Ecology 399:60–67.

Coble, D. W. 1975. Smallmouth Bass. Pages 21–33 in H. Clepper, ed. Black bass biology and management. Sport Fishing Institute, Washington, DC.

Cummins, K. W. 1974. Structure and function of stream ecosystems. BioScience 24:631–641.

Cummins, K. W., and M. J. Klug. 1979. Feeding ecology of stream invertebrates. Annual Review of Ecology and Systematics 10:147–172.

Cummins, K. W., M. A. Wilzbach, D. M. Gates, J. B. Perry, and W. B. Taliaferro. 1989. Shredders and riparian vegetation: leaf litter that falls into streams influences communities of stream invertebrates. BioScience 39:24–30.

Fisher, S. G., and G. E. Likens. 1973. Energy flow in Bear Brook, New Hampshire: an integrative approach to stream metabolism. Ecological Monographs 43:421–439.

Flecker, A. S. 1992. Fish trophic guilds and the structure of a tropical stream: weak direct vs. strong direct effects. Ecology 73:927–940.

Flecker, A. S. 1996. Ecosystem engineering by a dominant detritivore in a diverse tropical stream. Ecology 77:1845–1854.

Gardner, T. J. 1993. Grazing and the distribution of sediment particle sizes in artificial stream systems. Hydrobiologia 252:127–132.

Gelwick, F. P., and W. J. Matthews. 1992. Effects of an algivorous minnow on temperate stream ecosystem properties. Ecology 73:1630–1645.

Gelwick, F. P., and W. J. Matthews. 1997. Effects of algivorous minnows (*Campostoma*) on spatial and temporal heterogeneity of stream periphyton. Oecologia 112:386–392.

Gelwick, F. P., M. S. Stock, and W. J. Matthews. 1997. Effects of fish, water depth, and predation risk on patch dynamics in a north-temperate river ecosystem. Oikos 80:382–398.

Gido, K. B. 2002. Interspecific comparisons and the potential importance of nutrient excretion by benthic fishes in a large reservoir. Transactions of the American Fisheries Society 131:260–270.

Gido, K. B., and W. J. Matthews. 2001. Effects of water column minnows in experimental streams. Oecologia 126:247–253.

Gido, K. B., K. N. Bertrand, J. N. Murdock, W. K. Dodds, and M. R. Whiles. 2010. Disturbance-mediated effects of fishes on stream ecosystem processes: concepts and results from highly variable prairie streams. Pages 593–617 in K. B. Gido and D. A. Jackson, eds. Community ecology of stream fishes: concepts approaches, and techniques. American Fisheries Society Symposium 73. American Fisheries Society, Bethesda, MD.

Gilbert, B., T. D. Tunney, K. S. McCann, et al. 2014. A bioenergetic framework for the temperature dependence of trophic interactions. Ecology Letters 17:902–914.

Gillette, D. P. 2007. Trophic spatial ecology of invertivorous stream fishes. PhD dissertation, University of Oklahoma, Norman, OK.

Grimm, N. B. 1988. Feeding dynamics, nitrogen budgets, and ecosystem role of a desert stream omnivore, *Agosia chrysogaster* (Pisces: Cyprinidae). Environmental Biology of Fishes 21:143–152.

Hall, C. A. S. 1972. Migration and metabolism in a temperate stream ecosystem. Ecology 53: 585–604.

Hargrave, C. W. 2005. Effects of fish density, identity, and species richness on stream ecosystems. PhD dissertation, University of Oklahoma, Norman, OK.

Hargrave, C. W. 2006. A test of three alternative pathways for consumer regulation of primary productivity. Oecologia 149:123–132.

Hargrave, C. W. 2009. Effects of fish species richness and assemblage composition on stream ecosystem function. Ecology of Freshwater Fish 18:24–32.

Hargrave, C. W., S. Hamontree, and K. P. Gary. 2010. Direct and indirect food web regulation of microbial decomposers in headwater streams. Oikos 119:1785–1795.

Hargrave, C. W., R. Ramirez, M. Brooks, M. A. Eggleton, K. Sutherland, R. Deaton, and H. Galbraith. 2006. Indirect food web interactions increase growth of an algivorous stream fish. Freshwater Biology 51:1901–1910.

Harvey, B. C. 1991. Interactions among stream fishes: predator-induced habitat shifts and larval survival. Oecologia 87:29–36.

Hata, H., and M. Kato. 2004. Monoculture and mixed-species algal farms on a coral reef are maintained through intensive and extensive management by damselfishes. Journal of Experimental Marine Biology and Ecology 313:285–296.

Hrbacek, J., M. Dvorakova, V. Korinek, and L. Prochazkova. 1961. Demonstration of the effect of the fish stock on the species composition of zooplankton and the intensity of metabolism of the whole plankton assemblage. Verhandlungen der Internationalen Vereinigung für Theoretische und Angewandte Limnologie 14:192–195.

Hubbs, C. 1987. Summary of the symposium. Pages 265–267 in W. J. Matthews and D. C. Heins, eds. Community and evolutionary ecology of North American stream fishes. University of Oklahoma Press, Norman, OK.

Hutchinson, G. E. 1957–1993. A treatise on limnology. 4 vols. John Wiley and Sons, New York, NY.

Hynes, H. B. N. 1970. The ecology of running waters. University of Toronto Press, Toronto, ONT.

Hynes, H. B. N. 1975. The stream and its valley. Verhandlungen der Internationalen Vereinigung für Theoretische und Angewandte Limnologie 19:1–15.

Juday, C., W. H. Rich, G. I. Kemmerer, and A. Mann. 1932. Limnological studies of Karluk Lake, Alaska, 1926–1930. Bulletin of the Bureau of Fisheries 47:407–436.

Keast, A. 1968. Feeding of some Great Lakes fishes at low temperatures. Journal of the Fisheries Research Board of Canada 25:1199–1218.

Kohler, T. J., J. N. Murdock, K. B. Gido, and W. K. Dodds. 2011. Nutrient loading and grazing by the minnow *Phoxinus erythrogaster* shift periphyton abundance and stoichiometry in mesocosms. Freshwater Biology 56:1133–1146.

Likens, G. E., and F. H. Bormann. 1974. Linkages between terrestrial and aquatic ecosystems. BioScience 24:447–456.

Marsh-Matthews, E., and W. J. Matthews. 2000a. Spatial variation in relative abundance of a widespread, numerically dominant fish species and its effect on fish assemblage structure. Oecologia 125:283–292.

Marsh-Matthews, E., and W. J. Matthews. 2000b. Geographic, terrestrial and aquatic factors: which most influence the structure of stream fish assemblages in the midwestern United States? Ecology of Freshwater Fish 9:9–21.

Marsh-Matthews, E., W. J. Matthews, and N. R. Franssen. 2011. Can a highly invasive species re-invade its native community? the paradox of the Red Shiner. Biological Invasions 13: 2911–2924.

Marsh-Matthews, E., J. Thompson, W. J. Matthews, A. Geheber, N. R. Franssen, and J. Barkstedt. 2013. Differential survival of two minnow species under experimental sunfish predation: implications for re-invasion of a species into its native range. Freshwater Biology 58:1745–1754.

Martin, E. C., K. B. Gido, N. Bello, W. K. Dodds, and A. Veach. 2016. Influence of fish richness on ecosystem properties of headwater prairie streams. Freshwater Biology 61:887–898.

Matthews, W. J. 1985. Distribution of midwestern fishes on multivariate environmental gradients, with emphasis on *Notropis lutrensis*. American Midland Naturalist 113:225–237.

Matthews, W. J. 1986. Fish faunal structure in an Ozark stream: stability, persistence, and a catastrophic flood. Copeia 1986:388–397.

Matthews, W. J. 1998. Patterns in freshwater fish ecology. Chapman and Hall, New York, NY.

Matthews, W. J., and L. G. Hill. 1980. Habitat partitioning in the fish community of a southwestern river. Southwestern Naturalist 25:51–66.

Matthews, W. J., and E. Marsh-Matthews. 2007. Extirpation of Red Shiner in direct tributaries of Lake Texoma (Oklahoma-Texas): a cautionary case history from a fragmented river-reservoir system. Transactions of the American Fisheries Society 136:1041–1062.

Matthews, W. J., and E. Marsh-Matthews. 2011. An invasive fish species within its native range: community effects and population dynamics of *Gambusia affinis* in the central United States. Freshwater Biology 56:2609–2619.

Matthews, W. J., and E. Marsh-Matthews. 2016. Dynamics of an upland stream fish community over 40 years: trajectories and support for the loose equilibrium concept. Ecology 97:706–719.

Matthews, W. J., K. B. Gido, G. P. Garrett, F. P. Gelwick, J. G. Stewart, and J. Schaefer. 2006. Modular experimental riffle-pool stream system. Transactions of the American Fisheries Society 135:1559–1566.

Matthews, W. J., B. C. Harvey, and M. E. Power. 1994. Spatial and temporal patterns in the fish assemblages of individual pools in a midwestern stream (U.S.A.). Environmental Biology of Fishes 39:381–397.

Matthews, W. J., E. Marsh-Matthews, R. C. Cashner, and F. Gelwick. 2013. Disturbance and trajectory of change in a stream fish community over four decades. Oecologia 173: 955–969.

Matthews, W. J., M. E. Power, and A. J. Stewart. 1986. Depth distribution of *Campostoma* grazing scars in an Ozark stream. Environmental Biology of Fishes 17:291–297.

Matthews, W. J., W. D. Shepard, and L. G. Hill. 1978. Aspect of the ecology of the Duskystripe Shiner, *Notropis pilsbryi* (Cypriniformes, Cyprinidae) in an Ozark stream. American Midland Naturalist 100:247–252.

Matthews, W. J., A. J. Stewart, and M. E. Power. 1987. Grazing fishes as components of North American steam ecosystems: effects of *Campostoma anomalum*. Pages 128–135 in W. J. Matthews and D. C. Heins, eds. Community and evolutionary ecology of North American stream fishes. University of Oklahoma Press, Norman, OK.

McCann, K. S. 2000. The diversity-stability debate. Nature 405:228–233.

McCann, K. S., A. Hastings, and G. R. Huxel. 1998. Weak trophic interactions and the balance of nature. Nature 395:794–798.

Miller, R. L., and H. W. Robison. 2004. Fishes of Oklahoma. University of Oklahoma Press, Norman, OK.

Minshall, G. W. 1988. Stream ecosystem theory: a global perspective. Journal of the North American Benthological Society 7:263–288.

Morin, P. J. 2011. Community ecology, 2nd ed. Wiley-Blackwell, West Sussex, UK.

Murdock, J. N., W. K. Dodds, K. B. Gido, and M. R. Whiles. 2011. Dynamic influences of nutrients and grazing fish on periphyton during recovery from flood. Journal of the North American Benthological Society 30:331–345.

Murdock, J. N., K. B. Gido, W. K. Dodds, K. N. Bertrand, and M. R. Whiles. 2010. Consumer return chronology alters recovery trajectory of stream ecosystem structure and function following drought. Ecology 91:1048–1062.

Oliver, D. R. 1971. Life history of the chironomidae. Annual Reviews of Entomology 16:211–230.

Partridge, W. D. 1991. Effects of *Campostoma anomalum* on the export of various size fractions of particulate organic matter in streams. MS thesis, University of Oklahoma, Norman, OK.

Power, M. E. 1983. Grazing responses of tropical freshwater fishes to different scales of variation in their food. Environmental Biology of Fishes 9:103–115.

Power, M. E. 1984a. Habitat quality and the distribution of algae-grazing catfish in a Panamanian stream. Journal of Animal Ecology 53:357–374.

Power, M. E. 1984b. The importance of sediment in the grazing ecology and size class interactions of an armored catfish, *Ancistrus spinosus*. Environmental Biology of Fishes 10:173–181.

Power, M. E. 1990a. Resource enhancement by indirect effects of grazers: armored catfish, algae, and sediment. Ecology 71:897–904.

Power, M. E. 1990b. Effects of fish in river food webs. Science 250:811–814.

Power, M. E., and W. J. Matthews. 1983. Algae-grazing minnows (*Campostoma anomalum*), piscivorous bass (*Micropterus* spp.), and the distribution of attached algae in a small prairie-margin stream. Oecologia 60:328–332.

Power, M. E., W. J. Matthews, and A. J. Stewart. 1985. Grazing minnows, piscivorous bass and stream algae: dynamics of a strong interaction. Ecology 66:1448–1456.

Power, M. E., A. J. Stewart, and W. J. Matthews. 1988. Grazer control of algae in an Ozark Mountain stream: effects of short-term exclusion. Ecology 69:1894–1898.

Rooney, N., and K. S. McCann. 2012. Integrating food web diversity, structure and stability. Trends in Ecology and Evolution 27:40–48.

Ross, S. T., and W. J. Matthews. 2014. Evolution and ecology of North American freshwater fish assemblages. Pages 1–49 in M. L. Warren Jr. and B. M. Burr, eds. Freshwater fishes of North America. Vol. 1. Petromyzontidae to Catostomidae. Johns Hopkins University Press, Baltimore, MD.

Ruttner, F. 1963. Fundamentals of limnology, 3rd ed. D. G. Frey and F. E. J. Frey, trans. University of Toronto Press, Toronto, ONT.

Shuter, B. J., A. G. Finstad, I. P. Helland, I. Zweimuller, and F. Hoker. 2012. The role of winter phenology in shaping the ecology of freshwater fish and their sensitivities to climate change. Aquatic Sciences 74:637–657.

Stewart, A. J. 1987. Responses of stream algae to grazing minnows and nutrients: a field test for interactions. Oecologia 72:1–7.

Vannote, R. L., G. W. Minshall, K. W. Cummins, J. R. Sedell, and C. E. Cushing. 1980. The river continuum concept. Canadian Journal of Fisheries and Aquatic Sciences 37:130–137.

Vaughn, C. C., 2010. Biodiversity losses and ecosystem function in freshwaters: emerging conclusions and research directions. BioScience 60:25–35.

Vaughn, C. C., F. P. Gelwick, and W. J. Matthews. 1993. Effects of algivorous minnows on production of grazing stream invertebrates. Oikos 66:119–128.

Vaughn, C. C., S. J. Nichols, and D. E. Spooner. 2008. Community and foodweb ecology of freshwater mussels. Journal of the North American Benthological Society 27:409–423.

Wesner, J. S. 2010a. Trophic connections between stream and terrestrial food webs. PhD dissertation, University of Oklahoma, Norman, OK.

Wesner, J. S. 2010b. Seasonal variation in the trophic structure of a spatial prey subsidy linking aquatic and terrestrial food webs: adult aquatic insects. Oikos 119:170–178.

Wesner, J. S. 2010c. Aquatic predation alters a terrestrial prey subsidy. Ecology 91:1435–1444.

Wesner, J. S. 2012. Predator diversity effects cascade across an ecosystem boundary. Oikos 121:53–60.

Wesner, J. S. 2013. Fish predation alters benthic, but not emerging, insects across whole pools of an intermittent stream. Freshwater Science 32:438–449.

Wesner, J. S. 2016. Contrasting effects of fish predation on benthic versus emerging prey: a meta-analysis. Oecologia 180:1205–1211.

Wotton, R. S., and B. Malmqvist. 2001. Feces in aquatic ecosystems. BioScience 51:537–544.

CHAPTER 10

A Critical Synthesis

In a paper titled "Are There General Laws in Ecology?" Lawton (1999) made the statement that "community ecology is a mess." He noted that there were what could be considered "laws" or broad generalizations at both smaller and larger scales of study, but at the intermediate scale of community ecology, laws were evasive. Ecological studies at reductionist scales clearly indicate that ecological processes are constrained by the general laws of physics and chemistry and are generally predictable with respect to responses to natural or artificial selection. At the other extreme of scale, macroecology, Lawton (1999) noted that general patterns can emerge because of the lack of consideration of details of local processes. Community ecology, however, is studied at the scale where outcomes largely rely on context, thus imposing contingency (to a "mindboggling degree," in the words of Lawton 1999) on outcomes and predictions at this scale and making the discovery of general laws unlikely.

Despite this caveat, in this book we have attempted to analyze our past studies and to revisit old data with new approaches to look for patterns and identify processes of community dynamics in our study systems. We acknowledge from the outset that our outcomes are in fact contingent and may not be generalizable beyond the types of systems we have studied, that is, warm-water streams with fish communities largely dominated by minnows or sunfish. But given the prevalence of these types of systems in North America, our conclusions may in fact relate to a large number of stream fish communities and make contingency somewhat more manageable (despite Lawton's caveat!).

A Brief Review of Our Approach

As we have noted throughout the book (and emphasized in chaps. 1 and 2), a major strength of our approach is the fact that our study systems were all sampled using small-meshed seines by one or both of the authors (with the exception of a few samples made by students or colleagues with whom WJM had worked closely and who followed the same sampling regimen). The fish communities in some of these streams were sampled at multiple sites (our global communities), and in other streams only one site was sampled (our local communities). With the exception of our broad survey of sites in the lower Great Plains (which were sampled twice), all sites were sampled at least four times (and some many more).

The repeated sampling of our sites is another strength of our approach. Our use of the word "community" in this book has embraced temporal variation as an important component of community structure, and it is this temporal variation that we have used to evaluate community dynamics. Thus, by our definition, a stream fish community is integrated over time. The spatial aspect of our study communities was defined by the initial sampling regimen relative to the number of sites in a given stream. For streams with multiple sites, samples from all individual sites were pooled to construct a global community sample during a given survey. Defining the community at this scale relies on the assumption that although all community members may not interact at a given time due to spatial constraints, they do have the potential to interact over time as individuals and species move within the system. For streams with a single site, the local community at that site at any given time is clearly a subset of what might occur within the entire system. Repeated sampling of the local community, however, provides a "bigger picture" of what species can occur at that site over time. Throughout the book, we have separately examined global and local dynamics, and we have repeatedly found differences in community structure and dynamics at the different scales (more on this below).

Although all of our study systems are warm-water streams, they represent a diverse array of streams that vary in size, hydrology, and environmental harshness, and in the composition and complexity of the fish community. Because the taxonomic composition of our study communities was largely a function of biogeographic factors, we chose to compare communities with respect to a number of emergent properties. This approach was similar to the one we used in a paper published in *Ecology of Freshwater Fish* (Marsh-Matthews and Matthews 2000), where we asked which factors best explained spatial variation in fish community composition across 65 sites in the lower Great Plains. The answer was that taxonomic composition was mostly explained by the zoogeographic patterns in species distributions (which was not surprising) but that emergent properties of communities such as species richness were better predicted by local conditions.

In the analyses presented in this book (chap. 3), the emergent properties examined include those descriptive of the entire community over all times sampled, including number of families represented, total number of species captured, and number of core species (those occurring in all collections). We also used emergent metrics for each collection to calculate average number of species captured, average diversity, average evenness, average percentage of the community composed of core species, and coefficients of variation in these metrics. Our community descriptors were therefore designed to incorporate temporal variability in community structure.

After characterizing each community relative to emergent properties, we used principal components analysis (PCA) to place communities in emergent property space. This allowed us to examine correlations among emergent properties as well as assess similarity of our study systems with respect to these properties. We were also able to use PCA plots to superimpose convex hulls representing community characteristics or environmental factors that we thought might affect community dynamics, and PCA scores to examine community dynamics as a function of several continuous quantitative environmental variables.

We used other multivariate approaches to describe community variation over time and space and in response to extreme environmental events (chaps. 6–8). Specifically, we employed the Morisita-Horn similarity index as a tool to make pairwise comparisons of community structure over time or space and used this index in ordinations of communities using nonmetric multidimensional scaling (NMDS). Use of a density-invariant index like the Morisita-Horn index allowed us to compare structure of communities despite different community sample sizes.

To evaluate community dynamics with respect to the loose equilibrium concept (Collins 2000; Matthews et al. 2013; Matthews and Marsh-Matthews 2016), we examined trajectories of the communities in multivariate (NMDS) space for number of returns toward the centroid (average community composition) and relative lengths of time-to-time "steps." We were thus able to classify dynamics of our study systems using the system outlined by Matthews et al. (2013) and Matthews and Marsh-Matthews (2016), also described in chapter 7 in this book.

We also examined species traits (chap. 4) and biotic interactions (chap. 5) that potentially affect community dynamics, particularly in our study systems, where environmental conditions are often highly variable or harsh. In chapter 9, we reviewed effects of fishes on ecosystem dynamics and examined (or speculated upon) the effects that changes in community composition can have on those ecosystem properties.

What Have We Learned about Factors Affecting Stream Fish Community Dynamics?

As we point out in subsequent discussion, some of our findings provided support for the "conventional wisdom" of fish ecology, while others pointed out "contingencies" by which the conventional wisdom should be modified. Of those that support the conventional wisdom, the importance of the scale of analysis is perhaps the most "expected" of our conclusions (and one that we treated almost as an assumption from the outset).

Scale of Analysis

The importance of scale in ecological analyses has been recognized for a long time (Levin 1992). In our analyses of stream fish community dynamics, 9 of our study systems were analyzed at a global scale, for which collections at multiple sites within a watershed were pooled to construct the fish community. In contrast, 31 communities were sampled only from a given reach (single site) within a watershed. This difference in scale influences multiple factors that affect the fish communities. The global scale is physically much larger, includes more types of habitats, and encompasses more variation in stream dynamics (e.g., those related to longitudinal variation in flow, temperature, etc.). Not surprisingly, we found that our global communities had more total species, more core species, higher diversity, and less temporal variation in composition than randomly chosen local communities from the same watershed. Overall, temporal similarity in community structure (measured using the Morisita-Horn similarity index) was greater in global communities.

Global and local communities differed in additional ways that may have important implications for conservation strategies. Across all of our global communities, for those with more total species detected, the greater number of species was attributable to an increased number of core species (those present in every collection) relative to one-time species (those collected only once), but the opposite was true for local communities. For those local communities with greater numbers of species, the increased number of species was largely attributable to one-time taxa. There are current conservation initiatives, such as to establish Native Fish Conservation Areas (Williams et al. 2011), that target entire communities rather than individual taxa. For such efforts to be effective, it will be important to define the community at a scale for which community composition is predictable. Our findings throughout the book suggest that conservation targets defined at the global scale would be preferable to those at the local scale.

Environmental Factors Affecting Community Dynamics

We examined environmental factors ranging from geography to stream morphology to climate variables for effects on the dynamics of our study systems. For both global and local communities, there was no effect of river basin on community dynamics, with the exception that local communities in the Neosho–Verdigris drainage in northeastern Oklahoma were more variable that those elsewhere. The geographic effect is confounded by the fact that all of these communities were also dominated by Western Mosquitofish (*Gambusia affinis*) and may be related to the family composition of these communities, as discussed below. There were also no differences among dynamics of communities related to whether streams were classified as upland or lowland or according to primary land use of the surrounding area. The apparent lack of land-use effects may reflect the fact that most of our stream study systems were not heavily affected by human activity.

We did find that some aspects of stream morphology were correlated with community dynamics. Specifically, stream size was related to community variability at both global and local scales: communities from smaller streams (assessed by stream width and depth) tended to be more variable. This finding confirms the conventional wisdom in fish ecology regarding stream size and variability (Schlosser 1987). Some aspects of stream morphology, however, showed different relationships to community variability at the different spatial scales. Our PCA analyses showed that at the global scale, fish communities from higher-elevation and higher-gradient streams were more variable, but at the local scale the opposite was true. This conclusion for local communities was confirmed using time-to-time similarity assessed with the Morisita-Horn index, which also showed that higher-gradient streams were less variable (in some cases, such as Crooked Creek, because they had simple communities with only one to three species).

Hydrologic variation (intermittency) was associated with community variation at both local and global scales, such that fish communities from intermittent streams were more variable. This finding is consistent with the conventional wisdom that greater environmental variation is associated with higher community variation. Our comparative analyses of the fish communities in Brier Creek and Piney Creek supported this idea as well. These two systems were the two longest studied in our analyses and differed in a number of ways, one of which was the degree of environmental variation. Brier Creek is a prairie-margin stream in southern Oklahoma that occasionally dries in summer to the degree that flow ceases and water is confined to isolated pools. In severely dry periods, long reaches of the creek may be dry, and pools may shrink to an extremely small size. Water temperatures in the creek may reach a lethal level for some fishes during those drought events. Piney Creek, on the other hand, is a spring-fed upland stream in the Ozark region of north-central Arkansas. Flow and water temperature vary far less than in Brier Creek (except during the occasional flood), and environmental conditions are benign compared to those in Brier Creek. Our comparisons of community dynamics in these two systems showed that the fish community of Piney Creek is less variable than that of Brier Creek.

Floods and Droughts

The extremes of hydrologic variation during floods and droughts can affect community structure and result in displacement of the fish community in multivariate space (chap. 6). But for both types of disturbances, community structure in our study systems tended to recover toward a predisturbance state, although pre- and postflood communities tended to be more similar than pre- and postdrought communities (chap. 6). Timing of a disturbance can also affect the magnitude of community change. For two major floods in Piney Creek, for example, community return toward the preflood community was faster for a winter flood than for a flood that occurred in springtime (Matthews et al. 2014).

Intrinsic Community Properties Affecting Community Dynamics

As mentioned above, we found that local communities dominated by Western Mosquitofish (Family Poeciliidae) occupied regions of PCA space that were characterized by higher variation in several emergent properties. These communities also happened to be located within the same watershed, and all were characterized by harsh and variable environmental conditions, but we feel the family composition of these communities contributes to the explanation for their higher level of variation in community composition. Western Mosquitofish are livebearers with short gestation time and well-documented tolerances for harsh environmental conditions (Pyke 2008). These species traits in combination are probably responsible for the boom-and-bust dynamics of mosquitofish populations that dominate these communities, which would contribute to the high variation in evenness. Most of the other fish communities in our studies were dominated by minnows (in combination with other taxa in some), and there was no apparent effect of family on community dynamics of these.

Variation in community structure was not related to total number of species detected for either the global or local systems. Although PCA analyses revealed negative correlations between number of species and coefficients of variation in several emergent properties, the more direct measure of time-to-time community similarity using the Morisita-Horn index showed no relationship to overall species richness. We did find that the simplest communities were often the least variable, likely owing to the combination of community domination by one or a few highly tolerant species (as noted above for Western Mosquitofish) and the fact that our community comparisons did not include times when no fishes were collected at a site. In these intermittent habitats, when fishes were present, it was always the same species dominating an extremely simple community.

Loose Equilibrium as the Null Model for Community Dynamics

The long-standing assumption of equilibrium communities in ecology has gradually given way to the concept of loose equilibrium (May 1973; DeAngelis et al. 1985; Collins 2000; Matthews and Marsh-Matthews 2016) as the model best explaining community dynamics. For all of the communities we examined for fit to the model of loose equilibrium, all but one conformed to the model, and the one that did not was from a site that had been extremely altered by prolonged drought. We therefore suggest that loose equilibrium (LE) should indeed stand as the null model for dynamics of fish communities from unimpacted systems.

Here we offer a simple overview of the components of this null model (fig. 10.1). The figure summarizes the essential components of the LE model, and those components provide a framework with which to dissect the processes driving community dynamics. This depiction is overly simple but has utility in that it illustrates the important components that contribute to the community dynamics that result in LE. These are (1) the boundary, (2) the trajectory arrows, and (3) the "exit" arrow. Understanding the factors that contribute to each of these is essential to dissecting the processes that drive community dynamics.

Community Boundary

Loose equilibrium is characterized by displacement and return of the community (depicted as smaller circles) within some "bounded" community composition space

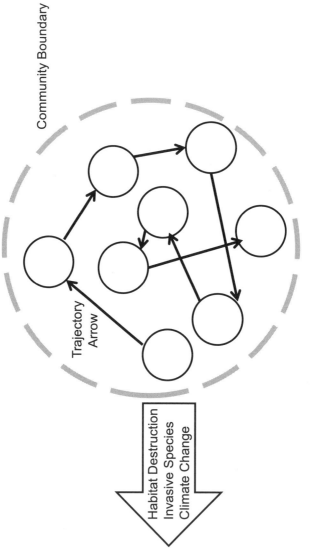

Community Boundary

Trajectory Arrow

Habitat Destruction
Invasive Species
Climate Change

FIG. 10.1 Schematic depiction of the components of the loose equilibrium model of community dynamics.

(gray dashed circle), such as that resulting from ordination analyses. Although this space in figure 10.1 is two dimensional, higher dimensions could apply. But in our analyses using NMDS (chaps. 6–8) with the Morisita-Horn as the similarity index, two-dimensional solutions were often the most appropriate. Note that the boundary in figure 10.1 is depicted as "perforated" to suggest that it may be breached, but if so, community structure displaced outside the boundary may return to a state within the normal range of variation defined by the boundary.

Factors that determine this boundary for a given community are perhaps the most difficult to characterize because the limits of community change that do not result in a permanently altered state are typically not known. Nonetheless, there are factors that likely interact to determine the boundary. As suggested by numerous authors (see Matthews 1998 for a review), characteristics of the environment should limit the species that can inhabit the community based on abilities of species to tolerate local environmental vagaries and extremes. This is the phenomenon that has been called "environmental filtering." As an example, our analyses of western range limits for species in Oklahoma based on temperature and oxygen tolerance (chap. 4) illustrate this phenomenon. Not all species that can potentially inhabit a community are found therein, however. Interspecific interactions (competition, predation) can result in exclusion (or elimination) of potential community members. In our overview of biotic interactions in our study systems (chap. 5), evidence for competitive exclusion in our study systems was mixed because patterns in some cases could be explained by habitat preferences. Analyses of predation in our study systems suggested that predation pressure in the community was influenced by (or at least correlated with) environment conditions: predator:prey ratios were highest in intermittent or low-gradient streams with deep pools and low current speeds. Thus the importance of predation in our study systems is in part predictable from environmental factors and may have a role in defining the community boundary. Numerous other factors could affect the community boundary as well, particularly those that can give rise to negative feedback dynamics of populations comprising the community. These would include resource limitation (either seasonal or periodic), life-history trade-offs, predatory switching behavior, and the like. Although all of these factors are known to affect community dynamics (see below), determining their role in setting the community boundary will require creative experiments and/or modeling.

Community Trajectory Arrows

Within the community boundary, community composition is expected to change in response to the collective forces driving changes in relative abundances of species in the community. These collective forces are summarized by the "community trajectory arrows" that determine the direction and distance of community change. These community trajectory arrows integrate all of the factors that affect community change, including factors with opposing effects on species' populations. The trajectory illustrated by the arrows represents the net change in community composition overall.

Environmental factors most certainly affect community change. For example, we found in our drought experiments (chap. 6) that environmental perturbation does not affect all species in the community equally. In our simulated droughts in meso-cosms, some species in our experimental communities suffered higher mortality than others, and some were able to recover better than others after drought. So, differential susceptibility of species within the community to environmental perturbations will drive change in community composition. Similarly, environmental variation may drive stochastic changes in species abundances (unrelated to traits) that in turn affect overall community structure.

Intraspecific interactions that could affect community change may vary with abundance of a given species. For example, reproductive success depends on finding a mate, and low population densities that decrease the chance of doing so may result in a critical decline in recruitment. Low population density of a given species may drive hybridization between that species and a more abundant congener. For example, hybrids have been found between introduced Red Shiners (*Cyprinella lutrensis*) and other *Cyprinella* species that are native to the invaded community at multiple sites in the central and southeastern United States (Page and Smith 1970; Wallace and Ramsey 1982; Walters et al. 2008; Blum et al. 2010). Not only does hybridization threaten the genetic integrity of the native species, but it also alters population dynamics of both species involved and can hence shift community structure.

Likewise, interspecific interactions may vary with community structure such that effects on species vary with species abundances. A well-known example of such an interspecific interaction is predator switching behavior (which may actually impose negative feedback on community change; Murdoch 1969). Competitive interactions are also density dependent and are expected to vary with productivity. Productivity, in turn, is related to both abiotic factors (temperature, rainfall, etc.) and the effects of fishes (and other organisms) on primary productivity or nutrient flux (chap. 9). Collectively, interspecific interactions are the most complex factors potentially driving community change and are certainly the least understood. The role of competition in stream fish community dynamics is very much contingent and still in need of carefully designed experiments.

The factors that most affect the trajectory arrows for a given community are also likely to differ among communities. Environmental stability, frequency of disturbance, and presence of keystone species (as a few examples) would all be expected to affect which factors are most likely to strongly impact community change. The lengths of trajectory arrows may also vary in a predictable manner, such that arrows are longer or more variable in length in harsh environments; our analyses of LE of the Piney and Brier Creek global communities showed this (chap. 7).

Even compiling the list of factors that may affect the trajectory arrows is an arduous (if not impossible) task, but recognition that the shift in community composition from time to time is in response to multiple factors is important. And it is likely that the dominant factors affecting change of a given community (or community type) will "pop up" on any list compiled by a student of the system.

Exit Arrow

The final component of figure 10.1 is a large "exit arrow" (leaving the bounded space) that summarizes the types of forces that may drive the community beyond the boundary of its LE space and result in change of the community to a permanently altered state. Types of perturbations that would force a community to an alternate state include habitat alteration (such as well-known examples of permanent shifts from "stream" to "lake" species following impoundment; Matthews et al. 2004), introductions of invasive species (particularly apex predators that eliminate prey species, such as the tragic case of introduction of Nile Perch into Lake Victoria; Goldschmidt et al. 1993), and climate change.

With respect to climate change, the only community for which we detected directional change rather than LE was from the Borrow Pits of southeastern Oklahoma, which, across the time span of our assessment, was exposed to prolonged drought (chap. 7). But we may expect that other fish communities in our area of study may shift beyond their historical boundaries of LE as continued changes in hydrology occur. Many of the species from sandbed prairie streams have declined precipitously or even disappeared from communities where they were once among the dominant species. Two species are of particular concern in central Oklahoma and elsewhere on the southern Great Plains: Arkansas River Shiner, *Notropis girardi* (Federal Register 1998, 2005), and Plains Minnow, *Hybognathus placitus* (Jelks et al. 2008, pers. obs.; K. B. Gido, pers. comm.).

Determining the precise factors that may force a community to an altered state outside of its LE boundary may be difficult, but if those can be identified, creative solutions to reverse or mitigate those factors may be possible. For example, Perkin and Gido (2012) found that fragmentation of riverine systems in the southern Great Plains has resulted in changes for many fish communities, but also that fragment length was an important predictor of community integrity. Using these data, they were able to predict which fragments would result in the most restoration of communities and make recommendations to fisheries managers. In Oklahoma, it may be possible to restore altered communities (if environmental conditions are also restored) because many species known from the earliest comprehensive collections (Ortenburger and Hubbs 1926; Hubbs and Ortenberger 1929a, 1929b) are still present in the state and in the watersheds sampled by those early expeditions (Matthews and Marsh-Matthews 2015).

Have We Learned Anything New?

The questions in ecology are not new: they date to before the beginning of the discipline as a defined subset of biology. They are the fundamental questions of why organisms are found where they are (e.g., Jackson et al. 2001), what factors or processes determine that distribution, and how those processes act at numerous levels (and integrate over those various levels) to determine distribution and abundance of organisms. What is new, or what has progressed over time in ecology, is the refinement with which we can answer those questions or attempt to answer those questions, by using new mathematical constructs and tools and the development of theory that allows us to pose testable hypotheses and find reasonable answers for the particular question at hand. Our ability to answer questions has evolved over the course of the century and a half—nearly two centuries—during which ecology has been a self-identified discipline. But those questions have never been fully answered and thus continue to recur, becoming the focal questions for a given generation of ecologists, or at least a cohort. Since we have been practitioners of the study of ecology (starting in the early 1970s), we have seen the change in focus in community ecology, from a time when competitive interactions and resource partitioning were considered to be the most important driving forces in structuring communities—only to have that focus displaced by concern with disturbance events, pulse and press disturbances, abiotic factors, and habitat selection—to a time when, in the last decade, competition has re-emerged as an important alternative driver within the context of the focus on the relative importance of phylogenic relatedness versus environmental filtering as processes driving community assembly (Webb et al. 2002). We also witnessed that this turnover in focal topics in ecology, at least to us, seems to have a two-student-generation turnover time, because as students ourselves, it is now our graduate students 40 years later who are coming back to the study of competition. There are many other issues that similarly have undergone this phenomenon of "coming back around on the guitar" (with apologies to Arlo Guthrie), but at each juncture the questions can be more refined and the answers can be more precise until, as in the past, they either become intractable with the current tools and methods, or they are displaced by something that is considered a "hotter" or "more cutting-edge" topic, which is likely another resurrection of a different process. All that said, we are asking if what we have learned is new. The answer is "not really," but in the face of the contingency in community ecology, we hope that we have been able to contribute some refinement to some answers in community ecology. We still lack precision in our answers, however, and in some aspects of community dynamics, "contingency" or "context" is the answer.

The Future of Studies in Community Dynamics

The most important element for progress in our understanding of community dynamics is the continuation of long-term studies (Matthews 2015). Long-term studies provide the essential data for understanding community dynamics, responses to disturbance, conservation needs, and projected changes in communities or ecosystem effects under global change scenarios. Few freshwater stream fish communities have been studied in the very long term (over several decades), but the number of studies at the scale of decades is growing (Ross 2013). As these studies move into the future, it would be ideal for younger researchers to collaborate with older investigators or actually inherit their long-term study sites. In fact, our colleagues Ginny and Reid Adams at the University of Central Arkansas have worked with us in Piney Creek, and they have begun their own (long-term, it is hoped) studies, continuing to sample all of our long-term sites. Such long-term studies are challenging, partly owing to the lack of support by granting agencies because they are viewed as "incremental" rather than "transformative" science. But it is only with the increments of information about community change that we will really understand community dynamics and the factors that drive those.

The future of our understanding of community dynamics also lies in the continued study of basic biology of community members (Matthews 2015). We know a lot about a few species in freshwater stream communities. There are countless studies of game species or those subject to management (e.g., Bluegill, *Lepomis macrochirus*; Largemouth Bass, *Micropterus salmoides*), but the basic tolerances, diets, habitat requirements, and reproductive strategies of many species are little studied or basically unknown. The "fishes of" books abound with comments like "reproductive biology is unknown" (Matthews 2015). Perhaps worse, traits of closely related species are simply assumed to be identical (or similar) to one another. But many traits vary across closely related species (e.g., unlike most other sunfish, Redear Sunfish, *Lepomis microlophus*, are molluscivores), and many traits vary within species across populations or even within the same individual under different environmental conditions (e.g., egg size in Orangethroat Darter, *Etheostoma spectabile*; Marsh 1984, 1986). Yet it is these traits that may have important impacts on populations and hence community dynamics. It may be unreasonable to think that studies of basic biology and species traits will proceed at a rate necessary to provide the resolution needed to predict community dynamics, but it is our hope that these studies continue, if only in the context of larger, more "fundable" efforts.

Future studies of community dynamics will also benefit from the development of new tools, both empirical and theoretical. Genetic approaches may allow major strides in understanding communities. The use of eDNA to take a "roll call" of species present has great promise for detecting rare species or early invasives (Rees et al. 2014), but its use for quantitative assays of community composition is still uncertain, particularly in flowing streams (Rees et al. 2014). At the level of detecting expression of "traits" characterizing the community at a given time or under given environmental conditions, the tools of genomics (also transcriptomics and proteomics) may prove highly informative and even far more practical and cost effective than traditional approaches (Mehinto et al. 2012). Likewise, the continued refinement of current analytical methods (e.g., papers cited in Magurran and McGill 2011) and development of new approaches may provide tools that will allow finer resolution of community dynamics in future studies. And the growing application of meta-analyses (such as those undertaken by study groups at the National Center for Ecological Analysis and Synthesis) will undoubtedly accelerate understanding of dynamics of many types of communities and thus provide a framework within which stream fish communities may be compared with those of other taxa.

In looking to the future of studies in stream fish community dynamics, it is clear that these studies will become even more important as conservation strategies are developed. Freshwater fishes are among the worst affected in the current biodiversity crisis (Williams et al. 2011), and it is imperative that we work to understand the dynamics of these communities if we have any chance of saving them. But that understanding takes time and study, and in the meantime steps must be taken to conserve these systems for future recovery and management. In the United States, since the enactment of the Endangered Species Act, efforts to conserve biodiversity have focused on individual species and (sometimes) their critical habitats. According to data on freshwater fishes from the US Fish and Wildlife Service (http://ecos.fws.gov/tess_public/reports/delisting-report), some of these efforts have been successful (e.g., Modoc Sucker, *Catostomus microps*), but others have not (Amistad Gambusia, *Gambusia amistadensis*). Currently, conservation efforts for native fishes are expanding to target entire communities in whole watersheds (Williams et al. 2011) with the establishment of Native Fish Conservation Areas. Our understanding of community dynamics will therefore become all the more critical and important.

Closing Thoughts and a Retrospective

During our lifetimes, stream fish ecology has gone from almost nonexistent to the vibrant, energetic discipline it is today. Some years ago, on WJM's fiftieth birthday, he went to the university library and looked through paper copies (there was no Google-Scholar then) of all the journals that were likely to have fish ecology papers in 1946, the year he was born. There were few. The year 1946 probably saw a low publication rate in many scientific disciplines, as the world was still reeling from the effects of World War II, which had only ended the year before. But even taking that factor into account, there was little published about stream fishes.

An entire day was spent in the university library, searching all journals related to fish, ecology, general natural history, and state or regional academies of science that were then available. The following journals did not even exist in 1946: *Journal of Fish Biology*, *Environmental Biology of Fishes*, *Ecology of Freshwater Fish*, *Oecologia*, *Oikos*, and *Ecological Applications*. No papers on fish were published in the journal *Ecology* in 1946. This search revealed a total of 39 papers in fish ecology in other journals, which included not only papers on stream fish per se, but also general papers on evolution, zoogeography, genetics, pond fish production, or other aspects of fish ecology. The most common themes of these 39 papers included 6 on growth or development of fishes, 5 on surveys and distributions or regional ichthyology, and 4 on fish production or population sizes. Three each were on "foods of," general ecological theory with some information on fishes, and aquatic ecology with reference to fishes. None of the papers were on community structure, competition or predation, temporal changes in fish communities, or effects of fish in ecosystems.

A great many other things have changed in stream fish ecology. Today, women are important leaders in the field, whereas in the 1940s, or even when EMM was in graduate school (1974–1980), women studying field ecology of fishes were few. Now at any professional meeting about fishes, the gender balance, especially for younger participants, is approximately equal.

From this review and synthesis of our 40 years or so of studying stream fish, we hope we have convinced you that the study of stream fish community dynamics has made great progress since V. E. Shelford (1911) first contemplated spatial and temporal changes in fish communities in small tributaries to Lake Michigan and William Starrett (1951) provided insight into dynamics of fish in large prairie rivers. But we hope we have also convinced you that much more needs to be known about the ways that traits of individual species, or interactions among species, affect stream fish community dynamics, and that much more needs to be known about how stream fish affect the dynamics of ecosystems. And we certainly hope we have convinced you of the critical importance of long-term studies of stream fish communities, continuing those that have been ongoing, or, especially for graduate students or younger faculty, the importance of beginning what will become long-term studies.

In that light, we offer a few suggestions. First, if you begin a long-term study of a fish community, think of it as a personal commitment that you might continue regardless of funding sources. Many well-meaning "long-term" sampling initiatives have been undertaken by various government entities during our 40 years of experience, but many have ultimately stopped as administrations, emphasis of funding agencies, or general problems with logistics or personnel have developed over time. If you are lucky, you may find yourself in a situation in which you can inherit or join in a long-term database from an ongoing effort. But if you start a long-term study in your own favorite stream(s), plan that over years and decades you may have to just get in your own truck, recruit volunteers to help, and go "do it." If you do begin such a study, take advantage of the many resources that were not available 40 years ago, or that we might have overlooked by chance when we began our work on Piney and Brier Creeks. First, if possible, plan such studies where there have been and are likely to be reliable US Geological Survey (USGS) gages (yes, the USGS historically spells it "gage," because that is the way their founder, John Wesley Powell, spelled the word). This is not easy in many parts of the country, because USGS gages tend to be on larger streams, and it is hard, at least in Oklahoma, to find long-term gages on small streams. But without reliable data on discharge in "your" stream, you will be left, like us, to frequent trips to document in person the effects of floods or drought, to interview landowners, or to form correlations between rainfall and discharge.

In addition to stream gages, it also would be advisable to plan for your long-term study to be in reasonable proximity to an active weather station site. In Oklahoma, we are fortunate now to have the Mesonet, with weather stations in every county of the state providing real-time information on rainfall and an array of other meteorological measures, with details that go back more than 20 years. Before that time weather information was spotty, however, often originating from the recordings made by "weather observers" who provided visual daily observations of manual rain gauges in their own backyards. If we were planning to begin a long-term stream fish community study today, we would probably select a site on a reasonably sized, seinable stream that did have a USGS gage and was near a Mesonet site.

Lastly, our research has been almost entirely by seining. While this has worked well for us, in a new study you should give careful consideration to how you will sample. By seining (which two people can do, with a third helpful but not essential)? By electroshocking (which requires a larger crew and wearing of waders even in hot weather)? A combination of the two? Should you use block nets? (We do not, as natural breaks like ends of pools or riffles seem to allow us to trap fish.) Should you standardize sampling by numbers of members of the crew, by time in the water, by distance sampled, or by time of day, or simply do the best you can across the vagaries of years and opportunities?

Should you use full randomization of site selection, partial randomization, or simply set sites where access is available? Are you going to be more successful in the long run by establishing sites on public land (where changes in policy might kick you out) or on private land (which is great when landowners are friendly, as so many have been to us, but where changes in ownership can create challenges or at a minimum a need to educate the new owners of the importance of your long-term study)? And do you realize that most long-term study is incremental, with each survey building on the one before and offering opportunities to see the effects of natural or human factors against the baseline of your previous surveys? Long-term studies become transformative only as you find creative new questions as the data set lengthens and as you think of new questions that go beyond your initial hypotheses when the study was initiated.

These, and many other considerations, are important if you are contemplating a long-term ecological study, not just of stream fish, but probably also for any other research you might consider over years or decades in streams. Our long-term studies of Piney and Brier Creeks simply developed, almost by chance, and at the time of the first surveys of those streams there was little thought given that there would be sampling at the same sites in the future. But it is almost impossible to predict what course your career may take, and as we were sampling Piney Creek in July 2012, almost 40 years to the day when it was first sampled in July 1972, it was easy to look back and think of all the things WJM *didn't* think of back then. We hope you will plan ahead for your long-term study in more ways than WJM did. But we close by suggesting that even with all the variability in our studies, the data have been useful for helping to clarify what stream fish communities really do and say something about their dynamics over relatively long periods of ecological time. With publication of this book, we will release all of our global and local data sets for use by any interested parties who may envision analytical or modeling studies with the data that we have not thought of. We would appreciate acknowledgment of the source of the data and knowing about any future uses of the data, but they are for readers to use as they see fit. We wish you well with using our data or gathering your own and hope that the next generation of fish ecologists enjoys the work as much as we have. We have had great experiences in the field over the last 40 years and have enjoyed collecting and analyzing data and thinking about what it all means. And we're not done: we plan to continue our work and to publish using our existing data sets or new ones we may gather (but we also plan to spend more time playing with our five grandkids).

References

Blum, M. J., D. M. Walters, N. M. Burkhead, B. J. Freeman, and B. A. Porter. 2010. Reproductive isolation and the expansion of an invasive hybrid swarm. Biological Invasions 12:2825–2836.

Collins, S. L. 2000. Disturbance frequency and community stability in native tallgrass prairie. American Naturalist 155:311–325.

DeAngelis, D. L., J. C. Waterhouse, W. M. Post, and R. V. O'Neill. 1985. Ecological modelling and disturbance evaluation. Ecological Modelling 29:399–419.

Federal Register. 1998. 50 CFR Part 17. Endangered and threatened wildlife and plants: final rule to list the Arkansas River Basin population of the Arkansas River Shiner (*Notropis girardi*) as threatened. November 23, 1998.

Federal Register. 2005. 50 CFR Part 17. Endangered and threatened wildlife and plants: final designation of critical habitat for the Arkansas River Basin population of the Arkansas River Shiner (*Notropis girardi*). November 23, 2005.

Goldschmidt, T., F. Witte, and J. Wanink. 1993. Cascading effects of the introduced Nile Perch on the Detritivorous/Phytoplanktivorous species in the sublittoral areas of Lake Victoria. Conservation Biology 7:686–700.

Hubbs, C. L., and A. I. Ortenburger. 1929a. Further notes on the fishes of Oklahoma with descriptions of new species of Cyprinidae. Publications of the University of Oklahoma Biological Survey 1:17–43.

Hubbs, C. L., and A. I. Ortenburger. 1929b. Fishes collected in Oklahoma and Arkansas in 1927. Publications of the University of Oklahoma Biological Survey 1:47–112.

Jackson, D. A., P. R. Peres-Neto, and J. D. Olden. 2001. What controls who is where in freshwater fish communities—the roles of biotic, abiotic, and spatial factors. Canadian Journal of Fisheries and Aquatic Sciences 58:157–170.

Jelks, H. L., S. J. Walsh, N. M. Burkhead, et al. 2008. Conservation status of imperiled North American freshwater and diadromous fishes. Fisheries 33:372–407.

Lawton, J. H. 1999. Are there general laws in ecology? Oikos 84:177–192.

Levin, S. A. 1992. The problem of pattern and scale in ecology: the Robert H. MacArthur Award lecture. Ecology 73:1943–1967.

Magurran, A. E., and B. J. McGill. 2011. Biological diversity: frontiers in measurement and assessment. Oxford University Press, Oxford, UK.

Marsh, E. 1984. Egg size variation in central Texas populations of *Etheostoma spectabile* Pisces: Percidae). Copeia 1984:291–301.

Marsh, E. 1986. Effects of egg size on offspring fitness and maternal fecundity in *Etheostoma spectabile* (Pisces: Percidae). Copeia 1986:18–30.

Marsh-Matthews, E., and W. J. Matthews. 2000. Geographic, terrestrial and aquatic factors: which most influence the structure of stream fish assemblages in the midwestern United States? Ecology of Freshwater Fish 9:9–21.

Marsh-Matthews, E., and W. J. Matthews. 2010. Proximate and residual effects of exposure to simulated drought on prairie stream fishes. Pages 461–486 in K. B. Gido and D. A. Jackson, eds. Community ecology of stream fishes: concepts, approaches, and techniques. American Fisheries Society Symposium 73. American Fisheries Society, Bethesda, MD.

Matthews, W. J. 1987. Physicochemical tolerance and selectivity of stream fishes as related to their geographic ranges and local distributions. Pages 111–120 in W. J. Matthews and D. C. Heins, eds. Community and evolutionary ecology of North American stream fishes. University of Oklahoma Press, Norman, OK.

Matthews, W. J. 1998. Patterns in freshwater fish ecology. Chapman and Hall, New York, NY.

Matthews, W. J. 2015. Basic biology, good field notes, and synthesizing across your career. Copeia 103:495–501.

Matthews, W. J., and E. Marsh-Matthews. 2015. Comparison of historical and recent fish distribution patterns in Oklahoma and western Arkansas. Copeia 103:170–180.

Matthews, W. J., and E. Marsh-Matthews. 2016. Dynamics of an upland stream fish community over 40 years: trajectories and support for the loose equilibrium concept. Ecology 97: 706–719.

Matthews, W. L., K. B. Gido, and F. P. Gelwick. 2004. Fish assemblages of reservoirs, illustrated by Lake Texoma (Oklahoma-Texas USA) as a representative system. Lake and Reservoir Management 20:219–239.

Matthews, W. J., E. Marsh-Matthews, G. L. Adams, and S. R. Adams. 2014. Two catastrophic floods: similarities and differences in effects on an Ozark stream fish community. Copeia 102:682–693.

Matthews, W. J., E. Marsh-Matthews, R. C. Cashner, and F. Gelwick. 2013. Disturbance and trajectory of change in a stream fish community over four decades. Oecologia 173: 955–969.

May, R. M. 1973. Stability in randomly fluctuating versus deterministic environments. American Naturalist 107:621–650.

Mehinto, A. C., C. J. Martyniuk, D. J. Spade, and N. D. Denslow. 2012. Applications of next-generation sequencing in fish ecotoxicogenomics. Frontiers in Genetics 3:1–10.

Murdoch, W. W. 1969. Switching in general predators: experiments on predator specificity and stability of prey populations. Ecological Monographs 39:335–354.

Ortenburger, A. I., and C. L. Hubbs. 1926. A report on the fishes of Oklahoma, with descriptions of new genera and species. Proceedings of the Oklahoma Academy of Science 6: 123–141.

Page, L. M., and R. L. Smith. 1970. Recent range adjustments and hybridization of Notropis lutrensis and Notropis spilopterus in Illinois. Transactions of the Illinois Academy of Science 63:264–272.

Perkin, J. S., and K. B. Gido. 2012. Fragmentation alters stream fish community structure in dendritic ecological networks. Ecological Applications 22:2176–2187.

Pyke, G. H. 2008. Plague minnow or mosquito fish? a review of the biology and impacts of introduced Gambusia species. Annual Review of Ecology, Evolution, and Systematics 39: 171–191.

Rees, H. C., B. C. Maddison, D. J. Middleditch, J. R. M. Patmore, and K. C. Gough. 2014. The detection of aquatic animal species using environmental DNA—a review of eDNA as a survey tool in ecology. Journal of Applied Ecology 51:1450–1459.

Ross, S. T. 2013. Ecology of North American freshwater fishes. University of California Press, Berkeley, CA.

Schlosser, I. J. 1987. A conceptual framework for fish communities in small headwater streams. Pages 17–24 in W. J. Matthews and D. C. Heins, eds. Community and evolutionary ecology of North American stream fishes. University of Oklahoma Press, Norman, OK.

Shelford, V. E. 1911. Ecological succession: I. stream fishes and the method of physiographic analysis. Biological Bulletin 21:9–34.

Starrett, W. C. 1951. Some factors affecting the abundance of minnows in the Des Moines River, Iowa. Ecology 32:13–27.

Wallace, R. K., and S. J. Ramsey. 1982. A new cyprinid hybrid, *Notropis lutrensis* and *N. callitaenia*, from the Apalachicola drainage in Alabama. Copeia 1982:214–217.

Walters, D. M., M. J. Blum, B. Rashleigh, B. J. Freeman, B. A. Porter, and N. M. Burkhead. 2008. Red Shiner invasion and hybridization with Blacktail Shiner in the upper Coosa River, USA. Biological Invasions 10:1229–1242.

Webb, C. O., D. D. Ackerly, M. A. McPeek, and M. J. Donoghue. 2002. Phylogenies and community ecology. Annual Review of Ecology and Systematics 33:475–505.

Williams, J. E., R. N. Williams, R. F. Thurow, et al. 2011. Native fish conservation areas: a vision for large-scale conservation of native fish communities. Fisheries 36:267–277.

INDEX

abiotic factors, 2, 76, 77, 113, 269, 280, 304, 306; continuous environmental variables, 66–70; harsh environments, 77–87, 113

abundance: as basis for naming communities, 43; changes and community trajectory arrows, 303–4; changes in species dominance, 200–202; core species, 48–50; effects of complementarity or species repulsions, 114–19, 141; effects of habitat loss, 92–93; effects of intraspecific and interspecific interactions, 304; effects of resource partitioning, 120–23; evenness measure, 51–53; facilitation effects, 112, 136–37; family composition, 57–59; flood effects, 159–60, 169–72; MH index based on, 173, 190–91, 203–6, 212, 242–43, 250; minnows, 82–85, 88, 114–22, 138, 141, 169, 200–202, 220, 226; percent similarity index, 186, 221; predation effects, 125–36; relation to ecosystem effects, 272–82, 288; relation to species richness, 46–48, 50, 138–39; represented by sampling, 5, 8, 9–10, 12, 15; temporal dynamics, 187–90, 196, 198, 224; temporal variation in spatial patterns, 220–22, 231–33, 253. *See also* numerical dominance

Agosia chrysogaster (Longfin Dace), 259

algae: ash-free dry mass, 263, 264, 266; assemblages of, 267, 283; Baron Fork, Illinois River, 88–89, 100–101, 261, 264–66; Brier Creek, 25, 261, 263, 264, 266, 269, 278, 283; downstream effects of fish, 269–71; effect of Largemouth Bass presence, 135, 143, 267, 290; filamentous, 260, 261, 263, 264, 269, 270, 280, 283; fish effects on algal productivity, 263, 264, 267–69, 273–75, 282, 289–90; Gila River, 120; insectivore effects, 266–67, 282, 283–84, 287; nutrient translocation and algal growth, 259, 263, 267, 268, 273, 275, 278, 280, 283, 284, 287, 289–90; Piney Creek, 280; Red Shiner effects, 285; removal by scrapers, 259; stoneroller grazing, 12, 14, 89, 100–101, 259, 260–66, 275, 283, 285, 289

algivores/algivory, 14, 259, 266–67, 268, 271, 283, 284, 288; Brier Creek, 261, 263, 264,

266, 276, 280, 283; damselfish, 266; ecosystem effects, 281, 282, 283; Kings Creek, 267, 282, 283–84; Ozark Minnow, 266, 289; Piney Creek, 275–77, 280; Southern Redbelly Dace, 266–67, 275, 282, 287, 289; stonerollers, 12, 14, 88–89, 100–101, 259, 260–66

Alum Creek, AR, 19, 30, 36; beta diversity, 225, 226; community characteristics, 44; environmental characteristics, 60, 66, 67; intermittency, 62; land use, 66; MH values for communities, 204, 205, 206, 241, 243–45, 252; numerically dominant minnow species, 201, 202; patterns of species presence-absence, 197; predator:prey ratios, 127; sampling, 30; South Fork, 19, 30, 62, 127; temporal variation in spatial patterns, 235–36, 252

Ameiurus: melas (Black Bullhead), 161; *natalis* (Yellow Bullhead), 12, 161, 162, 196; spp. (bullhead catfishes), 10, 101, 156

Amistad Gambusia *(Gambusia amistadensis),* 308

apex predators, 131–32, 143, 287, 305

Aphredoderus sayanus (Pirate Perch), 126, 128

Aplodinotus grunniens (Freshwater Drum), 161, 198

Arkansas Darter *(Etheostoma cragini),* 92, 97

Arkansas River, OK, 31, 32

Arkansas River basin, OK, 20, 80, 189, 233, 251

Arkansas River Shiner *(Notropis girardi),* 80, 82, 201, 305

ash-free dry mass (AFDM), 263, 264, 266

Asian carp, 140

assemblages of fish, 2; data sets, 3; downstream effects, 288; facilitation effects, 137; flood effects, 160; Kiamichi River, 25; naming, 43; niche separation, 120; Ouachita National Forest streams, 30; relation of land use to variations, 66; resource partitioning, 184; spatial variation in temporal dynamics, 224; taxonomic composition, 41; temporal changes, 186, 219, 221; using rarefaction to adjust number of species detected, 46. *See also* communities of fish

association coefficients, 218

Grass Pickerel (*Esox americanus*), 42, 70–71, 128, 198

Gravel Chub (*Erimystax x-punctatus*), 84

grazing on algae by stonerollers, 12, 14, 89, 100–101, 259, 260–66, 275, 283, 285, 289; Largemouth Bass effects, 135, 143, 267, 290

Great Blue Heron, 101

Great Plains streams, 3, 18, 33–34, 170, 297, 305; environmental tolerance related to east-west gradient, 77–84, 104; MH values of communities, 231, 250; nonnative species, 139, 140; sampling, 3, 8, 297

Greenside Darter (*Etheostoma blennioides*), 12, 97, 242

Green Sunfish (*Lepomis cyanellus*), 12, 91; abundance, 126, 278; Brier Creek, 49, 93, 126, 161, 162, 177, 240, 244, 249, 252, 278; Crutcho Creek, 243; ecosystem effects, 268, 270, 273, 274, 275, 278, 284, 290; flood effects, 133, 158; group size, 101; Piney Creek, 275; predation, 126, 128, 134; reproduction, 102, 103; thermal tolerance, 157

Greenthroat Darter (*Etheostoma lepidum*), 123, 233

group size, 100–101

Gulf Killifish (*Fundulus grandis*), 203–4

habitat loss, 77, 92–93, 105

habitat selectivity, 77, 87–90, 114, 123, 157, 168, 306

habitat specificity, 90–92, 184, 231

harsh environment, 77–78, 113, 297, 298, 300–301, 304; Alum Creek, 202; Bread Creek, 235; Brier Creek, 25, 78, 86, 173, 177, 188–89, 202, 210, 238, 240, 244, 250, 252, 300; Crutcho Creek, 202; drought effects, 160–68, 170, 173–77; extreme temperature effects, 156–57; flood effects, 157–60, 173–77; Grand River tributaries, 31, 176; habitat selectivity effects, 87–90, 105, 168; MH values of communities, 240, 244; mosquitofish, 59, 71, 177, 301; nestedness, 225; Peckarsky harsh–benign gradient, 23, 77–78, 113, 185; South Canadian River, 32, 129; species dominance, 202; species tolerance, 77–87, 91, 104; temporal changes in communities, 185, 186, 188–89, 224, 225, 238

Hauani Creek, OK, 20, 42, 45, 46, 61, 127, 205, 206

heat death of fish, 151, 156

Hesperoleucas symmetricus (California Roach), 259

Hickory Creek, OK, 20, 42, 45, 61, 127, 129, 201, 205, 206

Highland Stoneroller (*Campostoma spadiceum*), 12, 91, 227, 229, 242, 243, 282

homogenization, 92, 189, 190

Hornyhead Chub (*Nocomis c.f. biguttatus*), 169

House Creek, OK, 130, 163

Hybognathus placitus (Plains Minnow), 82, 156, 165, 202, 305

Hybopsis amblops (Bigeye Chub), 84, 169

hybridization, 304

hydrologic subregions, 55–57, 71

hydrologic unit codes (HUCs), 55–56, 190

Hypentelium nigricans (Northern Hog Sucker), 12, 100

Hypophthalmichthys: molitrix (Silver Carp), 140; *nobilis* (Bighead Carp), 140

Ictaluridae, 41, 126

Ictalurus: furcatus (Blue Catfish), 41; *punctatus* (Channel Catfish), 101

Illinois River, OK, 20, 33, 42, 45, 61, 127, 129, 201, 202, 206. *See also* Baron Fork, Illinois River

Imhof Creek, OK, 140

individual traits. *See* species traits

insectivores/insectivory, 41, 267; with algivores, 283, 284; benthic, 268, 271, 277–79; Brier Creek, 277–79, 282; darters, 41, 141, 268, 271, 277; downstream effects, 269, 271; ecosystem effects, 271–72, 273, 275–80, 287, 288; madtom catfishes, 41; minnows, 41, 100, 104, 115, 116, 141, 200–201, 267, 268, 269, 273, 275–80, 287, 288; Piney Creek, 275–80, 287; sunfish, 126, 142, 276; surface-feeding, 100, 268, 271, 273, 276–80, 282, 287; water column, 41, 100, 104, 115, 200–201, 268, 271, 273–74, 276–80

in-stream pens, 12–14, 283; arrowhead-shaped, Baron Fork, 89, 264, 265; Kings Creek, 267, 283

interbrood interval, 102

intermittency, 57, 59–64, 67, 70, 71, 104; Brier Creek, 78, 300; effects on temperature or oxygen tolerance, 86, 168; experimental flood effects, 177; Piney Creek, 78, 300; relation to community variation, 300, 301; relation to MH values of communities, 204–5, 213; relation to predator:prey ratios, 126, 128, 142, 143, 303; Roanoke River tributary, 86; Sister Grove Creek, 160

interquartile range (IQR) of MH values, 204–6, 213

interspecific competition. *See* competition, interspecific

metabolic rates, 90; shoaling, 89, 100–101, 131, 260, 261. *See also specific species*

Strawberry River, AR, 115, 219

stream gradient, 59, 60–62, 66, 67, 69, 70, 71, 78, 79; Brier Creek, 25; Crooked Creek, 42; Crutcho Creek, 31; Kiamichi River, 26; Mustang Creek, 42; Piney Creek, 22; relation to beta dynamics, 226; relation to MH values of communities, 205, 213; relation to predator:prey ratios, 126, 128, 129, 142; relation to spatiotemporal dynamics, 219, 221, 231; Roanoke River, 28

Streamline Chub (*Erimystax dissimilis*), 191

stream morphology, 40, 299–300. *See also* environmental variables

stream width, 59, 60–62, 66–69, 71, 299; relation to MH values of communities, 205, 213, 245–46, 252

Striped Bass (*Morone saxatilis*), 124, 139

Striped Shiner (*Luxilus chrysocephalus*), 12, 94, 95, 122–23, 201, 243, 268, 273, 274, 289

study systems, 18; Brier Creek, 23–25; Crutcho Creek and North Canadian River tributaries, 31–32; Grand River tributaries, 30–31; Kiamichi River, 25–27; map, 21; other OK streams, 33; Ouachita National Forest streams, 30; Piney Creek, 21–23; Red River basin, 34; Roanoke River, 27–30; South Canadian River, 32–33; southern Great Plains streams, 33–34; spatial scales of communities, 34–36; streams included, 19–20

Suckermouth Minnow (*Phenacobius mirabilis*), 82

suckers (Catostomidae): drought and flood effects, 169–70, 178; ecosystem effects, 100, 284, 285; group size, 101; Roanoke River, 158; sampling, 10, 12; temporal dynamics, 186. *See also specific species*

Sunburst Darter (*Etheostoma mihileze*), 84, 92, 97

sunfish, 41, 43; Brier Creek, 132, 137, 161, 172, 267, 272, 277, 278, 280; cold effects, 157; drought effects, 161, 163, 166, 178; ecosystem effects, 272, 277, 278, 280, 287; facilitative interactions with Largemouth Bass, 137, 143; flood effects, 172; group size, 101; numerical dominance of, 42, 70, 296; predator effects, 85, 93, 126, 129, 130, 131, 132–34, 142, 267–68, 278, 287; reproduction, 103. *See also specific species*

surface-feeding species, 268, 271, 273, 276–80, 282, 287

survivorship after drought, 166–68

Tallgrass Prairie Preserve, OK, 156–57

Tar Creek, OK, 30–31, 173, 175

taxonomic composition, 40–46, 70–71, 297

Telescope Shiner (*Notropis telescopus*), 116, 169, 191, 200

temperature, air, 59, 60–62, 67, 71, 78; effects of extreme heat or cold, 151, 153–54, 156–57; fish ecosystem effects, 286, 288; heat death of fish, 151, 156; mean annual temperature, 62, 67, 71; reproductive effects, 102, 176

temperature, water, 77, 79, 83, 299; Brier Creek, 25, 78, 188, 212, 239, 300; effects of extreme heat or cold, 156–57; habitat selectivity effects, 87–88, 90; habitat specificity effects, 91; Piney Creek, 78, 212, 240; predation effects, 129, 130, 143

temperature extremes, 151, 153–54, 156–57, 177. *See also* thermal tolerance; winter

temporal dynamics, 103, 183–90; assessment methods, 190–92; Brier Creek communities, 186, 187–89, 210–13, 220, 221–22, 224, 304; changes in species dominance, 200–202; changes in species presence-absence, 191, 195–99; community trajectories in multivariate space, 206–10; contrasting Brier Creek and Piney Creek, 210; detecting loose equilibrium, 192–95; MH index comparisons, 190–91, 203–6; Piney Creek communities, 185–89, 191, 195, 197, 198, 200–202, 204, 207–13, 286, 300; summary, 210–14. *See also* spatiotemporal dynamics

tetragnathid spiders, 272

Texas Hill Country streams, 82, 138, 139, 142, 187

Texas Shiner (*Notropis amabilis*), 233

thermal selectivity, 88

thermal tolerance, 12, 77, 78–86, 103–5, 156–57, 168, 303

Threadfin Shad (*Dorosoma petenense*), 139

Tilapia, 139

timber harvest, 30

tolerance: of harsh environments, 77–87, 91, 104; oxygen, 77, 80, 84, 86, 168, 303; thermal, 12, 77, 78–86, 103–5, 156–57, 168, 303

trajectories of change, 185, 186, 303–5; Borrow ditches, 209–10, 213; Brier Creek, 188, 206–8, 213, 247–50, 251, 252; community trajectory arrows, 302–5; evaluation for fit with loose equilibrium, 192–95, 209, 212, 251, 298, 301; exit arrow, 302, 305; in multivariate space, 206–10; NMDS biplots, 191–95, 206–9, 212, 247–48; Piney Creek, 22, 207–8, 213, 247; spatial differences, 247–50; types, 192–93, 208–9, 213

trophic cascades, 113, 135, 259–60; Brier Creek, 23, 267, 278

trophic groups, 77, 99, 200, 269, 280, 287, 290

Tyner Creek, OK, 20, 33, 45, 61, 127, 201, 202, 206, 261

University of Oklahoma Aquatic Research Facility, 13

University of Oklahoma Biological Station (UOBS), 5, 13, 23, 25, 263, 267, 268, 269–70, 271

University of Oklahoma FishLab, 189, 269

upland streams, 43, 59–61, 62, 64–65, 70, 71, 299; habitat selectivity, 87–88; habitat specificity, 91–92; Kiamichi River, 85; minnow species dominance, 202; Ouachita National Forest, 41, 42; Piney Creek, 23, 78, 202, 300; predator:prey ratios, 126, 129, 142; Red River basin, 34; Roanoke River, 152, 202; stoneroller numbers, 282; Strawberry River, 219; tolerance of harsh environments, 78, 79, 82–85

US Fish and Wildlife Service, 308

US Geological Survey (USGS): hydrologic unit codes, 55–56, 190; stream gages, 27, 33, 309, 310

Warmouth (*Lepomis gulosus*), 126

Washita River, OK-TX, 25, 57, 132, 273–75, 285

water willow (*Justicia americana*), 27

weather stations, 62, 309–10

Wedgespot Shiner (*Notropis greenei*), 84, 119, 191

West Cache Creek, OK, 165–66

Western Mosquitofish (*Gambusia affinis*), 47, 91; Crutcho Creek, 243; drought and flood effects, 160, 164, 165, 176, 178; ecosystem effects, 268, 271, 273, 274, 275, 276, 286, 288, 289, 290; lack of expected biotic interactions, 138–39, 143; numerical dominance, 42, 70, 71, 176, 205, 210; Piney Creek, 198, 213, 245, 286; population dynamics, 59, 71, 102, 125, 275, 288, 299, 300–301; predation, 125, 126, 128; reproduction, 102, 176; tolerance of harsh environments, 59, 156, 177

Western Starhead Topminnow (*Fundulus blairae*), 91, 210

White Crappie (*Pomoxis annularis*), 161, 196

White River, AR, 22, 158, 284

White Shiner (*Luxilus albeolus*), 88, 120, 201, 202

White Sucker (*Catostomus commersoni*), 84

Whitetail Shiner (*Cyprinella galactura*), 116–17, 201

winter: ecosystem effects, 285–86; effect on habitat selectivity, 88–90; Kiamichi River stage rises, 27; Piney Creek sampling, 238, 276; reproduction, 103; Roanoke River sampling, 28, 30; survival, 139, 153–54, 156–57, 177

Yanubbe Creek, OK, 209

Yellow Bullhead (*Ameirus natalis*), 12, 161, 162, 196

zonation, longitudinal, 116–18, 141, 218–20, 224, 250